Molec ... y and Function of Carrier Proteins

Society of General Physiologists Series • Volume 48

Molecular Biology and Function of Carrier Proteins

Society of General Physiologists • 46th Annual Symposium

Edited by
Luis Reuss
The University of Texas Medical Branch at Galveston
John M. Russell, Jr.
Medical College of Pennsylvania
and
Michael L. Jennings
The University of Texas Medical Branch at Galveston

Marine Biological Laboratory
Woods Hole, Massachusetts

10–13 September 1992

© The Rockefeller University Press
New York

Contents

Preface

This volume is a collection of the written contributions of the invited speakers to the 46th Annual Symposium of the Society of General Physiologists, focusing on molecular biology and function of carrier proteins, held in Woods Hole, Massachusetts, 10–13 September 1992. In addition to the keynote address and the talks included in the five official plenary sessions of the conference, the symposium was broadened for the first time this year by two sessions on "New Ideas, New Faces," designed to focus attention on the work of junior investigators. Judging by the number of posters submitted, close to 150, the symposium topic greatly interested scientists in the field. Abstracts appeared in the December 1992 issue of *The Journal of General Physiology.*

The 46th Symposium owes much to the explosive growth of molecular biology. The present volume makes it clear that the powerful methods of molecular biology have allowed investigators of transport proteins to address new kinds of questions, as well as to ponder new kinds of answers. Although we are far from a detailed understanding of the structure and structure–function relationships of membrane carriers, it must also be recognized that during the last few years progress has been enormous. The symposium presentations and the contributions in this volume represent an effort to bring together examples of the best structural and functional approaches to the experimental study of membrane carriers. We feel that during the symposium there was a significant degree of interdisciplinary communication, and we hope that this is reflected in this book.

We are grateful to Ron Kaback and Ernie Wright for their wise advice throughout the planning stages of the meeting, and to the Council of the Society of General Physiologists for continued support of our efforts. Jane Leighton and Sue Lahr, in Woods Hole, and Olwen Hooks, in Galveston, kept everything on track before, during, and after the meeting.

Thanks are due to the Rockefeller University Press and, in particular, to Jeanne Bye for her untiring and reliable help in getting the proceedings published.

This symposium would not have been possible without the financial support of the Office of Naval Research, the United States Army, the National Science Foundation, SmithKline Beecham Pharmaceuticals, Wyeth-Ayerst Company, and President Thomas N. James of The University of Texas Medical Branch at Galveston, to all of whom the organizers express their thanks.

Finally, we organizers recognize that the heart and soul of the meeting resided in the speakers' participation. Coming from various parts of the world, each brought enthusiasm, a spirit of cooperation, and a breadth of knowledge that made the symposium a true exchange of camaraderie and ideas which was profitable for those attending the meeting, and which will continue to be profitable for those who read the printed pages of the proceedings.

Luis Reuss
John M. Russell, Jr.
Michael L. Jennings

Keynote Address

Chapter 1

The Lactose Permease of *Escherichia coli*: A Paradigm for Membrane Transport Proteins

Miklós Sahin-Tóth, Rhonda L. Dunten, and H. R. Kaback

Howard Hughes Medical Institute, Departments of Physiology and Microbiology & Molecular Genetics, Molecular Biology Institute, University of California Los Angeles, Los Angeles, California 90024-1570

Molecular Biology and Function of Carrier Proteins © 1993 by The Rockefeller University Press

Little is known regarding the mechanism of energy transduction in biological membranes at the molecular level. Thus, although it has become apparent over the past 20 years that the immediate driving force for a wide range of seemingly unrelated phenomena (e.g., secondary active transport, oxidative phosphorylation, and rotation of the bacterial flagellar motor) is a bulk-phase, transmembrane electrochemical ion gradient, the mechanism by which free energy stored in such gradients is transduced into work or into other forms of chemical energy (i.e., ATP) remains enigmatic. To gain insight into the general problem, studies in this laboratory have concentrated on the lactose (lac) permease of *Escherichia coli* as a paradigm.

Accumulation of β-galactosides against a concentration gradient in *E. coli* occurs via lac permease, a hydrophobic polytopic cytoplasmic membrane protein that catalyzes the coupled translocation of β-galactosides and H^+ with a stoichiometry of unity (i.e., β-galactoside/H^+ symport or cotransport; see Kaback, 1989, 1992, and Kaback et al., 1990 for reviews). Under physiological conditions, where the proton electrochemical gradient across the cytoplasmic membrane ($\Delta\bar{\mu}_{H^+}$) is interior negative and/or alkaline, lac permease utilizes free energy released from downhill translocation of H^+ to drive accumulation of β-galactosides against a concentration gradient. In the absence of $\Delta\bar{\mu}_{H^+}$, the permease catalyzes the converse reaction, utilizing free energy released from downhill translocation of β-galactosides to drive uphill translocation of H^+ with generation of a $\Delta\bar{\mu}_{H^+}$, the polarity of which depends upon the direction of the substrate concentration gradient.

Lac permease is encoded by the *lacY* gene, the second structural gene in the *lac* operon, and it has been cloned into a recombinant plasmid (Teather et al., 1978) and sequenced (Büchel et al., 1980). By combining overexpression of *lacY* with the use of a highly specific photoaffinity probe (Kaczorowski et al., 1980) and reconstitution of transport activity in artificial phospholipid vesicles (i.e., proteoliposomes; Newman and Wilson, 1980), the permease was solubilized from the membrane, purified to homogeneity (Newman et al., 1981; Foster et al., 1982; Viitanen et al., 1986; also see Wright et al., 1986), and shown to catalyze all the translocation reactions typical of the β-galactoside transport system in vivo with comparable turnover numbers (Matsushita et al., 1983; Viitanen et al., 1984). Therefore, a single gene product, the product of *lacY,* is solely responsible for all of the translocation reactions catalyzed by the β-galactoside transport system.

Structure

Circular dichroic measurements on purified lac permease indicate that the protein is > 80% helical in conformation, an estimate consistent with the hydropathy profile of the permease, which suggests that ~ 70% of its 417 amino acid residues are found in hydrophobic domains with a mean length of 24 ± 4 residues (Foster et al., 1983). Based on these findings, a secondary structure was proposed in which the permease is composed of a hydrophilic NH_2 terminus followed by 12 hydrophobic segments in α-helical conformation that traverse the membrane in zigzag fashion connected by hydrophilic domains (loops) with a 17-residue COOH-terminal hydrophilic tail (Fig. 1). Support for the general features of the model and evidence that both the NH_2 and COOH termini of the permease are exposed to the cytoplasmic face of the membrane was obtained subsequently from laser Raman spectroscopy (Vogel et al.,

1985),[1] immunological studies (Carrasco et al., 1982, 1984*a,b;* Seckler et al., 1983, 1986; Herzlinger et al., 1984, 1985; Seckler and Wright, 1984; Danho et al., 1985), limited proteolysis (Goldkorn et al., 1983; Stochaj et al., 1986), and chemical modification (Page and Rosenbusch, 1988). However, none of these approaches differentiates between the 12-helix structure and other models containing 10 (Vogel et al., 1985) or 13 (Bieseler et al., 1985) putative transmembrane helices.

Calamia and Manoil (1990) have provided elegant and exclusive support for the topological predictions of the 12-helix model by analyzing an extensive series of lac permease–alkaline phosphatase (*lacY-phoA*) chimeras. Alkaline phosphatase is

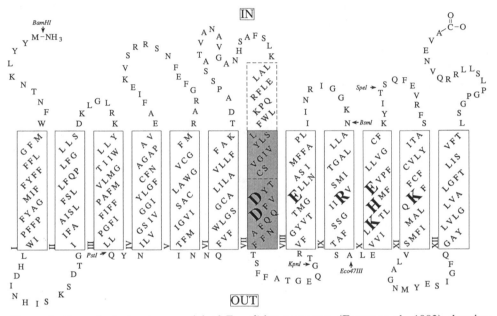

Figure 1. Secondary structure model of *E. coli* lac permease (Foster et al., 1983) showing modification suggested by King et al. (1991). Positions of putative intramembrane charged residues (Asp237, Asp240, Glu269, Arg302, Lys319, His322, Glu325, and Lys358) are highlighted. The single-letter amino acid code is used, and hydrophobic transmembrane helices are shown in boxes. The shaded box contains transmembrane helix VII as proposed by King et al. (1991), and the box enclosed by the dashed line defines putative helix VII as described originally (Foster et al., 1983). Also shown are the approximate positions of the restriction sites encoded by the cassette version of the *lacY* gene (EMBL-X56095).

synthesized as an inactive precursor in the cytoplasm of *E. coli* with a short signal sequence that directs its secretion into the periplasmic space where it dimerizes to form active enzyme. If the signal sequence is deleted, the enzyme remains in the cytoplasm in an inactive form. When alkaline phosphatase devoid of the signal sequence is fused to the COOH termini of fragments of a cytoplasmic membrane

---

[1] In addition to circular dichroic and laser Raman spectroscopy, Fourier-transform infrared studies also show that purified *lac* permease is largely helical (Roepe, P. D., H. R. Kaback, and K. J. Rothschild, unpublished work).

protein in vivo, enzyme activity reflects the ability of the NH_2-terminal portions of the chimeric polypeptides to translocate alkaline phosphatase to the outer surface of the membrane (Manoil and Beckwith, 1986; see Boyd et al., this volume). Alkaline phosphatase activity in cells independently expressing each of 36 *lacY-phoA* fusions exclusively favors the topological predictions of the model with 12 transmembrane domains.

In addition, Calamia and Manoil (1990) demonstrated that approximately half of a transmembrane domain is needed to translocate alkaline phosphatase to the external surface of the membrane. Thus, the alkaline phosphatase activity of fusions engineered at every third amino acid residue in putative helices III and V (Fig. 1) increases as a step function as the fusion junction proceeds from the 8th to the 11th residue of each of these transmembrane domains. Furthermore, when fusions are constructed at each amino acid residue in putative helices IX and X of the permease, the data obtained are in good agreement with the model (Fig. 1; Ujwal, M. L., E. Bibi, C. Manoil, and H. R. Kaback, unpublished data).

Purified lac permease reconstituted into proteoliposomes exhibits a notch or cleft (Costello et al., 1984, 1987), an observation also documented by Li and Tooth (1987) using completely different techniques. The presence of a solvent-filled cleft in the permease may have important implications with regard to the mechanism of β-galactoside/H^+ symport, as the barrier within the permease may be thinner than the full thickness of the membrane. Therefore, the number of amino acid residues in the protein directly involved in translocation may be fewer than required for lactose and H^+ to traverse the entire thickness of the membrane.

The remainder of this discussion deals with a few selected recent observations with a specific membrane transport protein, lac permease; however, it should be emphasized that the application of molecular biological techniques has made it apparent that a huge number of proteins catalyzing similar transport reactions are present in virtually all biological membranes from archaebacteria to the mammalian central nervous system. On the other hand, detailed structural information is a prerequisite for determining the mechanism, but it is particularly difficult to crystallize hydrophobic membrane transport proteins (Deisenhofer and Michel, 1989).

Functional Interactions between Putative Intramembrane Charged Residues

Recently, King et al. (1991) found that lac permease mutants with Thr in place of Lys358 or Asn in place of Asp237 are defective with respect to active transport. Second-site suppressor mutations of K358T[2] yield neutral amino acid substitutions for Asp237 (Asn, Gly, or Tyr), while suppressors of D237N exhibit Gln in place of Lys358. Based on these findings, it was proposed that Asp237 and Lys358 interact via a salt bridge, thereby neutralizing each other. Replacement of either charged residue with a neutral residue creates an unpaired charge that causes a functional defect, while neutral substitutions for both residues do not cause inactivation. Consequently, the secondary structure model proposed for the permease was altered to accommo-

---

[2] Site-directed mutants are designated as follows: the one-letter amino acid code is used followed by a number indicating the position of the residue in wild-type *lac* permease. The sequence is followed by a second letter denoting the amino acid replacement at this position.

date a putative salt bridge between Asp237 and Lys358 in the low dielectric of the membrane by placing residues Thr235 to Phe247 in transmembrane helix VII rather than the hydrophilic domain between helices VII and VIII (Fig. 1).

As part of an extensive site-directed mutagenesis study with an engineered lac permease that is functional but devoid of Cys residues (C-less permease; van Iwaarden et al., 1991), putative intramembrane residues Asp237, Asp240, Glu269, Arg302, Lys319, His322, Glu325, and Lys358 were systematically replaced with Cys. Individual replacement of any of the residues essentially abolishes active lactose transport (Sahin-Tóth et al., 1992). By using the single Cys mutants D237C and K358C, a double mutant was constructed containing Cys replacements for both Asp237 and Lys358 in the same molecule. D237C/K358C transports lactose at about half the rate of C-less permease to almost the same steady-state level of accumulation. Moreover, replacement of Asp237 and Lys358, respectively, with Ala and Cys or Cys and Ala, or even interchanging Asp237 with Lys358 causes little change in activity. The observations provide a strong indication that Asp237 and Lys358 interact to form a salt bridge and that neither residue nor the salt bridge per se is important for transport. Despite the relatively high activity of the charge inversion mutant (D237K/K358D) and the mutants with neutral substitutions at positions 237 and 358, immunoblots reveal low levels of the polypeptides in the membrane, suggesting a role for the salt bridge in permease folding and/or stability (Dunten et al., 1993). The observations also raise the possibility that Asp237 and Lys358 may interact in a folding intermediate but not in the mature molecule. Remarkably, however, an inactive single mutant with Cys in place of Asp237 regains full activity upon carboxymethylation, which restores a negative charge at position 237. Therefore, it seems likely that the interaction between Asp237 and Lys358 is important for folding/stability and that the residues maintain proximity in the mature permease.

To test the possibility that other charged residues in transmembrane helices are neutralized by charge-pairing, 13 additional double mutants were constructed in which all possible interhelical combinations of negative and positively charged residues were replaced pair-wise with Cys (Sahin-Tóth et al., 1992). Of all the combinations of double mutants, only D240C/K319C exhibits significant transport activity. However, the functional interaction between Asp240 and Lys319 is different phenomenologically from Asp237-Lys358. Thus, D240C/K319C catalyzes lactose transport at about half the rate of C-less to a steady-state level of accumulation that is only ~25–30% of the control. Moreover, although significant activity is observed with the double Ala mutant or with the two possible Ala-Cys combinations, interchanging Asp240 and Lys319 completely abolishes active transport. Therefore, although neither Asp240 nor Lys319 per se is mandatory for active transport, the polarity of the interaction appears to be important for full activity. Finally, as opposed to the double mutants in D237 and K358, all of the D240/K319 mutants are found in the membrane in amounts comparable to the C-less control.

Importantly, Lee et al. (1992) have reached a similar conclusion regarding interaction between Asp240 and Lys319 by using a different experimental approach. These workers replaced Asp240 with Ala by site-directed mutagenesis and found little or no active sugar transport. Two second-site revertants were then isolated, one with Gln in place of Lys319 and the other with Val in place of Gly268. The double mutants also exhibit little or no accumulation of sugar, but manifest significant rates of lactose entry down a concentration gradient. Although suppression of D240A by

G268V is difficult to explain, the properties of the double mutant D240A/K319Q clearly indicate that there is a functional interaction between Asp240 and Lys319.

On the whole, the results suggest that charge-pairing between intramembrane charged residues is probably not a general feature of lac permease and may be exclusive to D237/K358 and D240/K319. However, it should be emphasized that the charge-pair neutralization approach is dependent upon permease activity and will not reveal charge-paired residues if they are essential for activity. In this regard, it is noteworthy that double Cys mutants involving residues suggested to be H-bonded and directly involved in lactose-coupled H^+ translocation and/or substrate recognition (i.e., Arg302, His322, and Glu325 [see Kaback, 1989, 1992, and Kaback et al., 1990], as well as Glu269, which has been shown to be essential [Hinkle et al., 1990; Ujwal, M. L., B. Persson, L. Patel, and H. R. Kaback, manuscript in preparation]) do not catalyze active lactose transport.

The modified secondary structure model of lac permease (Fig. 1) is based on a functional interaction between Asp237 and Lys358 and on the notion that the intramembrane charged residues are balanced. Despite the indication that Asp240 and Lys319 may also participate in a salt bridge, the evidence for the interactions is indirect, and other approaches are required to determine the location of the residues relative to the plane of the membrane and to demonstrate directly that the pairs are in close proximity. C-less permease mutants containing double Cys or paired Cys-Ala replacements will be particularly useful in this respect. Preliminary efforts to estimate the accessibility of Cys residues at positions 237 and 358 with water- or lipid-soluble sulfhydryl reagents suggest that Cys residues at the two positions are accessible to both types of reagents, although the lipid-soluble reagents are relatively more effective (Dunten et al., 1993). Other preliminary experiments with permease mutants specifically labeled at position 237 or 358 with paramagnetic or fluorescent probes, as well as a series of alkaline phosphatase fusions in helix VII, are consistent with placement of both residues near the membrane–water interface at the external surface of the membrane, rather than in the middle of helix VII (Fig. 1). Efforts to demonstrate disulfide bond formation directly by oxidation of appropriate double Cys mutants are also in progress. In any event, based on the evidence currently available, it is clear that Asp237-Lys358 and Asp240-Lys319 interact functionally, and it is reasonable to suggest that both pairs of residues may be in close proximity. It follows that putative helix VII (Asp237 and Asp240) may neighbor helices X (Lys319) and XI (Lys358) in the tertiary structure of the permease.

References

Bieseler, B., P. Heinrich, and C. Beyreuther. 1985. Topological studies of lactose permease of *Escherichia coli* by protein sequence analysis. *Annals of the New York Academy of Sciences.* 456:309–325.

Büchel, D. E., B. Gronenborn, and B. Müller-Hill. 1980. Sequence of the lactose permease gene. *Nature.* 283:541–545.

Calamia, J., and C. Manoil. 1990. *lac* permease of *Escherichia coli:* topology and sequence elements promoting membrane insertion. *Proceedings of the National Academy of Sciences, USA.* 87:4937–4941.

Carrasco, N., D. Herzlinger, R. Mitchell, S. DeChiara, W. Danho, T. F. Gabriel, and H. R. Kaback. 1984a. Intramolecular dislocation of the C-terminus of the *lac* carrier protein in reconstituted proteoliposomes. *Proceedings of the National Academy of Sciences, USA.* 81:4672–4676.

Carrasco, N., S. M. Tahara, L. Patel, T. Goldkorn, and H. R. Kaback. 1982. Preparation, characterization and properties of monoclonal antibodies against the *lac* carrier protein from *Escherichia coli. Proceedings of the National Academy of Sciences, USA.* 79:6894–6898.

Carrasco, N., P. Viitanen, D. Herzlinger, and H. R. Kaback. 1984*b.* Monoclonal antibodies against the *lac* carrier protein from *Escherichia coli.* I. Functional studies. *Biochemistry.* 23:3681–3687.

Costello, M. J., J. Escaig, K. Matsushita, P. V. Viitanen, D. R. Menick, and H. R. Kaback. 1987. Purified *lac* permease and cytochrome *o* oxidase are functional as monomers. *Journal of Biological Chemistry.* 262:17072–17082.

Costello, M. J., P. Viitanen, N. Carrasco, D. L. Foster, and H. R. Kaback. 1984. Morphology of proteoliposomes reconstituted with purified *lac* carrier protein from *Escherichia coli. Journal of Biological Chemistry.* 259:15570–15586.

Danho, W., R. Makofske, F. Humiec, T. F. Gabriel, N. Carrasco, and H. R. Kaback. 1985. Use of site-directed polyclonal antibodies as immunotopological probes for the *lac* permease of *Escherichia coli. In* Peptides, Structure and Function: Proceedings of the 9th American Peptide Symposium. Pierce Chemical Company, Rockford, IL. 59–62.

Deisenhofer, J., and H. Michel. 1989. The photosynthetic reaction center from the purple bacterium *Rhodopseudomonas viridis. Science.* 245:1463–1473.

Dunten, R.L., M. Sahin-Tóth, and H.R. Kaback. 1993. Role of the charge pair aspartic acid 237-lysine 358 in the lactose permease of *Escherichia coli. Biochemistry.* In press.

Foster, D. L., M. Boublik, and H. R. Kaback. 1983. Structure of the *lac* carrier protein of *Escherichia coli. Journal of Biological Chemistry.* 258:31–34.

Foster, D. L., M. L. Garcia, M. J. Newman, L. Patel, and H. R. Kaback. 1982. Lactose: Proton symport by purified *lac* carrier protein. *Biochemistry.* 21:5634–5638.

Goldkorn, T., G. Rimon, and H. R. Kaback. 1983. Topology of the *lac* carrier protein in the membrane of *Escherichia coli. Proceedings of the National Academy of Sciences, USA.* 80:3322–3326.

Herzlinger, D., N. Carrasco, and H. R. Kaback. 1985. Functional and immunochemical characterization of a mutant of *Escherichia coli* energy-coupled for lactose transport. *Biochemistry.* 24:221–229.

Herzlinger, D., P. Viitanen, N. Carrasco, and H. R. Kaback. 1984. Monoclonal antibodies against the *lac* carrier protein from *Escherichia coli.* II. Binding studies with membrane vesicles and proteoliposomes reconstituted with purified *lac* carrier protein. *Biochemistry.* 23:3688–3693.

Hinkle, P. C., P. V. Hinkle, and H. R. Kaback. 1990. Information content of amino acid residues in putative helix VIII of *lac* permease from *Escherichia coli. Biochemistry.* 29:10989–10994.

Kaback, H. R. 1989. Molecular biology of active transport: from membranes to molecules to mechanism. *Harvey Lectures.* 83:77–105.

Kaback, H. R. 1992. In and out and up and down with the lactose permease of *Escherichia coli. International Review of Cytology (A Survey of Cell Biology).* 137:97–125.

Kaback, H. R., E. Bibi, and P. D. Roepe. 1990. β-Galactoside transport in *Escherichia coli:* a functional dissection of *lac* permease. *Trends in Biochemical Sciences.* 15:309–314.

Kaczorowski, G. J., G. Leblanc, and H. R. Kaback. 1980. Specific labeling of the *lac* carrier protein in membrane vesicles of *Escherichia coli* by a novel photoaffinity reagent. *Proceedings of the National Academy of Sciences, USA.* 77:6319–6323.

King, S. C., C. L. Hansen, and T. H. Wilson. 1991. The interaction between aspartic acid 237 and lysine 358 in the lactose carrier of *Escherichia coli. Biochimica et Biophysica Acta.* 1062:177–186.

Lee, J.-I., P. P. Hwang, C. Hansen, and T. H. Wilson. 1992. Possible salt bridges between transmembrane a-helices of the lactose carrier of *Escherichia coli. Journal of Biological Chemistry.* 267:20758–20764.

Li, J., and P. Tooth. 1987. Size and shape of the *Escherichia coli* lactose permease measured in filamentous arrays. *Biochemistry.* 26:4816–4823.

Manoil, C., and J. Beckwith. 1986. A genetic approach to analyzing membrane protein topology. *Science.* 233:1403–1408.

Matsushita, K., L. Patel, R. B. Gennis, and H. R. Kaback. 1983. Reconstitution of active transport in proteoliposomes containing cytochrome *o* oxidase and *lac* carrier protein purified from *Escherichia coli. Proceedings of the National Academy of Sciences, USA.* 80:4889–4893.

Newman, M. J., D. L. Foster, T. H. Wilson, and H. R. Kaback. 1981. Purification and reconstitution of functional lactose carrier from *Escherichia coli. Journal of Biological Chemistry.* 256:11804.

Newman, M. J., and T. H. Wilson. 1980. Solubilization and reconstitution of the lactose transport system from *Escherichia coli. Journal of Biological Chemistry.* 255:10583–10586.

Page, M. G. P., and J. P. Rosenbusch. 1988. Topography of lactose permease from *Escherichia coli. Journal of Biological Chemistry.* 263:15906–15914.

Sahin-Tóth, M., R. L. Dunten, A. Gonzalez, and H. R. Kaback. 1992. Functional interactions between putative intramembrane charged residues in the lactose permease of *Escherichia coli. Proceedings of the National Academy of Sciences, USA.* 89:10547–10551.

Seckler, R., T. Möröy, J. K. Wright, and P. Overath. 1986. Anti-peptide antibodies and proteases as structural probes for the lactose/H+ transporter of *Escherichia coli:* a loop around amino acid residue 130 faces the cytoplasmic side of the membrane. *Biochemistry.* 25:2403–2409.

Seckler, R., and J. K. Wright. 1984. Sidedness of native membrane vesicles of *Escherichia coli* and orientation of the reconstituted lactose:H+ carrier. *European Journal of Biochemistry.* 142:269–279.

Seckler, R., J. K. Wright, and P. Overath. 1983. Peptide-specific antibody locates the COOH terminus of the lactose carrier of *Escherichia coli* on the cytoplasmic side of the plasma membrane. *Journal of Biological Chemistry.* 258:10817–10820.

Stochaj, V., B. Bieseler, and R. Ehring. 1986. Limited proteolysis of lactose permease from *Escherichia coli. European Journal of Biochemistry.* 158:423–428.

Teather, R. M., B. Müller-Hill, U. Abrutsch, G. Aichele, and P. Overath. 1978. Amplification of the lactose carrier protein in *Escherichia coli* using a plasmid vector. *Molecular and General Genetics.* 159:239–248.

van Iwaarden, P. R., J. C. Pastore, W. N. Konings, and H. R. Kaback. 1991. Construction of a functional lactose permease devoid of cysteine residues. *Biochemistry.* 30:9595–9600.

Viitanen, P., M. L. Garcia, and H. R. Kaback. 1984. Purified, reconstituted *lac* carrier protein from *Escherichia coli* is fully functional. *Proceedings of the National Academy of Sciences, USA.* 81:1629–1633.

Viitanen, P., M. J. Newman, D. L. Foster, T. H. Wilson, and H. R. Kaback. 1986. Purification, reconstitution and characterization of the *lac* permease of *Escherichia coli. Methods in Enzymology.* 125:429.

Vogel, H., J. K. Wright, and F. Jähnig. 1985. The structure of the lactose permease derived from Raman spectroscopy and prediction methods. *EMBO Journal.* 4:3625–3631.

Wright, J. K., R. Seckler, and P. Overath. 1986. Molecular aspects of sugar:ion cotransport. *Annual Reviews of Biochemistry.* 55:225.

Structure and Dynamics of Membrane Proteins

Chapter 2

Dimerization of Glycophorin A Transmembrane Helices: Mutagenesis and Modeling

D. M. Engelman, B. D. Adair, A. Brünger, J. M. Flanagan, J. F. Hunt, M. A. Lemmon, H. Treutlein, and J. Zhang

Department of Molecular Biophysics and Biochemistry, Yale University, New Haven, Connecticut 06511

Molecular Biology and Function of Carrier Proteins © 1993 by The Rockefeller University Press

Introduction

Our basic premise is that specific side-to-side interactions of transmembrane regions can have important roles in the formation of oligomers of membrane proteins, in the folding of polytopic membrane proteins, and in the functions associated with folding and oligomerization. In our studies of bacteriorhodopsin (BR) and glycophorin A (GpA), significant progress has been made toward showing that such interactions can provide both specificity and important interactive energy for folding and oligomerization.

Transmembrane Helices Can Be Independently Stable

Helical structure is known to be induced in polypeptides in nonaqueous environments (Singer, 1962, 1971), as is expected given the large free energy costs of transferring an unsatisfied hydrogen bond donor or acceptor from an aqueous to a nonpolar environment or of breaking such a bond in a nonpolar environment. It is therefore expected that hydrogen bonds must be satisfied in the transbilayer region. If a polypeptide has a sufficiently hydrophobic sequence of amino acid sidechains, the hydrophobic effect will favor its partition into the nonaqueous region of a lipid bilayer. Combining these concepts leads to the notion that nonpolar sequences will be stable as transbilayer structures in which backbone hydrogen bonds are systematically satisfied and the polar ends of the helix are in more polar environments outside the bilayer (Engelman and Steitz, 1981; Engelman et al., 1986; Popot and deVitry, 1990; Popot and Engelman, 1990). Experimental observation of many integral membrane proteins shows that, with the exception of some outer membrane proteins such as bacterial outer membrane porins, transmembrane proteins can be understood as having bundles of hydrophobic α-helices (Henderson and Unwin, 1975; Henderson, 1977; Henderson et al., 1990; Deisenhofer and Michel, 1991). Some studies have shown that subfragments of polytopic proteins retain helical structure when they are separately reconstituted into lipid bilayers, as would be predicted if their separate helices were stable (Popot et al., 1987). Thus, it is possible that single helices can be regarded as domains in the classical sense of being independently stable folding units.

Helix–Helix Association in Bilayers

If hydrogen bonding of the main chain and the hydrophobic effect are largely accounted for in forming independent transmembrane helices, then factors other than these may drive their association if they interact in tertiary folding and oligomerization. These factors may include polar interactions within the membrane, the constraints of interactions outside the bilayer (including helix–helix polypeptide links), and packing effects. These factors have been contemplated in the proposal that folding and oligomerization of membrane proteins involve two stages, one in which hydrophobic α-helices are established across the lipid bilayer and a second in which helices interact to form functional transmembrane structures (Popot and Engelman, 1990; Bormann and Engelman, 1992). Work with BR and with GpA constructs has supported the contentions of this two-stage model (see below).

BR Studies Show That Helix–Helix Interactions Dominate in Folding

In BR, the seven transmembrane helices of the structure interact very closely and some interactions of helices are also involved in forming the trimeric structure found in membranes (Henderson et al., 1990). BR can be regenerated from two chymotryptic fragments containing two and five helical transmembrane segments (Popot et al., 1987), and, more dramatically, from two separate helical transmembrane peptides plus the five helix chymotryptic fragment (Kahn and Engelman, 1992). Since the two separate helices were also studied as independent entities and found to be transbilayer helices (Hunt et al., 1991), the notions of the two-stage concept are supported. Calorimetric studies (Kahn et al., 1992) show that cleavage of the B-C link or the A-B link, or removal of the retinal all act to destabilize the structure somewhat but do so in independent ways and do not abolish the folding of the BR molecule. Since reformation of BR occurs from fragments first reconstituted as separate transmembrane helical structures, and then introduced into the same lipid bilayer, viewing each fragment as a subunit suggests that helix–helix interactions are a component of specific oligomerization in this system.

GpA Dimerization Is Driven by Specific Interactions between Transmembrane α-Helices

Oligomerization of transbilayer proteins in vivo is also thought to involve helix–helix interactions. The structures of the BR trimer (Henderson et al., 1990) and the photosynthetic reaction center (Diesenhofer and Michel, 1991) show many helix–helix interactions at oligomeric interfaces. Another example is the interaction between the T cell receptor α chain and CD3 δ (Manolios et al., 1990; Cosson et al., 1991) in the assembly of the T cell receptor complex. In this case, potentially charged groups appear to be important in the interaction; however, in other instances association of transmembrane domains without charged groups is also seen (Lemmon and Engelman, 1992). One of these is the human erythrocyte sialoglycoprotein glycophorin A (GpA), which forms dimers stable under the conditions of SDS-PAGE, and in which the interactions of its single transmembrane domain drive the association (Furthmayr and Marchesi, 1976; Bormann et al., 1989; Lemmon et al., 1992a). The properties of the GpA transmembrane domain were studied by genetically fusing it to the carboxy terminus of staphylococcal nuclease. The resulting chimera forms a dimer in SDS, which is disrupted upon addition of a peptide corresponding to the transmembrane domain of GpA, but not upon addition of transmembrane domains from the EGF receptor, *neu* oncogene, or BR. Deletion mutagenesis has been used to define the boundaries of the transmembrane domain responsible for dimerization, and site-specific mutagenesis has been used to explore the properties of the interactions in the dimer (Lemmon et al., 1992a,b). This work indicates that dimerization is driven largely by close association of the transmembrane portions of GpA in a parallel coiled-coil, and that the interactions are highly specific.

Similar interactions may be important in transmembrane signaling by growth factor receptors with a single transmembrane α-helix (Bormann and Engelman, 1992). For a number of these, most notably the epidermal growth factor receptor

(EGFR), specific interactions between transmembrane domains may provide part of the energy and specificity for dimerization. Indeed, a valine to glutamic acid mutation in the transmembrane domain of the *neu* oncogene product causes this EGFR-like molecule to become constitutively active (Bargmann et al., 1986), and increases the proportion of it found as dimer (Weiner et al., 1989).

In summary, the interactions of transmembrane helices can be highly specific and can have sufficient interaction energies in a lipid context to drive oligomerization.

SDS Gel Assay of Dimerization Using a Chimeric Construct

Specific, stable side-by-side interactions between transmembrane α-helices can occur in detergent environments that, to some extent, mimic the properties of lipid bilayers. We have developed a system for the genetic and biophysical analysis of such interactions (Lemmon et al., 1992a). The transmembrane domain of interest is fused to the COOH terminus of staphylococcal nuclease (SN). The resulting chimera can be expressed at high levels in *Escherichia coli* and is readily purified. Interactions can be studied by SDS-PAGE and other methods. We have used this approach to study the single transmembrane α-helix of GpA, which mediates the SDS-stable dimerization of this protein. Vectors and methods are described in Lemmon et al. (1992a,b).

The chimeric protein runs as a dimer in SDS gels, as shown in Fig. 1. Lane *1* in Fig. 1 *A* shows the presence of substantial amounts of dimer in the chimeric construct in SDS. In lanes *2–4*, the pure transmembrane peptide is added and shown to compete for binding at the dimer interface, resulting in a large proportion of heterodimer between the chimeric construct and the transmembrane peptide as shown in lane *4* and in Fig. 1 *B*. In lanes *5–8*, other transmembrane peptides are added, and are found not to disrupt the SN/GpA dimer. Thus, the dimerization arises from an interaction of the transmembrane part of the chimeric construct, and is sufficiently specific so that transmembrane domains from the *neu* oncogene, the EGFR, the HER-2 receptor, or helix A from BR do not compete. In Fig. 1 *B*, a fluorescently labeled peptide was used in the competition experiment, and shown to migrate at a molecular weight just above the monomer molecular weight of the chimeric construct. This establishes that the band is in fact a peptide-chimera heterodimer.

Mutational Analysis of GpA Dimerization

The behavior of the chimeric protein on SDS gels has permitted a mutational analysis of the requirements for dimerization. We explored COOH-terminal truncations and deletions of the extramembraneous region proximal to the fusion junction in order to define the part of the sequence responsible for the dimerization seen on SDS gels, and found that the dimerization potential depends on the transmembrane region from GpA and on the presence of a flexible linker with no particular sequence requirement.

Further experiments were conducted by saturation mutagenesis at each amino acid position in the transmembrane domain, using SDS gel migration as an assay (Lemmon et al., 1992b). The effects of sequence alterations were assigned to four categories: no effect, a modest effect with significant dimer remaining, a stronger

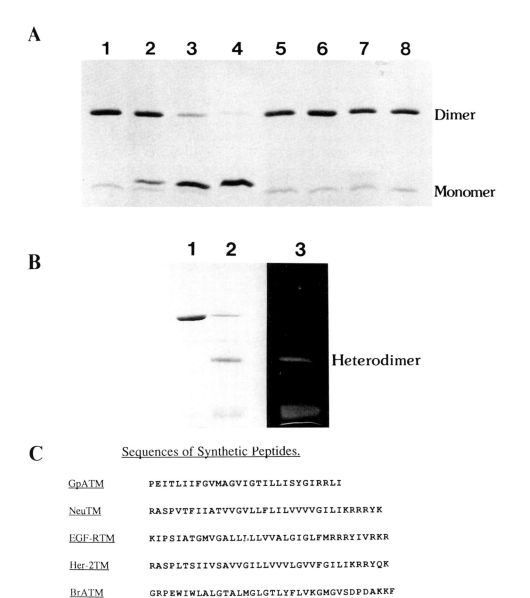

A

1 2 3 4 5 6 7 8

Dimer

Monomer

B

1 2 3

Heterodimer

C   <u>Sequences of Synthetic Peptides.</u>

<u>GpATM</u>   PEITLIIFGVMAGVIGTILLISYGIRRLI

<u>NeuTM</u>   RASPVTFIIATVVGVLLFLILVVVVGILIKRRRYK

<u>EGF-RTM</u>   KIPSIATGMVGALLLLLVVALGIGLFMRRRYIVRKR

<u>Her-2TM</u>   RASPLTSIIVSAVVGILLVVVLGVVFGILIKRRYQK

<u>BrATM</u>   GRPEWIWLALGTALMGLGTLYFLVKGMGVSDPDAKKF

Figure 1. (*A*) 12.5% polyacrylamide SDS Phastgel. Lane *1*, SN/GpA. Lanes *2–4* include a 1-, 5-, and 10-fold molar excess of GpATM, respectively. Lanes *5–8* contain a 10-fold molar excess of synthetic peptides of Neu, EGFR, HER-2, and BrA, respectively. Dimer and monomer positions are marked. (*B*) Competition experiment performed with GpA transmembrane peptide labeled with Dansyl. Lane *1* shows SN/GpA. Lane *2* shows SN/GpA with addition of 10-fold molar excess of Dansylated TM) peptide. Lane *3* is lane *2* under UV illumination. The diffuse fluorescent band at the bottom of the gel corresponds to free Dansylated peptide. (*C*) Sequences of peptides.

effect in which there remains relatively little dimer, or abolition of dimerization altogether resulting in the migration of all protein at the monomer position. Changes to charged amino acids produced strong effects, entirely abolishing all dimerization in almost all cases. Such effects could result from influences on the detergent micelle arrangement and are not easily interpreted. Terminations anywhere within the transmembrane region likewise abolish dimerization. The phenotypes of changes to moderately polar amino acids were more mixed. N, Q, and P substitutions generally abolished dimerization, but G, S, Y, and T were tolerated to some extent in some sequence locations; indeed, they are found in the native sequence.

More easily interpreted are the results of substitutions of less polar sidechains. Fig. 2 shows the current state of these mutagenesis studies. As will be readily

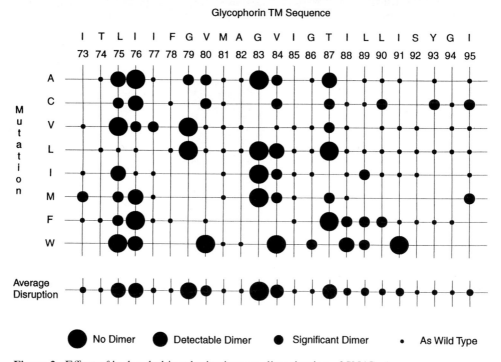

Figure 2. Effect of hydrophobic substitutions on dimerization of SN/GpA.

appreciated, this is a large body of work. Two striking messages emerge from the analysis. The first is that the dimeric interaction is extraordinarily specific, responding to modest changes in amino acid sidechain structure. The second is that this response varies along the sequence. Consideration, for example, of the three leucine residues in the transmembrane domain illustrates these points. At L75 all nonpolar substitutions disrupt the dimer. Many, including those to isoleucine or valine, have a large effect. However, the same substitutions at L89 and L90 are at most only slightly disruptive. Similarly, while all substitutions observed at G79 and G83 disrupt the dimer significantly, G86 and G94 will accommodate a variety of different residues with no significant effect. Thus, these results strongly suggest the intimate involvement of a subset of amino acid sidechains in the helix–helix interaction.

There is in general a periodicity in the ability of nonpolar sidechains to disrupt the dimer. The bottom of Fig. 2 shows the average degree of disruption by nonpolar substitutions at each position in the transmembrane domain. This shows that there are four clear regions in which the disruptive potential of substitutions is strong compared with other locations, namely, L75-76, G79-V80, G83-V84, and T87. A straightforward interpretation of this pattern would be that these residues comprise the interacting surfaces of α-helices in a supercoil. These putative sites of interaction are separated, on average, by almost four residues. This would be consistent with the dimer consisting of a pair of helices in a right-handed supercoil. Circular dichroism studies in lipid bilayers and in detergent environments show that both the isolated peptide and the transmembrane domain of the chimeric protein have strong helical character. Given the possibility that the interaction is between helices and that the interaction appears sensitive to small details in amino acid sidechain interactions, additional objectives arose. We wished to establish whether the interactions are of parallel helices, whether the interactions occur in lipid bilayers as well as in detergent environments, and whether theoretical descriptions can be created and tested against the experimental observations so that prediction of membrane protein folding and oligomerization can be facilitated. We have made progress in each of these directions.

The question of whether the dimers are parallel was addressed based on mutant complementation and on studies of the flexible link in the chimeric construct. When the link was shortened excessively, dimerization was abolished. Addition of alanines to replace the deleted residues progressively restored dimerization. This result can be explained if steric interference of the nuclease moiety in the chimera is acting to disrupt the dimer, which can only occur if it contains a parallel arrangement of helices. Further, we have recently found that dimerization can be restored in a mixture of L75V with I76A, in which dimerization was abolished for each substitution taken separately. The complementation suggests that Leu 75 and Ile 76 interact with one another, which is possible only in a parallel dimer (Lemmon et al., 1992*b*).

Simulated Annealing Approach to Model the Interaction

We have recently turned to studies of the dimerization of GpA using conformational searches by molecular dynamics searches and by simulated annealing techniques with restraints (Nilges and Brunger, 1991). Simulated annealing was implemented through molecular dynamic simulations with geometric (Brunger, 1991) and empirical energy functions (Karplus and Petsko, 1990). Our assumptions are that the transmembrane domains of glycophorin form dimers of α-helices with a parallel chain directionality and that the helices are in contact across the lipid bilayer. In the modeling, the helices were constrained with a half parabolic potential to prevent them from dissociating. Several initial conditions were chosen, including one in which the helices are straight and parallel. A search over a range of relative helix rotational angles resulted in the identification of two global energy minima (Treutlein et al., 1992). Modeling from initially parallel helices resulted in the surprising conclusion that the best structure is a right-handed supercoil. To check our proce-

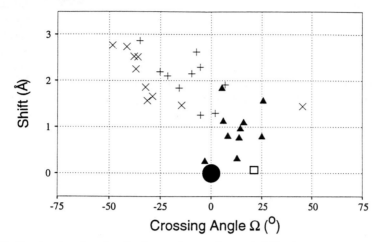

Figure 3. Scatterplot showing final crossing angles and shifts for models obtained starting from uncoiled dimers of ideal helices.

dure, we did a similar calculation with GCN4 (Jones and Fink, 1982; Pathak and Sigler, 1992), which is known to have a left-handed supercoiled dimer (O'Shea et al., 1991), and found that the modeling leads to the expected handedness (Treutlein et al., 1992).

Fig. 3 shows the result of a number of simulations using the glycophorin transmembrane helix and the GCN4 helix dimers, revealing a propensity for left-handed supercoiling (positive crossing angles) in the GCN4 and right-handed supercoiling (negative crossing angles) in the glycophorin cases. Fig. 4 shows a comparison of our best theoretical models with the mutagenesis data described above and in Lemmon et al. (1992b). The agreement for these right-handed supercoiled models is striking. We were unable to generate a left-handed coiled-coil of GpA helices that permitted good agreement with the mutagenesis data.

Our findings support the hope of predicting folded transmembrane protein structure in the hydrophobic region of the lipid bilayer. It is conceivable that a more

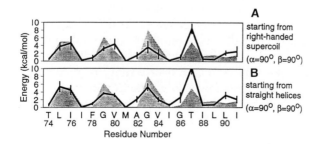

Figure 4. Comparison of average residue interaction energies with the mutational sensitivity profile of GpA.

complete modeling of the bilayer environment and more accurate empirical energy functions could improve the modeling, and we intend to move in these directions.

Conclusion

Our studies of the interaction of the transmembrane helices of GpA in a detergent environment suggest that dimerization occurs through highly specific interactions to form a right-handed supercoil of parallel helices. That such an interaction is sensitive to very small, normally conservative changes in the amino acid structures reveals the possibility of very high specificities in the interactions guiding membrane protein oligomerization and folding. Further, the finding that dimerization occurs via an extensive interaction surface in which the association is mainly through van der Waals contacts implies that such interactions can, on their own, provide sufficient energy for oligomerization. The fact that a single helix may act as a stable folding domain sets the conditions for the coordination of precise van der Waals contacts. Since the groups involved are held in a relatively fixed array in space, the coordination of interaction energies between two well-matched surfaces creates a strong summation of small energies that would not occur if the spatial correlations were not present. Thus, for example, contacts between groups in the lipid and groups on the protein would not be expected to have the same additivity. If the interaction is to be specific and sufficiently energetic, the surfaces must fit together well. Thus, GpA has provided insights concerning a set of mechanisms that must be considered in rationalizing the formation of higher order structure in biological membranes.

Acknowledgments

This work was supported by grants from the NIH, NSF, and National Foundation for Cancer Research. M. A. Lemmon is supported by a predoctoral fellowship from the Howard Hughes Medical Institute.

References

Bargmann, C. I., M.-C. Hung, and R. A. Weinberg. 1986. The neu oncogene encodes an epidermal growth factor receptor-related protein. *Nature.* 319:226–230.

Bormann, B. J., and D. M. Engelman. 1992. Intramembrane helix-helix association in oligomerization and transmembrane signalling. *Annual Review of Biophysics and Biophysical Chemistry.* 21:223–242.

Bormann, B.-J., W. J. Knowles, and V. T. Marchesi. 1989. Synthetic peptides mimic the assembly of transmembrane glycoproteins. *Journal of Biological Chemistry.* 264:4033–4037.

Brunger, A. T. 1991. Simulated annealing in crystallography. *Annual Review of Physical Chemistry.* 42:197–223.

Cosson, P., S. P. Lankford, J. S. Bonifacino, and R. D. Klausner. 1991. Membrane protein association by potential intramembrane charge pairs. *Nature.* 351:414–416.

Deisenhofer, J., and H. Michel. 1991. High-resolution structures of photosynthetic reaction centers. *Annual Review of Biophysics and Biophysical Chemistry.* 20:247–266.

Engelman, D. M., and T. A. Steitz. 1981. Insertion and secretion of proteins into and across membranes: the helical hairpin hypothesis. *Cell.* 23:411–422.

Engelman, D. M., T. A. Steitz, and A. Goldman. 1986. Identifying non-polar transbilayer helices in amino acid sequences of membrane proteins. *Annual Review of Biophysics and Biophysical Chemistry.* 15:321–353.

Furthmayr, H., and V. T. Marchesi. 1976. Subunit structure of human erythrocyte glycophorin A. *Biochemistry.* 15:1137–1144.

Henderson, R. 1977. The purple membrane from *halobacterium halobium. Annual Review of Biophysics and Bioengineering.* 6:87–109.

Henderson, R., R. Baldwin, T. A. Ceska, F. Zemlin, E. Beckman, and K. H. Downing. 1990. Model for the structure of bacteriorhodopsin based on high-resolution electron cryo-microscopy. *Journal of Molecular Biology.* 213:899–929.

Henderson, R., and P. N. Unwin. 1975. Three-dimensional model of purple membrane obtained by electron microscopy. *Nature.* 257:28–32.

Hunt, J. F., O. Bousche, K. Meyers, K. J. Rothschild, and D. M. Engelman. 1991. Biophysical studies of the integral membrane protein folding pathway. *Biophysical Journal.* 59:400*a*. (Abstr.)

Jones, E. W., and G. R. Fink. 1982. The Molecular Biology of the Yeast Saccharomyces: Metabolism and Gene Expression. J. N. Strathern, J. R. Broach, editors. Cold Spring Harbor Laboratory, Cold Spring Harbor, NY. 181–299.

Kahn, T. W., and D. M. Engelman. 1992. Bacteriorhodopsin can be refolded from two independently stable transmembrane helices. *Biochemistry.* 31:6144–6151.

Kahn, T. W., J. M. Sturtevant, and D. M. Engelman. 1992. Thermodynamic measurements of the contributions of helix-connecting loops and of retinal to the stability of bacteriorhodopsin. *Biochemistry.* 31:8829–8839.

Karplus, M., and G. A. Petsko. 1990. Molecular dynamics simulations in biology. *Nature.* 347:631–639.

Lemmon, M. A., and D. M. Engelman. 1992. Helix-helix interactions inside lipid bilayers. *Current Opinion in Structural Biology.* 2:511–518.

Lemmon, M. A., J. M. Flanagan, J. F. Hunt, B. D. Adair, B. J. Bormann, C. E. Dempsey, and D. M. Engelman. 1992*a*. Glycophorin A dimerization is driven by specific interactions between transmembrane α-helices. *Journal of Biological Chemistry.* 267:7683–7689.

Lemmon, M. A., J. M. Flanagan, H. R. Treutlein, J. Zhang, and D. M. Engelman. 1992*b*. Sequence specificity in the dimerization of transmembrane α-helices. *Biochemistry.* 31:12719–12725.

Manolios, N., J. S. Bonifacino, and R. D. Klausner. 1990. Transmembrane helical interactions and the assembly of the T cell receptor complex. *Science.* 249:274–277.

Nilges, M., and A. Brunger. 1991. Automated modeling of coiled coils: application to the GCN4 dimerization region. *Protein Engineering.* 4:649–659.

O'Shea, E. K., J. D. Klemm, P. S. Kim, and T. Alber. 1991. X-ray structure of the GCN4 leucine zipper, a two-stranded, parallel coiled coil. *Science.* 254:539–544.

Pathak, D., and P. B. Sigler. 1992. Updating structure-function relationships in the bZIP family of transcription factors. *Current Opinion in Structural Biology.* 2:116–123.

Popot, J.-L., and C. de Vitry. 1990. On the microassembly of integral membrane proteins. *Annual Review of Biophysics and Biophysical Chemistry.* 19:369–403.

Popot, J.-L., and D. M. Engelman. 1990. Membrane protein folding and oligomerization: the two-state model. *Biochemistry.* 29:4031–4037.

Popot, J.-L., S.-E. Gerchman, and D. M. Engelman. 1987. Refolding of bacteriorhodopsin in lipid bilayers, a thermodynamically controlled two-stage process. *Journal of Molecular Biology.* 198:655–676.

Singer, S. 1962. The properties of proteins in nonaqueous solvents. *Advances in Protein Chemistry.* 17:1–69.

Singer, S. 1971. The molecular organization of biological membranes. *In* Structure and Function of Biological Membranes. Academic Press, New York and London. 145–222.

Treutlein, H. R., M. A. Lemmon, D. M. Engelman, and A. T. Brünger. 1992. The glycophorin A transmembrane domain dimer: sequence-specific propensity for a right-handed supercoil of helices. *Biochemistry.* 31:12726–12732.

Weiner, D. B., J. Liu, J. A. Cohen, D. V. Williams, and M. L. Greene. 1989. A point mutation in the neu oncogene mimics ligand induction of receptor aggregation. *Nature.* 339:230–231.

Chapter 3

Gene Fusion Approaches to Membrane Protein Topology

Dana Boyd, Beth Traxler, Georg Jander, Will Prinz, and Jon Beckwith

Department of Microbiology and Molecular Genetics, Harvard Medical School, Boston, Massachusetts 02115

Molecular Biology and Function of Carrier Proteins © 1993 by The Rockefeller University Press

Introduction

The term topology has been used in the case of membrane proteins to describe the positioning of the different domains of those proteins relative to the lipid bilayer (Blobel, 1980). Topological analysis defines which segments of the protein span the membrane and which hydrophilic domains lie on either side of the membrane. Defining these aspects of the structure of membrane proteins is essential for assigning function to particular regions of the protein and for understanding the mechanism of membrane insertion and assembly of such proteins.

Many different approaches have been used in topological analysis of membrane proteins.[1] While we have reviewed these in more detail elsewhere (Traxler et al., 1993), here we briefly summarize each approach.

Prediction from Amino Acid Sequence

For certain proteins, topological models based only on sequence inspection have proved to be correct. Two major features of these proteins have proven sufficient to do topology predictions. First, runs of ~20 amino acids with a high hydrophobicity index are nearly always found to be membrane-spanning segments (Kyte and Doolittle, 1982). Second, hydrophilic domains that are localized to the cytoplasmic side of a membrane are enriched for basic amino acids compared with hydrophilic domains localized to the other side of the membrane (von Heijne, 1986; von Heijne and Gavel, 1988). Given a protein with an amino acid sequence that shows clear-cut membrane-spanning segments and an array of hydrophilic domains between them that alternate in their net positive charge, a topological model can be proposed that is highly likely to be correct.

However, many proteins do not provide such a user-friendly sequence. Hydrophobic regions may contain charged amino acids or may be considerably longer or shorter than 20 amino acids. Not all cytoplasmic domains are enriched for basic amino acids. Thus, one is not infrequently faced with sequences for which different possible topological models can be derived.

Physical Approaches

Both x-ray and electron diffraction studies have been used to determine the topology of membrane proteins. Presumably, NMR will be added to the list in the near future. X-ray analysis has provided a detailed picture of the arrangement in the membrane of the photosynthetic reaction center (Deisenhofer et al., 1985) from certain bacteria, including the topology of individual subunits. Electron diffraction was used to analyze the structure of bacteriorhodopsin (Henderson et al., 1990). While we can expect that more such structures will appear in the near future, these approaches have not been easy to extend to other proteins.

---

[1] For the purposes of this article, we are excluding from discussion the analysis of topology of outer membrane proteins of Gram-negative bacteria and other analogous proteins that may be found in mitochondria or chloroplasts of eukaryotic cells. Outer membrane proteins follow entirely different rules from other classes of membrane proteins.

Inherent Structural or Function Features of a Protein

Some aspects of topology can be deduced from inherent features of a protein. Such modifications as glycosylation and disulfide bond formation ordinarily take place on the extracytoplasmic side of the membrane, while protein phosphorylation is carried out in the cytoplasm. Determining the location of such modifications in a protein allows the placement of that particular domain within one or the other compartment. In addition, assignment to a specific domain of a function, which is known to operate in one or another cellular compartment, defines the location of that domain. Such an assignment may be effected by sequence comparisons or mutant analysis.

Chemical Modification

Treatment of cells with a chemical modifying agent under conditions where only the extracytoplasmic face of the membrane is exposed can reveal certain of the domains on that side of the membrane. This approach is limited by the nature of the agent and persistent questions about its membrane impermeability.

Proteolysis

Proteolysis of vesicles (inside-out or rightside-out) can be used to determine which domains of a protein are susceptible to degradation. Certain external and internal domains of proteins have been defined in this way, but to provide a complete picture, the approach requires that each hydrophilic domain be susceptible to the protease used. This is rarely the case.

Immunochemical Tagging of Membrane Protein Domains

Antibodies can be made to peptides corresponding to each of the hydrophilic domains of a membrane protein and then accessibility of the protein to antibody can be evaluated in different vesicle preparations. This approach has been used with impressive success in predicting a complete topological structure for the β-adrenergic receptor (Wang et al., 1989). However, for certain proteins this and the two previous approaches will be limited because of the small size of some of the hydrophilic domains.

Gene Fusion Techniques

When certain reporter proteins are fused to different domains of a membrane protein, the properties of the cells carrying the gene fusion can yield information on the location of the reporter protein. The reporter proteins are chosen on the basis that their phenotypic expression varies depending on their cellular location. Assuming that the reporter protein follows correctly the topological signals of the protein to which it is fused, phenotypic analysis of gene fusion strains can give a topological model for the protein. The limitations of this technique, which can be severe in some cases, are discussed below.

Gene Fusion Systems in Use for Studying Topology

The genes fusion systems that have been used in topological analysis are listed in Table I.

TABLE I
Gene Fusion Systems for Topological Analysis

Reporter protein	Phenotype	
	Fusion to external domain	Fusion to internal domain
Bacterial		
Alkaline phosphatase	Enzyme active	Enzyme inactive
β-Galactosidase	Enzyme inactive	Enzyme active
β-Lactamase	Ampicillin resistant	Ampicillin sensitive
Yeast		
Histidinol dehydroge-nase	No growth of HIS4 mutant on histidinol	Growth of HIS4 mutant on histidinol
Galactokinase	Glycosylated	Not glycosylated
Invertase	Glycosylated	Not glycosylated
Acid phosphatase	Glycosylated	Not glycosylated

   The most widely used is the bacterial alkaline phosphatase system (Manoil et al., 1990). The basis for this approach derives from the finding that alkaline phosphatase is enzymatically inactive when localized to the cytoplasm, but enzymatically active when translocated across the cytoplasmic membrane (Michaelis et al., 1983). The difference results from the absence of disulfide bond formation in the protein in the cytoplasm and its presence in the bacterial periplasm (Derman and Beckwith, 1991). The theory behind the approach is shown in Fig. 1.

   A second approach uses a reporter protein with opposite properties to alkaline phosphatase. β-Galactosidase is enzymatically active in the cytoplasm, but inactive when fused to a protein export signal (Froshauer et al., 1988). In the latter case, the

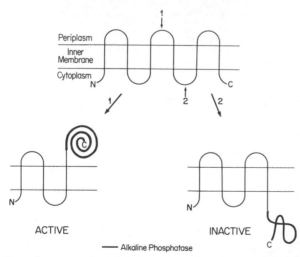

Figure 1. Theory of alkaline phosphatase fusion analysis of membrane protein topology. The membrane protein depicted is a hypothetical one. It is assumed in the theory that alkaline phosphatase will follow the same export signals that promote translocation across the membrane of external (periplasmic) domains. Thus, fusion to a periplasmic domain of the protein (*1*) will allow the export of alkaline phosphatase. Once localized to the periplasmic space, alkaline phosphatase will be enzymatically active, even though it is anchored to the membrane. It is also assumed that the same signals that cause retention of certain hydrophilic domains in the cytoplasm will also lead to the cytoplasmic localization of alkaline phosphatase. Thus, fusion to a cytoplasmic domain of the protein (*2*) will lead to a cytoplasmic location of alkaline phosphatase. In the cytoplasm, alkaline phosphatase is enzymatically inactive due to the absence of disulfide bonds (Derman and Beckwith, 1991).

enzyme becomes imbedded in the membrane where it cannot assembly properly. Thus, when fused to a cytoplasmic domain of a membrane protein, the β-galactosidase is active, while when fused to a periplasmic domain it is inactive.

When β-lactamase is fused to a periplasmic domain of a membrane protein, its export to the periplasm confers an ampicillin-resistant phenotype on cells (Broome-Smith et al., 1990). When it is fused to a cytoplasmic domain, its retention in the cytoplasm prevents its phenotypic expression and cells are ampicillin sensitive.

These three cases represent bacterial gene fusion systems for topological analysis. Several systems have also been used in the yeast *Saccharomyces cerevisiae*. Histidinol dehydrogenase is an enzyme that carries out the last step in histidine biosynthesis, converting histidinol to histidine. In yeast mutants missing the histidine pathway, the presence of the enzyme allows the cells to grow if they are fed histidinol. Only gene fusions in which the enzyme is retained in the cytoplasm, however, can satisfy this requirement (Senstag et al., 1990). In this way, cytoplasmic and extracytoplasmic domains of membrane proteins can, in principle, be distinguished.

Finally, fusion of a protein with a glycosylation site(s) to a membrane protein can help to distinguish the location of its hydrophilic domains. That is, the reporter protein will only be glycosylated if it is translocated into the endoplasmic reticulum. Several different reporter proteins have been used in this way (Ahmad and Bussey, 1988; Green et al., 1989; Deshaies and Schekman, 1990) (Table I).

There are two obvious assumptions that are made in using gene fusions to study topology. Since the gene fusion approach results in the replacement of carboxy-terminal segments of a protein by a reporter protein, correct predictions of topology would only result if the amino-terminal portion of the protein remaining in the fusion protein can assemble properly in the absence of carboxy-terminal sequences. Anomalies resulting from violations of this assumption will be discussed below. A second assumption is that the reporter protein is passive and does not itself influence topology. In general, this appears to be true, although differences in the behavior of β-lactamase and alkaline phosphatase fusions to be described below may indicate some exceptions.

The Experience with Alkaline Phosphatase Fusions

There are now a variety of techniques for generating alkaline phosphatase fusions to bacterial proteins or to proteins, the genes for which can be cloned into *Escherichia coli*, including eukaryotic proteins. A transposon, Tn*phoA,* can be used to obtain random fusions to membrane proteins in vivo (Manoil and Beckwith, 1985). Plasmids have been constructed that allow fusion of the *phoA* gene to a variety of different restriction sites (Hoffman and Wright, 1985; Ehrmann et al., 1990; Duchêne et al., 1992; Wilmes-Riesenberg and Wanner, 1992). Starting with the appropriate plasmids, polymerase chain reactions (Boyd et al., 1993) or oligonucleotide deletion mutagenesis (Boyd et al., 1987) have permitted the construction of fusions to specific sites within a gene. Using nuclease digestion with the appropriate starting plasmids, it is possible to obtain a nested set of *phoA* fusions to a gene for a membrane protein (Sugiyama et al., 1991; Yun et al., 1991).

There is enough experience now to show that the alkaline phosphatase fusion approach can give a complete and correct topological picture for certain membrane proteins. Proteins for which alkaline phosphatase fusions provided a topological model and for which there was preexisting or confirmatory evidence from other approaches include the family of chemotaxis receptors in certain Gram-negative bacteria (Manoil and Beckwith, 1986), the *E. coli* leader peptidase (San Millan et al., 1989), the *E. coli* MotB protein (a component of the flagellar apparatus; Chun and Parkinson, 1988), and the L subunit of the photosynthetic reaction center of *Rhodobacter sphaeroides* (Yun et al., 1991). These results show that one of the major assumptions of the approach holds true for at least these proteins. That is, amino-terminal portions of membrane proteins can assume the correct topological structure in the absence of carboxy-terminal sequences. Further, in the case of the L subunit of the photosynthetic reaction center, the protein assumes the same structure in the membrane on its own as its does in the multi-subunit complex (Deisenhofer et al., 1985).

Alkaline Phosphatase Fusions to the MalF Protein of *E. coli*

The MalF protein is part of a multi-subunit cytoplasmic membrane complex required for maltose transport in *E. coli* (Shuman and Silhavy, 1981; Davidson and Nikaido,

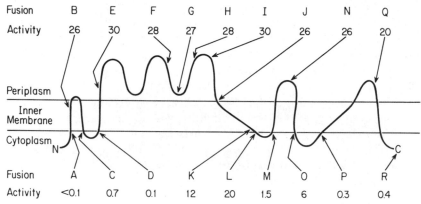

Figure 2. Alkaline phosphatase fusions to the MalF protein. The letters refer to the specific fusion at that site and in Tables II and III to the specific site of the fusion joint. The numbers represent the alkaline phosphatase assay levels for each fusion (Boyd et al., 1987).

1991). The complex also includes MalG, an integral membrane protein, and MalK, located peripherally to the cytoplasmic face of the membrane.

The sequence of MalF (Froshauer and Beckwith, 1984) allows the formulation of a topological model based on the rules outlined earlier in this paper (Fig. 2). Fusions of alkaline phosphatase to MalF were obtained using both Tn*phoA* and in vitro oligonucleotide deletion mutagenesis (Boyd et al., 1987). The properties of fusions to different positions of MalF give a topological picture consistent with the predictions from sequence alone.

However, there are anomalies in the results which, when first obtained, appeared to raise questions about the MalF model. Specifically, fusions K and L, with endpoints in cytoplasmic domains of MalF (Fig. 2), gave much higher levels of alkaline phosphatase units than were expected for cytoplasmic fusions. This finding led us to suspect that there were determinants within cytoplasmic domains that were important in assuring proper topology and these cytoplasmic determinants were missing in the K and L fusions. Based on the analysis of vonHeijne (1986) of the distribution of charged amino acids in hydrophilic domains of membrane proteins, the obvious candidates for such determinants were the basic amino acids of cytoplasmic domains (Fig. 3). Within the short hydrophilic domain in MalF that lies distal to the sites of the fusion joints of K and L are two arginines and a lysine. It was this possibility that led us to obtain the oligonucleotide mutagenesis-derived fusions in addition to those found with Tn*phoA*. A pattern emerged where, in each cytoplasmic

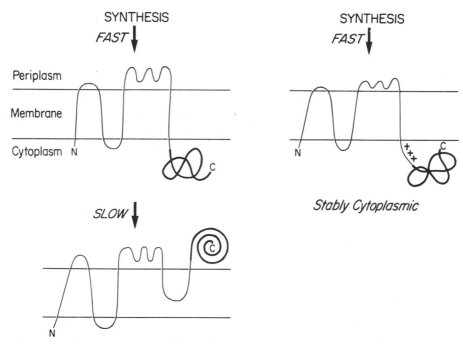

Figure 3. Membrane assembly of MalF-alkaline phosphatase fusion proteins. The fusion on the right-hand side of the figure (*M*) inserts rapidly in the membrane. The positive charges anchor alkaline phosphatase (*thick line*) stably in the cytoplasm. The fusion on the left-hand side (e.g., K or L) also inserts rapidly into the membrane, but is missing the basic amino acids. As a result, the alkaline phosphatase is not stably anchored in the cytoplasm and is slowly translocated across the membrane. This may occur either by the structure indicated or by complete translocation of the alkaline phosphatase *and* the hydrophobic membrane-spanning stretch preceding it.

domain, fusions to points where alkaline phosphatase was preceded by basic amino acids exhibited little alkaline phosphatase enzymatic activity, while those to points where the basic amino acids had been deleted exhibited considerably higher levels of activity (Fig. 2). Subsequently, we took the M fusion and mutated it so as to reduce the net positive charge (Boyd and Beckwith, 1989). In such constructs, the alkaline phosphatase activity was again higher than in the M fusion itself. These results demonstrated that the basic amino acids are important topological determinants, insuring stable anchoring of cytoplasmic domains.

These findings represent a case that violates one of our assumptions about the fusion approach. That is, the absence of the basic amino acids, which were carboxy-terminal to the fusion joint, resulted in an anomalous result that was not consistent with the overall topological picture. The contradictory results were resolved by the subsequent studies and they show, in fact, that anomalies can be exploited to deepen our understanding of how topology is achieved.

We have explored the properties of fusion strains missing basic amino acids even further. We have found that when alkaline phosphatase is fused to a periplasmic domain of MalF, it is translocated very rapidly across the membrane (Traxler et al., 1992). In contrast, in the K and L fusions, where it is fused to a cytoplasmic position, but missing the basic amino acids, it is translocated much more slowly. The ultimate amount found in the periplasm does not differ by more than a factor of two or three for all these fusions, but the kinetics of export differ dramatically. These findings suggest that ordinarily the MalF protein (and fusions of it) assemble very rapidly into their proper conformation in the membrane (Fig. 3). In the case of the MalF K and L fusions, the hybrid protein also assembles properly at first, but then the absence of basic amino acids preceding alkaline phosphatase in the cytoplasm results in an unstable arrangement with weak anchoring in the cytoplasm. Thus, the alkaline phosphatase is slowly exported across the lipid bilayer (Fig. 3).

We have further used MalF-AP fusions to assess topological signals in the MalF protein (McGovern et al., 1991). We do this by introducing deletions into the MalF portion and determining the effect on the topology of the protein. This is done both by measuring alkaline phosphatase activity and by proteolysis studies in spheroplasts, which give an indication of which parts of the protein are localized to the extracytoplasmic face of the membrane. For instance, we may delete a single membrane-spanning segment from a fusion and ask whether this causes an inversion in the topology of the protein. Extensive studies of this sort have led to the following conclusions: (*1*) the basic amino acids in cytoplasmic domains are the preeminent signals determining the orientation of membrane-spanning stretches; (*2*) periplasmic domains have no equivalent signals; (*3*) membrane-spanning segments (as others have found) have no information determining their orientation in the membrane; and (*4*) the strength of cytoplasmic domains as topogenic signals generally varies with the density of positive charges within them.

Alternative and Modified Gene Fusion Approaches to Topology

The studies with the K and L fusions to MalF and others like them show that anomalous results can be obtained in this type of analysis. We have developed new approaches to topology, particularly to avoid such anomalies, by taking advantage of our finding that alkaline phosphatase is exported much more slowly in the anomalous fusions K and L.

Fusions of a Biotinylatable Domain to Membrane Proteins

Cronan and colleagues have described a polypeptide fragment of acetylCoA carboxylase that is biotinylated in its native protein and can be biotinylated when fused to another protein (Cronan, 1990). They and we have shown that when this domain is fused to a rapidly secreted protein, very little biotinylation takes place. Apparently, this portion of the protein is exported rapidly enough so that it is not accessible to the biotinylating enzyme. However, if it is fused to a protein that is exported slowly, this domain is retained in the cytoplasm long enough to be biotinylated (Reed and Cronan, 1991; Jander, G., and J. Beckwith, unpublished results). This occurs, for example, when the rate of export of the secreted protein is slowed down by the presence of a mutation (*sec*) in the bacterial secretion machinery.

We reasoned that fusions of the biotinylatable domain to different positions in MalF should give differing degrees of biotinylation depending upon whether or for how long that domain was retained in the cytoplasm. Thus, we have constructed fusions of this domain to the same positions in MalF as alkaline phosphatase fusions such as H, I, J, K, L, and M (Jander, G., and J. Beckwith, unpublished results). We thought it possible that since the alkaline phosphatase in the K and L fusions was exported slowly, a biotinylatable domain fused at the same position would be retained in the cytoplasm long enough to be modified. This is, in fact, the case (Table II).

TABLE II
Properties of MalF Fusions to a Biotinylatable Domain

Site of fusion	Domain fused to	Biotinylation
H	Periplasmic	−
I	Periplasmic	−
J	Periplasmic	−
K	Cytoplasmic	+
L	Cytoplasmic	+
M	Cytoplasmic	+

A 79 amino acid biotinylatable domain from the *E. coli* acetylCoA carboxylase was fused to the same positions in MalF (Fig. 2) as the H-M alkaline phosphatase fusions. Cells were grown in tritiated biotin and extracts run on an SDS gel. An autoradiogram was prepared from the gel and the intensity of the radioactive bands in each of the fusion proteins observed. Fusions K, L, and M gave thick intense bands, while no radioactivity was seen in the bands corresponding to the H-J fusion proteins. The presence of all proteins in approximately equal amounts was verified on a gel using antibody to MalF.

These results, then, give a much more clear-cut picture in distinguishing fusions to cytoplasmic and periplasmic domains. Fusions to the former are biotinylated and fusions to the latter are not or only weakly so. We point out that in the process of constructing the fusions to the biotinylatable domains we introduced a basic amino acid close to the fusion joint. It is possible that this amino acid contributed to a more stable cytoplasmic location of this portion of the biotinylatable domain in fusions K and L. On the other hand, that charged amino acid did not interfere with the export of the same domain in fusions H, I, and J. We are currently removing the charged amino acid by mutagenesis to determine whether it contributed to the clear-cut pattern obtained. At any rate, this approach provides another way to look at topology and also a sensitive measure of kinetics of export.

Comparing Alkaline Phosphatase and β-Lactamase Fusions

We have also examined the properties of β-lactamase fusions to MalF at the same positions as a set of alkaline phosphatase fusions (Prinz, W., and J. Beckwith, unpublished results). Again, we find that the anomalies seen with alkaline phosphatase fusions disappear with β-lactamase fusions (Table III). That is, while alkaline phosphatase is exported slowly to the periplasm in the K and L cytoplasmic fusions, β-lactamase remains in the cytoplasm in the comparable fusions. We suspect that this difference in properties between the two types of fusions relates to the folding of the two proteins in the cytoplasm. Alkaline phosphatase does not fold properly in the cytoplasm; it is highly sensitive to protease, whereas the native form is not (Akiyama and Ito, 1989). The absence of the essential disulfide bonds may be the basis for this difference. In contrast, β-lactamase can fold into an active conformation without its disulfide bond. Thus, it may be that β-lactamase cannot be exported slowly in the fusions analogous to K and L because it has folded into a conformation that is incompatible with translocation. The relationship between translocation and protein folding is well established (Maher and Singer, 1986; Randall and Hardy, 1986; Lee et al., 1989; Nilsson and von Heijne, 1990).

If this explanation is correct, then it provides an example where the properties of the reporter protein do, in fact, interfere with the topological analysis. That is, the

TABLE III
Properties of MalF-β-Lactamase Fusions

Site of fusion	Domain fused to	MIC for ampicillin
J	Periplasmic	225
N	Periplasmic	225
Q	Periplasmic	125
K	Cytoplasmic	10
L	Cytoplasmic	10
M	Cytoplasmic	5

β-Lactamase, missing its signal sequence, was fused to MalF at the exact positions corresponding to the sites of alkaline phosphatase fusions indicated in Fig. 2. MIC stands for minimal inhibitory concentration of ampicillin necessary to kill the cells. MIC, in turn, reflects the amount of β-lactamase present in the periplasm.

failure of alkaline phosphatase to fold in the cytoplasm could be responsible for the anomalous results in fusions such as K and L. However, the β-lactamase fusions also contained a basic amino acid in the linker region, as did the fusions of the biotinylatable domain (see above). Further studies will be necessary to determine whether the folding of β-lactamase distinguishes it as a reporter protein from alkaline phosphatase.

Sandwich Fusions

The methods outlined above provide ways of avoiding the problems that might arise from certain cases of deletion of carboxy-terminal sequences in gene fusions. Another approach is to insert the reporter protein into an otherwise intact membrane protein. In this way, all sequences of the protein are present, and if they can interact with each other to allow proper assembly, such fusions could avoid some of the problems seen so far. We have tested this idea with several sandwich fusions, including ones derived from K and L (Ehrmann et al., 1990). In the sandwich fusions the anomalously high levels of alkaline phosphatase activity of K and L disappear. We suspect that this occurs because the presence of sequences both amino-terminal and carboxy-terminal to alkaline phosphatase that assemble properly in the membrane stabilizes the cytoplasmic location of the enzyme.

It is yet to be determined whether sandwich fusions can be used to resolve other questions that arise from ambiguous gene fusion results.

Constructing Fusions with the Right Endpoints

Another way to avoid the problems seen with fusions such as K and L can be gleaned from the data in Fig. 1. If one constructs fusions where the fusion joint follows the basic amino acids in the cytoplasmic domain, then there should be no problem in anchoring alkaline phosphatase in the cytoplasm if the topological model is correct. To test this approach, we have studied the topology of the MalG protein, the second integral membrane component of the maltose transport system (Boyd et al., 1993).

Inspection of the MalG sequence reveals six rather clear-cut membrane-spanning segments. Basic amino acids in the first and third hydrophilic domains suggest that they are localized to the cytoplasm. This analysis leads to the model in Fig. 4. Fusions of alkaline phosphatase to MalG were constructed with their

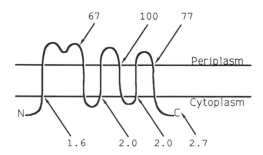

Figure 4. Alkaline phosphatase fusions to the MalG protein. The numbers represent the alkaline phosphatase activities of each fusion. The arrows point to the sites in MalG of the fusion joints. The activities are normalized to the rate of hybrid protein synthesis as measured in pulse-labeling experiments (Boyd et al., 1993).

endpoints at or near the carboxy terminus of each of the seven hydrophilic domains. The results obtained showed a very strong and consistent difference between the activities of fusions to putative cytoplasmic and periplasmic domains.

Shortcomings of Fusion Analysis and the Future

The two examples of topological analysis presented in some detail here, MalF and MalG, show proteins whose membrane-spanning segments are clearly hydrophobic and do not contain any charged amino acids. However, many membrane proteins require for their functioning charged amino acids within their membrane portions. Bacteriorhodopsin (Khorana et al., 1979; Ovchinnikov et al., 1979) and the lactose permease (LacY; Calamia and Manoil, 1992) are two examples. Membrane-spanning segments that contain charged amino acids may not be efficient export signals for alkaline phosphatase. Calamia and Manoil (1992) have shown that this is the case for LacY. A fusion of alkaline phosphatase to a domain of LacY that is known to be periplasmic from covalent labeling studies yielded relatively low levels of alkaline phosphatase. Mutational studies showed that an arginine in the membrane-spanning segment was responsible for the low efficiency of export. Again, an anomaly in the topological analysis may lead to a deepened understanding of membrane protein assembly. It may be that in this case the proper assembly of this portion of the protein requires interactions with carboxy-terminal portions of the protein. Gene fusion studies or other genetic approaches may reveal an explanation for this finding.

The examples described and referred to in this review all involve membrane proteins where the amino-terminal portion of the protein is localized to the cytoplasm. However, other proteins, such as the β-adrenergic receptor and the ProW protein of *E. coli,* have substantial amino-terminal sequences on the extracytoplasmic face of the membrane. Alkaline phosphatase fusions in the early regions of these proteins give results that are not consistent with the known topology of the proteins (LaCatena and Tocchini-Valentini, personal communication; Haardt and Bremer, personal communication). Again, the anomalies indicate unusual mechanisms to explore in the assembly of these proteins.

A final issue is a hypothetical one at this point. For those membrane proteins that act as channels, it might be expected that there would be extensive hydrophilic or amphipathic regions that are imbedded in the membrane. It seems possible that when fusion analysis is carried out on these, misleading results could be obtained. For instance, the MalF protein shows a 180 amino acid periplasmic domain by fusion analysis. Is it possible that this domain, in the intact protein, is imbedded in the membrane as a channel? In fusion strains, the disruption of the domain may result in its inability to assemble as a channel in the membrane and lead, rather, to its mislocalization to the periplasm.

Perhaps the most important lesson that has been learned from fusion studies is that the approach does work for a number of proteins. Even though the membrane protein is severely perturbed in the hybrid proteins (by deletion of a carboxy-terminal segment), the reporter molecule usually indicates the correct topology. These findings suggest that topology is determined, at least in part, by local signals, the hydrophobic stretches and the basic residues being the two known ones. Further study of cases in which fusion results conflict with results of other approaches to topology may reveal other classes of topogenic signals.

Acknowledgments

This research was supported by grants from the National Science Foundation to D. Boyd and from the National Institutes of Health to J. Beckwith. J. Beckwith is an American Cancer Society Research Professor.

References

Ahmad, M., and H. Bussey. 1988. Topology of membrane insertion *in vitro* and plasma membrane assembly *in vivo* of the yeast arginine permease. *Molecular Microbiology.* 2:627–635.

Akiyama, Y., and K. Ito. 1989. Export of *Escherichia coli* alkaline phosphatase attached to an integral membrane protein, SecY. *Journal of Biological Chemistry.* 264:437–442.

Blobel, G. 1980. Intracellular protein topogenesis. *Proceedings of the National Academy of Sciences, USA.* 77:1496–1500.

Boyd, D., and J. Beckwith. 1989. Positively charged amino acid residues can act as topogenic determinants in membrane proteins. *Proceedings of the National Academy of Sciences, USA.* 86:9446–9450.

Boyd, D., C. Manoil, and J. Beckwith. 1987. Determinants of membrane protein topology. *Proceedings of the National Academy of Sciences, USA.* 84:8525–8529.

Boyd, D., B. Traxler, and J. Beckwith. 1993. Analyzing the topology of a membrane protein using a minimum number of alkaline phosphatase fusions. *Journal of Bacteriology.* 175:553–556.

Broome-Smith, J. K., M. Tadayyon, and Y. Zhang. 1990. β-Lactamase as a probe of membrane protein assembly and protein export. *Molecular Microbiology.* 4:1637–1644.

Calamia, J., and C. Manoil. 1992. Membrane protein spanning segments as export signals. *Journal of Molecular Biology.* 224:539–543.

Chun, S., and J. S. Parkinson. 1988. Bacterial motility:membrane topology of the *Escherichia coli* MotB protein. *Science.* 239:276–278.

Collier, D. N., V. A. Bankaitas, J. B. Weiss, and P. J. Bassford, Jr. 1988. The antifolding activity of SecB promotes the export of the *E. coli* maltose-binding protein. *Cell.* 53:273–283.

Cronan, J. E., Jr. 1990. Biotination of proteins *in vivo*. A post-translational modification to label, purify, and study proteins. *Journal of Biological Chemistry.* 265:10327–10333.

Davidson, A. L., and H. Nikaido. 1991. Purification and characterization of the membrane-associated components of the maltose transport system from *Escherichia coli*. *Journal of Biological Chemistry.* 266:8946–8951.

Deisenhofer, J., O. Epp, K. Miki, R. Huber, and H. Michel. 1985. Structure of the protein subunits in the photosynthetic reaction centre of *Rhodopseudomonas viridis* at 3Å resolution. *Nature.* 318:618–624.

Derman, A. I., and J. Beckwith. 1991. *Escherichia coli* alkaline phosphatase fails to acquire disulfide bonds when retained in the cytoplasm. *Journal of Bacteriology.* 173:7719–7722.

Deshaies, R. J., and R. Schekman. 1990. Structural and functional dissection of Sec62p, a membrane-bound component of the yeast endoplasmic reticulum protein import machinery. *Molecular and Cellular Biology.* 10:6024–6035.

Duchêne, A. M., J. Patte, C. Gutierrez, and M. Chandler. 1992. A simple and efficient system for the construction of *phoA* gene fusions in Gram-negative bacteria. *Gene.* 114:103–107.

Ehrmann, M., D. Boyd, and J. Beckwith. 1990. Genetic analysis of membrane protein topology by a sandwich gene fusion approach. *Proceedings of the National Academy of Sciences, USA.* 87:7574–7578.

Froshauer, S., and J. Beckwith. 1984. The nucleotide sequence of the gene for MalF protein, an inner membrane component of the maltose transport system of *Escherichia coli*. *Journal of Biological Chemistry.* 259:10896–10903.

Froshauer, S., G. N. Green, D. Boyd, K. McGovern, and J. Beckwith. 1988. Genetic analysis of the membrane insertion and topology of MalF, a cytoplasmic membrane protein of *Escherichia coli*. *Journal of Molecular Biology.* 200:501–511.

Green, G. N., W. Hansen, and P. Walter. 1989. The use of gene-fusions to determine membrane protein topology in *Saccharomyces cerevisiae*. *Journal of Cell Science Supplement.* 11:109–113.

Henderson, R., J. M. Baldwin, T. A. Ceska, F. Zemlin, E. Beckmann, and K. H. Downing. 1990. Model for the structure of bacteriorhodopsin based on high-resolution electron cryo-microscopy. *Journal of Molecular Biology.* 213:899–929.

Hoffman, C., and A. Wright. 1985. Fusions of secreted proteins to alkaline phosphatase: an approach for studying protein secretion. *Proceedings of the National Academy of Sciences, USA.* 82:5107–5111.

Khorana, H. G., G. E. Gerber, W. C. Herlihy, C. P. Gray, R. J. Anderegg, K. Nijei, and K. Biemann. 1979. Amino acid sequence of bacteriorhodopsin. *Proceedings of the National Academy of Sciences, USA.* 76:5046–5050.

Kyte, J., and R. F. Doolittle. 1982. A simple method for displaying the hydropathic character of a protein. *Journal of Molecular Biology.* 157:105–132.

Lee, C., P. Li, H. Inouye, and J. Beckwith. 1989. Genetic studies on the inability of beta-galactosidase to be translocated across the *E. coli* cytoplasmic membrane. *Journal of Bacteriology.* 171:4609–4616.

Maher, P. A., and S. J. Singer. 1986. Disulfide bonds and the translocation of proteins across membranes. *Proceedings of the National Academy of Sciences, USA.* 83:9001–9005.

Manoil, C., and J. Beckwith. 1985. Tn*phoA:* a transposon probe for protein export signals. *Proceedings of the National Academy of Sciences, USA.* 82:8129–8133.

Manoil, C., and J. Beckwith. 1986. A genetic approach to analyzing membrane protein topology. *Science.* 233:1403–1408.

Manoil, C., J. J. Mekalanos, and J. Beckwith. 1990. Alkaline phosphatase fusions: sensors of subcellular location. *Journal of Bacteriology.* 172:515–518.

McGovern, K., M. Ehrmann, and J. Beckwith. 1991. Decoding signals for membrane protein assembly using alkaline phosphatase fusions. *EMBO Journal.* 10:2773–2782.

Michaelis, S., H. Inouye, D. Oliver, and J. Beckwith. 1983. Mutations that alter the signal sequence of alkaline phosphatase of *Escherichia coli. Journal of Bacteriology.* 154:366–374.

Nilsson, I., and G. von Heijne. 1990. Fine-tuning the topology of a polytopic membrane protein: role of positively and negatively charged amino acids. *Cell.* 62:1135–1141.

Ovchinnikov, Y. A., N. G. Abdulaev, M. Y. Feigina, A. V. Kiselev, and N. A. Lobanov. 1979. The structural basis of the functioning of bacteriorhodopsin:an overview. *FEBS Letters.* 100:219–224.

Randall, L. L., and S. J. S. Hardy. 1986. Correlation of competence for export with lack of tertiary structure of the mature species: a study *in vivo* of maltose-binding protein in *E. coli. Cell.* 46:921–928.

Reed, K. E., and J. E. Cronan, Jr. 1991. *Escherichia coli* exports previously folded and biotinated protein domains. *Journal of Biological Chemistry.* 266:11425–11428.

San Millan, J. L., D. Boyd, R. Dalbey, W. Wickner, and J. Beckwith. 1989. Use of *phoA* fusions to study the topology of the *Escherichia coli* inner membrane protein leader peptidase. *Journal of Bacteriology.* 171:5536–5541.

Senstag, C., C. Stirling, R. Schekman, and J. Rine. 1990. Genetic and biochemical evaluation of eukaryotic membrane protein topology: the multiple transmembrane domains of *S. cerevisiae* HMG-CoA reductase. *Molecular and Cellular Biology.* 10:672–680.

Shuman, H. A., and T. J. Silhavy. 1981. Identification of the *malK* gene product: a peripheral membrane component of the *E. coli* maltose transport system. *Journal of Biological Chemistry.* 256:560–562.

Sugiyama, J. E., S. Mahmoodian, and G. R. Jacobson. 1991. Membrane topology analysis of *Escherichia coli* mannitol permease by using a nested-deletion method to create *mtlA-phoA* fusions. *Proceedings of the National Academy of Sciences, USA.* 88:9603–9607.

Traxler, B., D. Boyd, and J. Beckwith. 1993. The analysis of membrane protein topology. *Journal of Membrane Biology.* In press.

Traxler, B., C. Lee, D. Boyd, and J. Beckwith. 1992. The dynamics of assembly of a cytoplasmic membrane protein in *Escherichia coli. Journal of Biological Chemistry.* 267:5339–5345.

von Heijne, G. 1986. The distribution of positively charged residues in bacterial inner membrane proteins correlates with the trans-membrane topology. *EMBO Journal.* 5:3021–3027.

von Heijne, G., and Y. Gavel. 1988. Topogenic signals in integral membrane proteins. *European Journal of Biochemistry.* 174:671–678.

Wang, H.-y., L. Lipfert, C. C. Malbon, and S. Bahouth. 1989. Site-directed anti-peptide antibodies define the topography of the β-adrenergic receptor. *Journal of Biological Chemistry.* 264:14424–14431.

Wilmes-Riesenberg, M. R., and B. L. Wanner. 1992. Tn*phoA* and Tn*phoA'* elements for making and switching fusions for study of transcription, translation, and cell surface localization. *Journal of Bacteriology.* 174:4558–4575.

Yun, C.-H., S. R. Van Doren, A. R. Crofts, and R. B. Gennis. 1991. The use of gene fusions to examine the membrane topology of the L-subunit of the photosynthetic reaction center and of the cytochrome *b* subunit of the *bc*$_1$ complex from *Rhodobacter sphaeroides. Journal of Biological Chemistry.* 266:10967–10973.

Chapter 4

The High-Resolution Three-dimensional Structure of a Membrane Protein

J. Deisenhofer

Howard Hughes Medical Institute and Department of Biochemistry, University of Texas Southwestern Medical Center, Dallas, Texas 75235-9050

Molecular Biology and Function of Carrier Proteins © 1993 by The Rockefeller University Press

The first integral membrane protein whose three-dimensional (3-D) structure could be determined at atomic resolution was the photosynthetic reaction center (RC) from the purple bacterium *Rhodopseudomonas viridis*. RCs of purple bacteria are complexes of proteins and pigments spanning the bacterial inner membrane. They catalyze the primary steps of the photosynthetic light reactions: light-driven charge separation across the membrane with a pair of bacteriochlorophylls acting as electron donor and quinones acting as acceptors.

The RC from *R. viridis* consists of the four protein subunits cytochrome (336 amino acids), M (323 amino acids), L (273 amino acids), and H (258 amino acids). The 14 cofactors associated with these protein subunits are: four bacteriochloro-phyll-b, two bacteriopheophytin-b, one menaquinone-9, one ubiquinone-9, one nonheme iron ion, one carotenoid, and the four heme groups. This complex, with a total molecular weight of $\sim 140,000$, was one of the first integral membrane proteins for which well-ordered 3-D crystals were obtained (Michel, 1982). The structure of these crystals was determined using x-ray crystallography in two steps: solution of the phase problem and building of an initial model at 3.0 Å resolution (Deisenhofer et al., 1984, 1985), followed by crystallographic refinement of this model at 2.3 Å resolution (Deisenhofer and Michel, 1989*a,b;* Deisenhofer, J., O. Epp., I. Sinning, and H. Michel, manuscript in preparation). The gene sequences of the protein subunits (Michel et al., 1985, 1986; Weyer et al., 1987) were essential for the structure analysis.

The RC complex has an elongated shape with a maximum dimension of 130 Å. Its central part consists of the subunits L and M, which, in spite of only 25% sequence homology, are folded very similarly and are arranged with an approximate twofold symmetry; the symmetry axis is perpendicular to the membrane plane. Each of these subunits has five hydrophobic helices as their membrane-spanning segments. The NH_2- and COOH-terminal polypeptide chain segments and the segments connecting the transmembrane helices are folded along the membrane surfaces. On either side of the membrane a peripheral subunit is attached to the L-M complex: the cyto-chrome on the periplasmic side, and the H subunit on the cytoplasmic side. The H subunit has a single membrane-spanning helix near its NH_2 terminus. An overall view of the model of the RC from *R. viridis* is shown in Fig. 1.

The cofactors located in the membrane-spanning region of the RC are arranged in two branches originating from a pair of bacteriochlorophylls and leading across the membrane to quinone binding sites. Only one of these branches is used for the initial charge transfer steps after the absorption of light energy. This strong prefer-ence for a unique electron pathway is in contrast to geometric symmetry between the two branches. Functional asymmetry within a molecular complex with significant geometric symmetry is also observed in the cytochrome subunit of the RC. Structural and functional considerations support the hypothesis that the L-M complex and its bound cofactors constitute a good model for the core of photosystem II from green plants and cyanobacteria (Michel and Deisenhofer, 1988).

The membrane-spanning region of the RC complex can be easily recognized by unusual structural features. Analysis of the molecular surface in the central region of the model reveals a hydrophobic zone 30 Å thick with sharp boundaries to the peripheral parts of the complex, which have a much more polar surface typical of proteins soluble in aqueous solutions. In the crystal this hydrophobic surface is covered by detergent (Roth et al., 1989). The membrane-spanning region of the RC is also completely free of charged residues and contains only a few tightly bound water molecules. Both charged residues and tightly bound waters are abundant in the peripheral regions of the RC.

The RC from the purple bacterium *Rhodobacter sphaeroides* lacks the bound four-heme cytochrome; its subunits M, L, and H form a complex very similar to the corresponding part of the RC from *R. viridis*. Several groups of investigators

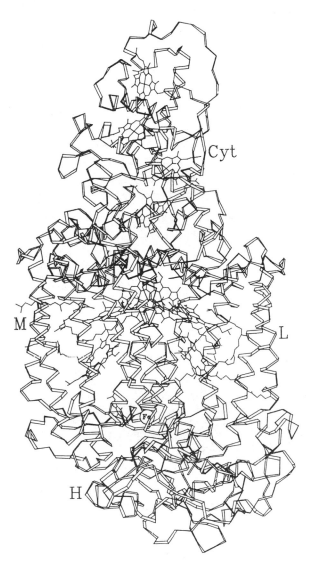

Figure 1. Schematic drawing of the photosynthetic reaction center from *R. viridis*. The protein subunits are represented as ribbons tracing the course of the polypeptide backbone; the cofactors are drawn with thin lines.

crystallized the RC from *Rb. sphaeroides* and solved the structure by molecular replacement, using the atomic coordinates of the *R. viridis* RC as a search model (Allen et al., 1986; Chang et al., 1986). More references to the crystallographic work on RCs from *Rb. sphaeroides* can be found, for example, in a recent review (Deisenhofer and Michel, 1991).

Recently, the 3-D structures of several other membrane proteins have been determined. The structure analysis of porin from *Rb. capsulatus,* based on x-ray diffraction data to 1.8 Å resolution, shows a molecular architecture consisting of 16 antiparallel strands of β-sheet per monomer (Weiss et al., 1991*a,b*). The same architecture was also observed for porin from *Escherichia coli* (Cowan et al., 1992). Electron crystallography of two-dimensional crystals of purple membrane was driven to high resolution and resulted in an atomic model of bacteriorhodopsin (Henderson et al., 1990). In studies at medium resolution (~6 Å) the structure of a plant light harvesting complex was determined by electron crystallography using two-dimensional crystals (Kühlbrandt and Wang, 1991).

References

Allen, J. P., G. Feher, T. O. Yeates, D. C. Rees, J. Deisenhofer, H. Michel, and R. Huber. 1986. Structural homology of reaction centers from *Rhodopseudomonas sphaeroides* and *Rhodopseudomonas viridis* as determined by x-ray diffraction. *Proceedings of the National Academy of Sciences, USA.* 83:8589–8593.

Chang, C.-H., D. M. Tiede, J. Tang, U. Smith, J. R. Norris, and M. Schiffer. 1986. Structure of *Rhodopseudomonas sphaeroides* R-26 reaction center. *FEBS Letters.* 205:82–86.

Cowan, S. W., T. Schirmer, G. Rummel, M. Steiert, R. Ghosh, R. A. Paupit, J. N. Jansonius, and J. P. Rosenbusch. 1992. Crystal structures explain functional properties of two *E. coli* porins. *Nature.* 358:727–733.

Deisenhofer, J., O. Epp, K. Miki, R. Huber, and H. Michel. 1984. X-ray structure analysis of a membrane protein complex: electron density map at 3 A resolution and a model of the chromophores of the photosynthetic reaction center from *Rhodopseudomonas viridis. Journal of Molecular Biology.* 180:385–398.

Deisenhofer, J., O. Epp, K. Miki, R. Huber, and H. Michel. 1985. Structure of the protein subunits in the photosynthetic reaction centre of *Rhodopseudomonas viridis* at 3 A resolution. *Nature.* 318:618–624.

Deisenhofer, J., and H. Michel. 1989*a.* The photosynthetic reaction centre from the purple bacterium *Rhodopseudomonas viridis. In* Les Prix Nobel 1988. T. Frangsmyr, editor. Nobel Foundation, Stockholm. 134–188.

Deisenhofer, J., and H. Michel. 1989*b.* The photosynthetic reaction centre from the purple bacterium *Rhodopseudomonas viridis. EMBO Journal.* 8:2149–2170.

Deisenhofer, J., and H. Michel. 1991. Structures of bacterial photosynthetic reaction centers. *Annual Review of Cell Biology.* 7:1–23.

Henderson, R., J. M. Baldwin, T. A. Ceska, F. Zemlin, E. Beckmann, and K. H. Downing. 1990. Model for the structure of bacteriorhodopsin based on high-resolution electron cryo-microscopy. *Journal of Molecular Biology.* 213:899–929.

Kühlbrandt, W., and D. N. Wang. 1991. Three-dimensional structure of plant light-harvesting complex determined by electron crystallography. *Nature.* 350:130–134.

Michel, H. 1982. Three-dimensional crystals of a membrane protein complex: the photosynthetic reaction centre from *Rhodopseudomonas viridis. Journal of Molecular Biology.* 158:567–572.

Michel, H., and J. Deisenhofer. 1988. Relevance of the photosynthetic reaction center from purple bacteria to the structure of photosystem II. *Biochemistry.* 27:1–7.

Michel, H., K. A. Weyer, H. Gruenberg, I. Dunger, D. Oesterhelt, and F. Lottspeich. 1986. The 'light' and 'medium' subunits of the photosynthetic reaction centre from *Rhodopseudomonas viridis:* isolation of the genes, nucleotide and amino acid sequence. *EMBO Journal.* 5:1149–1158.

Michel, H., K. A. Weyer, H. Gruenberg, and F. Lottspeich. 1985. The 'heavy' subunit of the photosynthetic reaction centre from *Rhodopseudomonas viridis:* isolation of the gene, nucleotide and amino acid sequence. *EMBO Journal.* 4:1667–1672.

Roth, M., A. Lewit-Bentley, H. Michel, J. Deisenhofer, R. Huber, and D. Oesterhelt. 1989. Detergent structure in crystals of a bacterial photosynthetic reaction centre. *Nature.* 340:659–662.

Weiss, M. S., U. Abele, J. Weckesser, W. Welte, E. Schiltz, and G. E. Schulz. 1991a. Molecular architecture and electrostatic properties of a bacterial porin. *Science.* 254:1627–1630.

Weiss, M. S., A. Kreusch, E. Schiltz, U. Nestel, W. Welte, J. Weckesser, and G. E. Schulz. 1991b. The structure of porin from *Rhodobacter capsulatus* at 1.8 Å resolution. *FEBS Letters.* 280:379–382.

Weyer, K. A., F. Lottspeich, H. Gruenberg, F. S. Lang, D. Oesterhelt, and H. Michel. 1987. Amino acid sequence of the cytochrome subunit of the photosynthetic reaction centre from the purple bacterium *Rhodopseudomonas viridis. EMBO Journal.* 6:2197–2202.

Chapter 5

Insights into Membrane Protein Folding and Translocation from the Behavior of Bacterial Toxins: Models for Membrane Translocation

Erwin London, Nancy D. Ulbrandt, Domenico Tortorella, Jean Xin Jiang, and Franklin S. Abrams

Department of Biochemistry and Cell Biology, State University of New York at Stony Brook, Stony Brook, New York 11794-5215

Molecular Biology and Function of Carrier Proteins © 1993 by The Rockefeller University Press

Introduction

Many bacterial toxins cross membranes in order to attack targets in the cytoplasm. Toxins face unique constraints as they must be able to take on both hydrophilic and membrane-inserting conformations and switch back and forth between these conformations at the appropriate times. How they accomplish this has important implications for how proteins insert into membranes. Bacterial toxins are also important models for protein translocation as they catalyze efficient self-translocation, and they represent much less complicated systems than the multiprotein complexes that catalyze translocation of ordinary proteins.

Studies of toxin behavior have begun to yield important clues as to how they accomplish such tasks. This report summarizes the recent studies that have shown that unfolding and refolding events play a critical role in the behavior of diphtheria toxin and *Pseudomonas* exotoxin A. Partial unfolding events, in which these toxins take on molten globule-like conformations, are closely linked to their membrane insertion. The mechanism by which these toxins translocate across membranes is also discussed. In addition, based on their behavior, a speculative model for the translocation of cholera toxin is proposed. Finally, new fluorescence methods that may allow us to refine our knowledge of the structure of membrane-inserted toxins and other membrane proteins are described.

Diphtheria Toxin Structure and Function

Diphtheria toxin (M_r 58,348) is secreted by *Corynebacterium diphtheriae*. It can be readily cleaved by proteolysis into A and B chains that remain joined by a single disulfide bond. Its native structure has been solved by x-ray crystallography (Choe et al., 1992), which has revealed that it is a three-domain protein.

The A chain is the NH$_2$-terminal third of the protein and is the catalytic (C) domain. It acts to inhibit protein synthesis by inactivation of elongation factor 2 (EF-2). This inactivation is due to ADP ribosylation of diphthamide, a modified His residue unique to EF-2. It has also been proposed that the toxin has a nuclease activity (Chang et al., 1989), but this idea is very controversial (Johnson, 1990; Wilson et al., 1990). The crystal structure shows the A chain to form a globular domain containing α-helices and β-sheets, with the sheets bordering its active site cleft.

The B chain comprises the remaining two-thirds of the protein. It is composed of two globular domains. Its NH$_2$-terminal T domain plays a critical role in the membrane insertion of the toxin. This domain is an α-helical region (Choe et al., 1992) that contains several long hydrophobic sequences that tend to be localized within hydrophobic helices. These sequences remain largely buried until the conversion of the B chain to a hydrophobic state (see below).

The remainder of the B chain, the COOH-terminal receptor binding (R) domain, contains, as its name implies, the receptor binding region. It is a β-sheet domain with a tertiary structure somewhat similar to that of an immunoglobulin (Choe et al., 1992). The receptor itself appears to be a specific small plasma membrane protein complexed with a second membrane protein (Iwamoto et al., 1991). The receptor cDNA encodes a protein of 185 residues with a single transmembrane domain and an extracellular domain corresponding to a heparin-binding epidermal growth factor (Naglich et al., 1992).

The possibility that the B chain has additional biological activities has not been ruled out (Kagan, 1991).

The Role of Endocytosis and Low pH in Diphtheria Toxin Entry

After the toxin binds to its receptor, it undergoes receptor-mediated endocytosis and enters acidic vacuoles called endosomes. Genetic, cell biological, and biochemical studies have shown that the exposure of the toxin to the acidic lumen of the endosome triggers its membrane penetration (see references in London, 1992*a*). Given the pH at which the toxin inserts into membranes (pH 5–5.5), late endosomes appear to be the location of membrane translocation, although one group has proposed that toxin entry might only be completed after recycling of endocytic vesicles to the cell surface (Hudson et al., 1988).

It is presumed that the receptor functions to carry the toxin to the endosomes. It is not yet clear whether it plays any additional role in translocation, or whether there are any endosomal proteins that might aid in the translocation process.

The Low pH-induced Conformational Change and Membrane Penetration Step: Molten Globule-like Behavior of Diphtheria Toxin

Exposure to low pH renders diphtheria toxin hydrophobic, as shown by the observation that it gains the abilities to bind to micelles of mild nonionic detergents and to insert into both model and biological membranes (see references in London, 1992*a*). This hydrophobic behavior is linked to major low pH–induced changes in its folding, which include exposure of previously buried proteolysis sites in the T domain (Dumont and Richards, 1988) and Trp residues (Blewitt et al., 1985). It appears that the low pH–induced exposure of the hydrophobic regions within the B chain, resulting from these conformational changes, is the central event triggering membrane insertion.

The conformational change at low pH can be best summarized as a partial unfolding event in which tertiary structure is disrupted (Blewitt et al., 1985; Zhao and London, 1986; Ramsay et al., 1989), but in which secondary structure, as shown by circular dichroism, remains largely intact (Hu and Holmes, 1984; Blewitt et al., 1985). This is supported by the similarity of toxin behavior at low pH to that under denaturing conditions such as high temperature and high pH (Zhao and London, 1986; Kieleczawa et al., 1990). The conformation of the toxin at low pH appears to be equivalent or very similar to the partly unfolded molten globule conformation. This conformation has now been recognized to exist for many proteins.

Low pH–induced partial unfolding probably involves alterations of electrostatic forces due to protonation of His, Asp, and Glu residues. Protonation could result in formation of buried charges, breakdown of salt bridges, and the strengthening of repulsions due to increased local or global positive charge, all of which favor unfolding. Local increases in hydrophobicity due to the loss of charge on anionic residues at low pH are likely to be another important factor in toxin hydrophobicity (Kieleczawa et al., 1990). The crystal structure of the toxin suggests that the protonation of the acidic residues at the tips of pairs of hydrophobic helices in the T domain renders these tips sufficiently hydrophobic to initiate membrane insertion of the helices (Choe et al., 1992).

Low pH can also induce changes within the isolated A chain that are remarkably similar to those involving the B chain seen in whole toxin. As in whole toxin, A chain unfolding is linked to an increase in its hydrophobicity and its ability to membrane insert (Zhao and London, 1988). The ability of the isolated A chain to insert is surprising, since unlike the B chain it has no sequences that appear to be hydrophobic. Furthermore, A chain unfolding may be quite different than that of the B chain, involving opening of a hinge that bisects its structural lobes (Choe et al., 1992). Another important difference between A chain and whole toxin behavior is that the changes within the A chain are readily reversible. This suggests that *transient* unfolding and insertion could have an important role in its translocation.

It is noteworthy that it is the low pH triggering of the change in toxin behavior that makes it such a useful model. It allows experimental control of the timing of membrane penetration using in vitro systems, and allows the dissection of the conformational change triggered by low pH from membrane interaction.

The Structure of Membrane-inserted Toxin

Membrane-inserted diphtheria toxin can be in one of several conformations. At lower temperatures the membrane-inserted toxin takes on conformations in which the A chain is folded, whereas at higher temperatures conformations in which the A chain is partially unfolded are observed (Jiang et al., 1991*a*). In addition, there is a change in toxin conformation after toxin that has inserted into model membranes at low pH is returned to neutral pH. After pH neutralization the toxin remains membrane bound, but is inserted into the membrane to a much lesser degree (Montecucco et al., 1985; Jiang et al., 1991*a*). Determination of the detailed structures of these different conformations would go a long way toward explaining the translocation process.

The crystal structure of diphtheria toxin provides important clues to the structure of its membrane-inserted conformations. Since the secondary structure of the native toxin appears to be similar (although not necessarily identical) to that at low pH, it is possible to identify several hydrophobic helices in the T domain that are likely to take on a transmembraneous orientation upon insertion (Choe et al., 1992). On the other hand, it appears that hydrophobic transmembrane α-helices cannot represent the whole story of toxin membrane insertion. The diphtheria toxin A chain and *Pseudomonas* exotoxin (see below) sequences do not show the presence of hydrophobic helices and yet exhibit hydrophobic behavior. What structure do such molecules have when membrane inserted?

One possibility for the structure of the inserted A chain is that some of its sequences form transmembrane β-sheets. Transmembrane β-sheets have been shown to exist in porins (Weiss et al., 1991), but may have often eluded detection in other cases up to now because they are very hard to detect from sequence data. This is true because transmembrane β-sheets are shorter than transmembrane helices (10 vs. 20 residues) and tend to have a pattern of alternating hydrophobic and hydrophilic residues that is easily obscured by one or two residues that break up the pattern. Inspection of the sequences encompassed by the β-sheets of the A chain show several that are long enough to be transmembranous and that contain hydrophobic residues in alternating positions. In particular, such a pattern overlaps β-sheets at positions 75–85, 88–96, 132–140, and 159–168. Therefore, diphtheria toxin could form a hybrid structure with both transmembrane T domain α-helices and A chain β-sheets. In this regard, it is interesting that the opening of the A domain hinge mentioned above could result in the formation of an even longer extended sheet structure that could play a role in translocation (Choe et al., 1992).

Another possibility is that other helices present in the toxin are more hydrophobic than believed, and come into contact with the lipid bilayer. This is possible because hydropathy scales that predict transmembrane helices are generally not adjusted for pH. At low pH, protonation of Asp and Glu residues renders sequences containing them more hydrophobic. In agreement with this idea, the unfolding of diphtheria toxin at low pH does result in more hydrophobic behavior than unfolding at high temperature or at high pH (Kieleczawa et al., 1990). Furthermore, as noted above, the anionic residues adjacent to the most hydrophobic sequences in diphtheria toxin have suggested that Asp and Glu protonation could strongly modulate hydrophobicity. Nevertheless, adjustment of hydropathy scales for pH does not suggest that additional hydrophobic transmembrane helices exist at low pH (unpublished observations).

Alternatively, a toxin could insert with some transmembrane helices less hydrophobic than those found in ordinary membrane proteins. Since the maintenance of bilayer integrity upon toxin insertion may not be important, it is possible that the interactions of toxin and lipid involve some contact of polar amino acid residues with hydrophobic acyl chains. A related possibility is that the toxin forms oligomers in membranes. In oligomers many transmembrane helices could be involved in internal protein–protein interactions and so would not have to be very hydrophobic (Papini et al., 1987). In the same way, contacts of toxin helices with a cellular transmembrane protein could substitute for contact with lipid (Stenmark et al., 1988).

A final possibility is that some new type of structure might be present. For example, there could be a hydrophobic surface formed by the spatial clustering of residues that are not contiguous in the primary sequence, analogous to the hydrophobic core of soluble proteins. Under conditions of partial unfolding the hydrophobic core of a toxin might remain largely intact and thus form a lipid interacting site at the tertiary structure level (Jiang and London, 1990).

The Translocation of Diphtheria Toxin across Model Membranes

The analysis of the effects of low pH on model membrane-inserted toxin has led to the proposal of various mechanisms for the translocation of diphtheria toxin. Observations showing that whole toxin and isolated B chain can both induce pore formation in membranes at low pH have raised the possibility that the A chain passes into the cytoplasm via this pore, or that the pore is formed by a discarded hydrophobic B chain wrapper which acts to shield the membrane-inserted A chain from contact with the bilayer (Kagan et al., 1981; Misler, 1984). However, it is still not clear whether the pore formed by the toxin is large enough to allow translocation of the A chain in vivo, as different pore sizes are found under different conditions. This may occur because toxin oligomers of various sizes form, with larger oligomers forming larger pores, as seen in the complement system (Malinski and Nelsestuen, 1989).

Studies demonstrating that the A chain will insert into lipid bilayers (Hu and

Holmes, 1984; Zalman and Wisnieski, 1984) even when isolated from the B chain (Montecucco et al., 1985; Zhao and London, 1988) have led to the proposal that there is transmembrane insertion of the A chain. In the latest version of such models a partially surrounding B chain acts to limit the degree of A chain contact with the lipid bilayer (Papini et al., 1987; Zhao and London, 1988). The B chain may also serve to properly orient the inserted A chain.

Another possible mechanism for translocation is that the toxin, which can destabilize and fuse bilayers, may escape endosomes by inducing membrane lysis. This seems unlikely in view of the limited size of the lesions formed by toxin in model membrane vesicles (Jiang et al., 1991*b*).

Translocation can be studied directly in model membrane vesicles. Exposure of toxin trapped within the lumen of such vesicles to low pH induces the insertion of the trapped toxin into the membrane and its exposure to the external solution (Gonzalez and Wisnieski, 1988; Jiang et al., 1991*b*). Interestingly, the degree of exposure appears to be greater for the high temperature conformation in which fragment A is unfolded (Jiang et al., 1991*b*). Subsequent reduction of the disulfide linking the A and B chains completes translocation, resulting in release of the A chain into the external solution (Jiang et al., 1991*b*).

Unlike exposure to the external face of the vesicles, the degree of A chain release is similar for toxin in its low and high temperature states (Jiang et al., 1991*b*). The explanation of the different dependencies of exposure and release upon conformation probably arises from the fact that some of the 100–250 toxin molecules trapped in each vesicle in these experiments were found to form nonspecific, protein-sized pores (Jiang et al., 1991*b*). A single pore in a vesicle could be sufficient to allow nonspecific release of many A chains that would otherwise not penetrate the membrane. This does not resolve whether a single toxin molecule trapped within a vesicle would form such nonspecific pores or translocate the A chain by a more specific process. Until this and the number of toxin molecules within endosomes under physiological conditions are more directly determined, it will be difficult to specify the relative roles of pore formation and A chain unfolding in translocation.

Pseudomonas Exotoxin A Structure and Function

Pseudomonas exotoxin A is a toxin that exhibits a number of fascinating similarities to and differences from diphtheria toxin. Like diphtheria toxin, the exotoxin catalyses the ADP ribosylation of EF-2. Crystallographic studies show the exotoxin to be a three-domain protein like diphtheria toxin (Allured et al., 1986). The catalytic domain (III) of the exotoxin is a globular region very similar in secondary and tertiary structure to the A chain of diphtheria toxin. It also shares significant sequence homology with the A chain around a cleft believed to form the active site (Carroll and Collier, 1988; Zhao and London, 1988). However, in the exotoxin the catalytic domain is the COOH-terminal domain rather than the NH_2-terminal domain as in diphtheria toxin. The NH_2-terminal domain (I) of the exotoxin is a β-sheet region that appears to be involved in receptor binding. The central domain (II) is an α-helical domain believed to be involved in membrane translocation, although it contains no strikingly hydrophobic sequences. Experimentally, all three domains of the exotoxin have been found to exhibit hydrophobic behavior under some conditions (Idziorek et al., 1990).

There appears to be a major proteolytic processing site in the exotoxin. However, unlike diphtheria toxin, it occurs not between domains but rather near the middle of the domain (II) (Ogata et al., 1990), which suggests division of the toxin into two disulfide-linked regions more reminiscent of other A-B toxins (see below). The idea of some sort of fundamental division at this point is reinforced by the switch in the relative frequency of different basic residues on either side of the proteolysis site. This type of switch in frequency has been correlated with subunit boundaries for several A-B type toxins (London and Luongo, 1989). Such behavior implies that there may be important rearrangements in the domain structure of the exotoxin at some point in the entry process.

Receptor-mediated Endocytosis of Exotoxin

The entry of the exotoxin into cells also shows parallels to and divergences from that of diphtheria toxin. The apparent receptor for the exotoxin is the α_2-macroglobulin receptor, a large plasma membrane protein (M_r 515,000; Kounnas et al., 1992). Like diphtheria toxin, the surface-bound exotoxin enters endocytic vacuoles via receptor-mediated endocytosis.

However, exotoxin may ultimately gain entry to a different set of organelles than diphtheria toxin. Studies have indicated that the exotoxin is transported at least to the Golgi (Morris, 1990). Furthermore, there is an increasing body of evidence that the COOH-terminal REDLG sequence of the exotoxin plays an important role in its translocation by targeting it to endoplasmic reticulum (ER). KDEL is the COOH-terminal targeting sequence recognized by a protein system that prevents escape of ER lumen proteins to the Golgi by a recovery process. The KDEL-like exotoxin COOH-terminus may be recognized by this system. It has been shown that except for the terminal Gly residue the REDLG sequence of the exotoxin is necessary for toxicity, and that substitution of KDEL for REDLG even enhances exotoxin toxicity (Chaudhary et al., 1990; Seetharam et al., 1991). One possibility is that this sequence helps the exotoxin reach the ER, where membrane translocation may take place.

The Role of Low pH and Other Factors in Exotoxin Membrane Insertion and Translocation

Cells with a defect in acidification are resistant to exotoxin (Moehring and Moehring, 1983), suggesting that, as in the case of diphtheria toxin, acidification plays an important role in exotoxin membrane translocation. However, it is not clear just what role low pH has in exotoxin entry. One possibility is that low pH induces membrane penetration, and in vitro it is possible to show that exotoxin becomes hydrophobic and can penetrate model membranes at low pH (Zalman and Wisnieski, 1985; Farabakhsh et al., 1986; Farabakhsh and Wisnieski, 1989; Jiang and London, 1990). As in the case of diphtheria toxin, this hydrophobic behavior is dependent on a low pH–induced partial unfolding event that results in the acquisition of molten globule-like behavior (Jiang and London, 1990; Tortorella, D., and E. London, unpublished observations).

However, it appears that the influence of low pH is only part of the story of exotoxin translocation. Isolated exotoxin shows two conformational transitions at low pH, and the first, near the endosomal pH of 5.5, does not induce the exotoxin to become very hydrophobic. Strong hydrophobicity is observed only after the second transition, which occurs below physiological pH, close to pH 4. This in itself does not rule out induction of hydrophobic behavior by low pH in vivo, as other factors, such as the presence of model membranes with a significant fraction of acidic lipids, can shift the hydrophilic-to-hydrophobic transition up to pH 5 (Jiang and London, 1990). However, it is suggestive of a process that is more complicated than a simple pH-dependent mechanism.

One additional process that may influence exotoxin behavior is proteolysis. The processing at the critical proteolytic site in the central domain of the exotoxin noted above appears to be necessary for exotoxin toxicity (Ogata et al., 1990). It is possible that the main effect of low pH is to increase the sensitivity of the exotoxin to proteolysis, as is found in vitro (Jiang and London, 1990). It is also possible that low pH and proteolysis combine to destabilize the native structure of the exotoxin and allow membrane insertion. Another possibility is that the exposure of exotoxin transported to the lumen of the ER to protein disulfide isomerase may result in the cleavage of stabilizing disulfide bonds within the exotoxin. A combination of low pH exposure, proteolysis, and disulfide reduction could then all combine in vivo to induce unfolding and insertion.

Analogous Role of Unfolding in the Action of Viral Membrane Penetration Proteins and Other Toxins

Partial unfolding events induced at low pH seem to play a role in more than just the cases of diphtheria toxin and *Pseudomonas* exotoxin A. For example, in the entry of viruses such as influenza virus there are striking parallels to diphtheria toxin behavior. In influenza virus entry the virus undergoes receptor-mediated endocytosis and enters endosomes. In the endosomes, a critical low pH–triggered conformational change is believed to occur in the viral hemagglutinin protein, resulting in fusion of the viral and endosomal membranes and release of the viral nucleocapsid into the cytoplasm. In this conformational change, low pH shifts a hydrophobic fusion sequence in the hemagglutinin protein from a buried to an exposed location, allowing its insertion into the endosomal membrane.

That this change is due to some limited unfolding event similar to those in diphtheria toxin is suggested by the observation that fusion can also be induced by thermal unfolding of the hemagglutinin (Wharton et al., 1986). Furthermore, using a series of mutant hemagglutinins it was found that mutants that form the fusion-competent conformation at a higher pH than the wild-type molecule also show decreased thermal stability at pH 7 relative to wild type (Ruigrok et al., 1986). In other words, these mutations simultaneously weaken the native structure toward low pH and high temperature effects. This implies that low pH and thermally induced conformational changes within the hemagglutinin must be closely related.

It should also be noted that diphtheria toxin and exotoxin A are not the only bacterial toxins in which low pH–induced partial unfolding is closely linked to membrane insertion. Similar behavior has recently been observed for colicins (Merrill et al., 1990; van der Goot et al., 1991; Benedetti et al., 1992).

Two key ideas coming from diphtheria toxin behavior that may be generally applicable are that the translocating polypeptide may be at least partly unfolded during the translocation process, and that there will be major rearrangements of the native tertiary structure throughout the protein due to its membrane insertion. Building upon these ideas, it is possible to propose a novel mechanism for translocation for cholera toxin (CT) and its close relative, the heat-labile toxin of *Escherichia coli* (LT).

Structure and Action of Cholera Toxin and Heat-labile Toxin

Both CT and LT are composed of A and B subunits. The crystal structure of LT shows that these toxins contain five B subunits forming a cylindrical ring, with a central "pore" filled by the A2 portion of the A subunit. The outside of the B subunits form a ring basically composed of β-sheets. The wall around the inner central pore is formed by α-helices of the B chains. The extended A2 chain is attached to a globular A1 domain which lies outside the B subunit ring (Sixma et al., 1991). The A subunit does not seem to possess classical transmembrane hydrophobic α-helices.

The B subunit binds this toxin to plasma membranes via interaction with ganglioside GM_1. The catalytic A subunit acts by ADP ribosylation of G_s proteins, which are located on the cytoplasmic side of membranes (Fishman, 1990).

Entry of CT and LT into Cells

The simplest models propose that CT and LT directly penetrate the plasma membrane to reach their targets (Fishman, 1990). Some studies of CT interaction with ganglioside-containing model membranes have suggested that the B subunits remain at the surface of the membrane or only insert slightly, suggesting that the A subunit penetrates the membrane by itself (Ribi et al., 1988; Fishman, 1990). The alternate suggestion that the unfolded A subunit would pass through a pore formed by membrane-inserted B chains (Gill, 1976) is not supported by such studies.

However, there is growing evidence that the translocation process is more complex than simple penetration of the A subunit directly into the plasma membrane. CT action shows a lag time often associated with a dependence of translocation upon endocytosis, and may well require transcytosis in order to act (Lencer et al., 1992). In addition, the cytoplasmic ADP ribosylation factor (ARF) protein that aids CT-catalyzed ADP ribosylation has been found to be concentrated at the Golgi (Serafini et al., 1991), not the plasma membrane. Furthermore, ARF is believed to play a role in ER–Golgi trafficking (Balch et al., 1992). Along these lines, it is fascinating that the A subunits have at their COOH terminus the KDEL signal for transport of lumenal proteins to the ER (Chaudhary et al., 1990). Given the important role a KDEL-like sequence has in exotoxin action, it seems reasonable that it will be important for CT and LT as well. Combining all these hints, it seems likely that CT and LT translocation is a complex process.

By analogy to other toxins, one would predict that an additional physiological factor might trigger a major change in CT and LT conformation. This seems all the more likely as the native structure of the A subunit does not have a structure that would obviously allow it to penetrate membranes. What is the physiological trigger? Low pH is one possibility, but despite observations showing hydrophobicity changes at low pH (De Wolf et al., 1987), a role for low pH in vivo is not certain. The presence of the KDEL sequence suggests parallels to exotoxin A behavior, and brings up the possibilities of proteolysis and disulfide bond reduction. However, nothing definite can be concluded in this regard at present.

A Model for the Conformation of the Inserted State and Translocation

There are several features in heat-labile toxin crystal structure that are suggestive of how A subunit translocation might occur. The ring of B subunits, thought to sit on the membrane surface upon GM_1 binding, has a smooth cylindrical shape and a β-sheet outer wall roughly resembling the structure of membrane-inserted porin (Weiss et al., 1991). The length of the B subunit cylinder is sufficient to span at least most of the hydrophobic core of a membrane if it were to insert. In addition, the spacial relationship of the B subunits to one another shows that they form a structure that looks like an iris (London, 1992b). A sliding action between B subunits could result in the opening of a large pore (London, 1992b). The unusual orientation of the side chains of Arg and Lys residues that line the interior wall formed by the central α-helices of the B subunits is also suggestive of a structure that might open. In the native state they are all pressed against the helices, lying parallel to the helix axis (Sixma et al., 1991), and are reminiscent of the ribs of a closed umbrella. Therefore, it is tempting to propose that these toxins undergo a major conformational change, linked to B subunit insertion, which results in the opening of the iris and passage of the A subunit through the pore thus formed.

There is one especially intriguing feature of this model. The interior of the open iris would have five clefts, each bounded by a β-sheet outer wall and by two α-helical lateral walls (London, 1992b). This corresponds to a structure that is known to bind peptide chains in an unfolded conformation, the cleft of the human lymphocyte antigen (HLA) protein (Bjorkman et al., 1987; Madden et al., 1991). Combined with the fact that the HLA cleft structure is already about the length that would span the hydrophobic core of a membrane, this suggests that a modification of this type of cleft could be adapted to protein translocation. In fact, it has been proposed that heat shock proteins, relatives of which play critical roles in protein translocation, may contain a fold very similar to that of HLA (Rippmann et al., 1991). Perhaps in CT and LT much of the A subunit polypeptide interacts with the analogous B subunit clefts in an unfolded form. The binding and release of the A subunit sequences from such clefts could be controlled by the regulation of the exact distance between B subunits. In this regard, it is noteworthy that HLA peptide binding can be regulated by pH, with the rate of peptide association and dissociation being much faster at low pH values corresponding to those in the endosomal lumen (Reay et al., 1992).

This model raises fascinating questions. Does translocation of ordinary cellular proteins involve a mechanism similar to that of CT and LT? If they all have a similar fold, what is the evolutionary relationship between HLA, heat shock proteins, and CT? However, it is certainly premature to consider the model as anything other than a spur to further experimentation. It does not solve the problem of understanding the entry pathway for CT and LT. And although it is interesting that a similar B subunit structure is seen with verotoxin-1, it is only large enough to span half a membrane (Stein et al., 1992). Could there be a dimer of verotoxin rings involved, or does verotoxin B subunit have a lesser role in translocation? Or is the model wrong to the extent that conformational changes occur that do not involve membrane insertion?

Parallels between Toxin Behavior and the Translocation of Ordinary Cellular Proteins

There is increasing evidence that the folding and translocation behavior of protein toxins is mirrored in the translocation of ordinary cellular proteins. The well-established idea that translocation of an ordinary cellular protein requires it to be held in a partly unfolded "translocation-competent" state (Eilers and Schatz, 1988) parallels the idea that the A chain of diphtheria toxin unfolds in order to be translocated. Very recently, we have found that there are also parallels between the membrane insertion of the translocation-promoting B chain of diphtheria toxin and cellular proteins that promote protein translocation.

In these experiments we have been examining the behavior of the SecA protein of *E. coli* in collaboration with the group of Dr. Donald B. Oliver (Wesleyan University, Middletown, CT). SecA is required for protein translocation in *E. coli*. It interacts with precursor proteins and with integral membrane proteins that are required for translocation, SecY and SecE. It also has an ATPase activity that is likely to be important in translocation (Oliver et al., 1990). Therefore, it is one of the central players in the translocation process. The observation that in *E. coli* SecA exists in both soluble and membrane-bound forms (Cabelli et al., 1991) prompted comparison to diphtheria toxin. Photolabeling and fluorescence revealed that partial unfolding induced insertion of SecA into the hydrocarbon region of model membrane vesicles and vice versa (Ulbrandt et al., 1992). Therefore, as for diphtheria toxin, the membrane insertion of SecA is closely linked to a partial unfolding event. Since the secondary structure of SecA remains largely intact during these changes, it appears that the conversion of some domain of SecA to a molten globule-like conformation is involved in insertion.

Based on the results with SecA, it seems likely that additional studies of toxin translocation will provide further clues to the mechanisms behind protein translocation in general.

Determining Toxin Structure at High Resolution: Use of Fluorescence Quenching

Up to now our knowledge of the structure of membrane-inserted toxins has come from indirect sources. To fully understand the entry of toxins into cells it will be necessary to more fully characterize their topography and tertiary structure when in the membrane-inserted state. This is a problem that can be attacked by a variety of methods including proteolytic, chemical labeling, and immunochemical methods. Our group has been developing a new fluorescence quenching approach which in combination with site-directed mutagenesis may reveal the structure of membrane-inserted toxins and other membrane proteins in great detail. This method allows the measurement of the transverse location (depth) of specific sites within a membrane at the angstrom level of resolution.

The approach, which we have named the parallax analysis, involves comparison of the amount of quenching of membrane-inserted proteins induced by phospholipids carrying spin-labeled quenchers at two different depths in the membrane (Chattopadhyay and London, 1987). Qualitatively, the method depends on the fact that fluorescence quenching is distance dependent. Therefore, if the amount of spin-label quenching of a fluorescent group, such as a Trp residue, is found to be greatest for a deep spin-label relative to a shallow one, the fluorophore must be closer in membrane depth to the deep quencher. Quantitatively, simple algebraic expressions that allow calculation of exact fluorophore depth from such experiments have been derived (Chattopadhyay and London, 1987).

Use of fluorescence quenching by spin-labeled lipids has several significant advantages: (*1*) Being a short range process it is sensitive to very small differences in distance. This is the basis for being able to do "high resolution" studies. (*2*) Spin-labels appear to be able to quench almost all types of fluorophores. Thus, with a single set of spin-label quenchers at calibrated depths it is possible (unlike the case of energy transfer) to use any type of fluorescent probe desired. This is a major advantage, allowing one to vary fluorophore properties over a tremendous range and to exploit any prior information gained about a particular fluorescence probe by other techniques. (*3*) Spin-labeled lipids have already been studied in great detail as ESR probes, and this has provided much information about their properties.

In a first series of studies it has been possible to calibrate the method and demonstrate its accuracy and sensitivity using fluorescent membrane probes. Comparison of the quenching properties of brominated and spin-labeled lipids has shown that the locations of the spin-label groups in the membrane can be defined, and that the parallax analysis provides accurate depths within the hydrocarbon chain region of the bilayer (Abrams and London, 1992). Studies of the depth of a series of anthroyloxy labeled fatty acids has shown that they take on a graded series of depths as the number of carbon atoms between the carboxyl and anthroyloxy groups is increased (Abrams et al., 1992). This is in agreement with the qualitative conclusion of earlier studies and further suggests the accuracy of the parallax analysis. (Recent studies in which we have used an additional spin-labeled lipid in which the spin-label is attached to the polar headgroup of a lipid extends this conclusion, suggesting that the distribution of anthroyloxy groups is in fact almost linearly related to their carbon attachment site [Abrams, F. S., and E. London, unpublished observations].) Just as important, it is possible to detect a 1–2-Å upward shift in the location of anthroyloxy-labeled fatty acids upon ionization of the carboxyl group, probably due to the tendency of the charged carboxyl to pull away from the nonpolar hydrocarbon region of the membrane. That such a small shift can be detected indicates that the method is very sensitive to small changes in depth.

It has also been possible to apply the method to proteins, localizing the depth of Trp residues. The experiments of Chattopadhyay and McNamee (1991) with the acetylcholine receptor have suggested a shallow Trp location in accordance with predictions based on sequence information. We have applied the analysis to membrane-inserted cytochrome b_5 (Abrams and London, 1992), diphtheria toxin (Jiang et al., 1991*a*), and the SecA protein of *E. coli* (Ulbrandt et al., 1992). These are all multi-Trp proteins, so only weighted average depths of Trp residues can be obtained. Nevertheless, it has been possible to derive useful conclusions based on quenching. For example, when diphtheria toxin or SecA bind to membranes, some Trp residues insert deeply into the membrane. This is in agreement with the deep insertion of the protein detected using hydrophobic photolabels (Hu and Holmes, 1984; Zalman and Wisnieski, 1984; Montecucco et al., 1985). In the case of cytochrome b_5, the depth for its three clustered Trp residues obtained by applying the parallax analysis to spin-label quenching agrees closely with that obtained previously by brominated lipid quenching (Abrams and London, 1992).

Such studies demonstrate that it will be practical to use spin-label quenching experiments on membrane proteins to obtain information about depth. However, to obtain high resolution information it will be necessary to prepare proteins with only a single fluorescent group. This can be accomplished with site-directed mutagenesis. The two most obvious strategies are to introduce single Cys residues that can be labeled with fluorescent groups and to introduce single Trp residues. In either case, Cys and Trp residues in the wild-type protein must be removed. This may not be as large a problem as it might seem. In the case of the lactose permease it has been shown that the protein remains active after replacement of all Trp by Phe residues (Menezes et al., 1990). Although the quenching method has not yet been applied to proteins with single Trp residues, it has been used by Chung et al. (1992) to investigate the behavior of a synthetic hydrophobic peptide in which a single Trp residue was introduced over a whole range of positions. For model membrane-inserted peptide the results cleanly demonstrated a helical periodicity in depth consistent only with an orientation parallel to the membrane surface, thus demonstrating the potential of the method.

It should also be noted that there are at present several other limitations on the use of the parallax analysis. The distribution and motions of fluorophores and quenchers can complicate analysis of depth (Abrams and London, 1992). In many cases, such effects will average out, but this may be dependent on the nature of the identity of the fluorescent probe used for labeling. How the presence of a protein influences the depth of the quenching probes is another potential problem. These questions will have to be faced in future studies. Nevertheless, it appears that the parallax analysis will be a powerful method for characterizing membrane protein structure in many cases.

Acknowledgments

This work was supported by NIH grant GM-31986.

References

Abrams, F. S., A. Chattopadhyay, and E. London. 1992. Determination of the location of fluorescent probes attached to fatty acids using the parallax analysis of fluorescence quenching. *Biochemistry*. 31:5322–5327.

Abrams, F. S., and E. London. 1992. Calibration of the parallax fluorescence quenching method for determination of membrane penetration depth. *Biochemistry*. 31:5312–5322.

Allured, V. S., R. J. Collier, S. F. Carroll, and D. B. McKay. 1986. Structure of exotoxin A of *Pseudomonas aeruginosa* at 3.0-angstrom resolution. *Proceedings of the National Academy of Sciences, USA*. 83:1320–1324.

Balch, W. E., R. A. Kahn, and R. Schwaninger. 1992. ADP-ribosylation factor is required for vesicular trafficking between the endoplasmic reticulum and the cis-Golgi compartment. *Journal of Biological Chemistry*. 267:13053–13061.

Benedetti, H., R. Lloubes, C. Lazdunski, and L. Letellier. 1992. Colicin A unfolds during its translocation in *Escherichia coli* cells and spans the whole cell envelope when its pore has formed. *EMBO Journal*. 11:441–447.

Bjorkman, P. J., M. A. Saper, B. Samraoui, W. S. Bennett, J. L. Strominger, and D. C. Wiley.

1987. Structure of the human class I histocompatibility antigen, HLA-A2. *Nature.* 329:512–518.

Blewitt, M. G., L. A. Chung, and E. London. 1985. The effect of pH upon the conformation of diphtheria toxin and its implications for membrane penetration. *Biochemistry.* 24:5458–5464.

Cabelli, R. J., K. M. Dolan, L. Qian, and D. B. Oliver. 1991. Characterization of membrane-associated and soluble states of SecA protein from wild-type and *SecA51(TS)* mutant strains of *Escherichia coli. Journal of Biological Chemistry.* 266:24420–24427.

Carroll, S. F., and R. J. Collier. 1988. Amino acid sequence homology between enzymic domain of diphtheria toxin and *Pseudomonas aeruginosa* exotoxin A. *Molecular Microbiology.* 2:293–296.

Chang, M. P., R. L. Baldwin, C. Bruce, and B. J. Wisnieski. 1989. Second cytotoxic pathway of diphtheria toxin suggested by nuclease activity. *Science.* 246:1165–1168.

Chattopadhyay, A., and E. London. 1987. Parallax method for direct measurement of membrane penetration depth by fluorescence quenching. *Biochemistry.* 26:39–45.

Chattopadhyay, A., and M. G. McNamee. 1991. Average membrane penetration depth of tryptophan residues of the nicotinic acetylcholine receptor by the parallax method. *Biochemistry.* 30:7159–7164.

Chaudhary, V. K., Y. Jinno, D. FitzGerald, and I. Pastan. 1990. Pseudomonas exotoxin contains a specific sequence at the carboxyl terminus that is required for cytotoxicity. *Proceedings of the National Academy of Sciences, USA.* 87:308–312.

Choe, S., M. J. Bennett, G. Fujii, P. M. G. Curmi, K. A. Kantardjieff, R. J. Collier, and D. Eisenberg. 1992. The crystal structure of diphtheria toxin. *Nature.* 357:216–221.

Chung, L. A., J. D. Lear, and W. F. DeGrado. 1992. Fluorescence studies of the secondary structure and orientation of a model ion channel peptide in phospholipid vesicles. *Biochemistry.* 31:6608–6616.

De Wolf, M. J. S., G. A. F. Van Dessel, A. R. Lagrou, H. J. J. Hilderson, and W. S. H. Dierick. 1987. pH-induced transitions in cholera toxin conformation: a fluorescence study. *Biochemistry.* 26:3799–3806.

Dumont, M. E., and F. M. Richards. 1988. The pH-dependent conformation change of diphtheria toxin. *Journal of Biological Chemistry.* 263:2087–2097.

Eilers, M., and G. Schatz. 1988. Protein unfolding and the energetics of protein translocation across biological membranes. *Cell.* 52:481–483.

Farabakhsh, Z. T., R. L. Baldwin, and B. J. Wisnieski. 1986. Pseudomonas exotoxin A: membrane binding, insertion and traversal. *Journal of Biological Chemistry.* 261:11404–11408.

Farabakhsh, Z. T., and B. J. Wisnieski. 1989. The acid-triggered entry pathway of *Pseudomonas* exotoxin A. *Biochemistry.* 28:580–585.

Fishman, P. H. 1990. ADP-ribosylating Toxins and G-Proteins: Insights into Signal Transduction. J. Moss and M. Vaughan, editors. American Society of Microbiology, Washington, DC. 127–140.

Gill, D. M. 1976. The arrangement of subunits in cholera toxin. *Biochemistry.* 15:1242–1248.

Gonzalez, J. E., and B. J. Wisnieski. 1988. An endosomal model for acid triggering of diphtheria toxin translocation. *Journal of Biological Chemistry.* 263:15257–15259.

Hu, V. W., and R. K. Holmes. 1984. Evidence for direct insertion of fragments A and B of diphtheria toxin into model membranes. *Journal of Biological Chemistry.* 259:12226–12233.

Hudson, T. H., J. Scharff, M. A. G. Kimak, and D. M. Neville, Jr. 1988. Energy requirements for diphtheria toxin translocation are coupled to the maintenance of a plasma membrane potential and a proton gradient. *Journal of Biological Chemistry.* 263:4773–4781.

Idziorek, T., D. FitzGerald, and I. Pastan. 1990. Low-pH induced changes in *Pseudomonas* exotoxin and its domains: increased binding of Triton X-114. *Infection and Immunity.* 58:1415–1420.

Iwamoto, R., H. Senoh, Y. Okada, T. Uchida, and E. Mekada. 1991. An antibody that inhibits the binding of diphtheria toxin to cells revealed the association of a 27-kDa membrane protein with the diphtheria toxin receptor. *Journal of Biological Chemistry.* 266:20463–20469.

Jiang, J. X., F. S. Abrams, and E. London. 1991a. Folding changes in membrane-inserted diphtheria toxin that may play important roles in its translocation. *Biochemistry.* 30:3857–3864.

Jiang, J. X., L. A. Chung, and E. London. 1991b. Self-translocation of diphtheria toxin across model membranes. *Journal of Biological Chemistry.* 266:24003–24010.

Jiang, J. X., and E. London. 1990. Involvement of denaturation-like changes in Pseudomonas exotoxin A hydrophobicity and membrane penetration determined by characterization of pH and thermal transitions. *Journal of Biological Chemistry.* 265:8636–8641.

Johnson, V. G. 1990. Does diphtheria toxin have nuclease activity? *Science.* 250:832–834.

Kagan, B. L. 1991. Inositol 1,4,5-trisphosphate directly opens diphtheria toxin channels. *Biochimica et Biophysica Acta.* 1069:145–150.

Kagan, B. L., A. Finkelstein, and M. Colombini. 1981. Diphtheria toxin fragment forms large pores in phospholipid bilayer membranes. *Proceedings of the National Academy of Sciences, USA.* 78:4950–4954.

Kieleczawa, J., J.-M. Zhao, C. L. Luongo, L.-Y. D. Dong, and E. London. 1990. Effect of high pH upon diphtheria toxin conformation and membrane penetration. *Archives of Biochemistry and Biophysics.* 282:214–220.

Kounnas, M. Z., R. E. Morris, M. R. Thompson, D. J. FitzGerald, D. K. Strickland, and C. B. Saelinger. 1992. The α_2-macroglobulin receptor/low density lipoprotein receptor-related protein binds and internalizes *Pseudomonas* exotoxin A. *Journal of Biological Chemistry.* 267:12420–12423.

Lencer, W. I., C. Delp, M. R. Neutra, and J. L. Madara. 1992. Mechanism of cholera toxin action on a polarized human intestinal epithelial cell line. Role of vesicular traffic. *Journal of Cell Biology.* 117:1197–1209.

London, E. 1992a. Diphtheria toxin: membrane interaction and membrane translocation. *Biochimica et Biophysica Acta.* 1113:25–51.

London, E. 1992b. How bacterial toxins enter cells. *Molecular Microbiology.* 6:3277–3282.

London, E., and C. L. Luongo. 1989. Domain-specific bias in arginine/lysine usage by protein toxins. *Biochemical and Biophysical Research Communications.* 160:333–339.

Madden, D. R., J. C. Gorga, J. L. Strominger, and D. C. Wiley. 1991. The structure of HLA-B27 reveals nonamer self-peptides bound in an extended conformation. *Nature.* 353:321–325.

Malinski, J. A., and G. L. Nelsestuen. 1989. Membrane permeability to macromolecules mediated by the membrane attack complex. *Biochemistry.* 28:61–70.

Menezes, M. E., P. D. Roepe, and H. R. Kaback. 1990. Design of a membrane transport protein for fluorescence spectroscopy. *Proceedings of the National Academy of Sciences, USA.* 87:1638–1642.

Merrill, A. R., F. S. Cohen, and W. A. Cramer. 1990. On the nature of the structural change of the colicin E1 channel peptide necessary for its translocation-competent state. *Biochemistry.* 29:5829–5836.

Misler, S. 1984. Diphtheria toxin fragment channels in lipid bilayer membranes. Selective sieves or discarded wrappers? *Biophysical Journal.* 45:107–109.

Moehring, J. M., and T. J. Moehring. 1983. Strains of CHO-K1 cells resistant to Pseudomonas exotoxin A and cross-resistant to diphtheria toxin and viruses. *Infection and Immunity.* 41:998–1009.

Montecucco, C., G. Schiavo, and M. Tomasi. 1985. pH-dependence of the phospholipid interaction of diphtheria-toxin fragments. *Biochemical Journal.* 1985. 231:123–128.

Morris, R. E. 1990. Trafficking of Bacterial Toxins. C. B. Saelinger, editor. CRC Press, Inc., Boca Raton, FL. 49–70.

Naglich, J. G., J. E. Metherall, D. W. Russell, and L. Eidels. 1992. Expression cloning of a diphtheria toxin receptor: identity with a heparin-binding EGF-like growth factor receptor. *Cell.* 69:1051–1061.

Ogata, M., V. K. Chaudhary, I. Pastan, and D. J. FitzGerald. 1990. Processing of *Pseudomonas* exotoxin by a cellular protease results in the generation of a 37,000 Da toxin fragment that is translocated to the cytosol. *Journal of Biological Chemistry.* 265:20678–20685.

Oliver, D. B., R. B. Cabelli, and G. P. Jarosik. 1990. SecA protein: autoregulated initiator of secretory precursor protein translocation across the *E. coli* plasma membrane. *Journal of Bioenergetics and Biomembranes.* 22:311–336.

Papini, E., G. Schiavo, M. Tomasi, M. Colombatti, R. Rappuoli, and C. Montecucco. 1987. Lipid interaction of diphtheria toxin and mutants with altered fragment B. *European Journal of Biochemistry.* 169:637–644.

Ramsay, G., D. Montgomery, D. Berger, and E. Freire. 1989. Energetics of diphtheria membrane insertion and translocation: calorimetric characterization of the acid pH induced transition. *Biochemistry.* 28:529–533.

Reay, P. A., D. A. Wettstein, and M. M. Davis. 1992. pH dependence and exchange of high and low responder peptides binding to a class II MHC molecule. *EMBO Journal.* 11:2829–2839.

Ribi, H. O., D. S. Ludwig, K. L. Mercer, G. K. Schoolnik, and R. D. Kornberg. 1988. Three-dimensional structure of cholera toxin penetrating a lipid membrane. *Science.* 239:1272–1276.

Rippmann, F., W. R. Taylor, J. B. Rothbard, and N. M. Green. 1991. A hypothetical model for the peptide binding domain of hsp70 based on the peptide binding domain of HLA. *EMBO Journal.* 10:1052–1059.

Ruigrok, R. W. H., S. R. Martin, S. A. Wharton, J. J. Skehel, P. M. Bayley, and D. C. Wiley. 1986. Conformational changes in the hemagglutinin of influenza virus which accompany heat-induced fusion of virus with liposomes. *Virology.* 155:484–497.

Seetharam, S., V. K. Chaudhary, D. FitzGerald, and I. Pastan. 1991. Increased cytotoxic activity of *Pseudomonas* exotoxin and two chimeric toxins ending in KDEL. *Journal of Biological Chemistry.* 266:17376–17381.

Serafini, T., L. Orci, M. Amherdt, M. Brunner, R. A. Kahn, and J. E. Rothman. 1991.

ADP-ribosylation factor is a subunit of the coat of Golgi-derived COP-coated vesicles: a novel role for a GTP-binding protein. *Cell.* 67:239–253.

Sixma, T. K., S. E. Pronk, K. H. Kalk, E. S. Wartna, B. A. M. van Zanten, B. Witholt, and W. G. J. Hol. 1991. Crystal structure of a cholera toxin-related heat-labile enterotoxin from *E. coli. Nature.* 351:371–377.

Stein, P. E., A. Boodhoo, G. J. Tyrrell, J. L. Brunton, and R. J. Read. 1992. Crystal structure of the cell-binding B oligomer of verotoxin-1 from *E. coli. Nature.* 355:748–760.

Stenmark, H., S. Olsnes, and K. Sandvig. 1988. Requirement of specific receptors for efficient translocation of diphtheria toxin A fragment across the plasma membrane. *Journal of Biological Chemistry.* 263:13449–13455.

Ulbrandt, N. D., E. London, and D. B. Oliver. 1992. Deep penetration of a portion of the Escherichia coli SecA protein into model membranes is promoted by anionic phospholipids and by partial unfolding. *Journal of Biological Chemistry.* 267:15184–15192.

van der Goot, F. G., J. M. Gonzalez-Manas, J. H. Lakey, and F. Pattus. 1991. A 'molten-globule' membrane-insertion intermediate of the pore forming domain of colicin A. *Nature.* 354:408–410.

Weiss, M. S., U. Abele, J. Weckesser, W. Welte, E. Schiltz, and G. E. Schulz. 1991. Molecular architecture and electrostatic properties of a bacterial porin. *Science.* 254:1627–1630.

Wharton, S. A., J. J. Skehel, and D. C. Wiley. 1986. Studies of influenza virus haemagglutinin-mediated membrane fusion. *Virology.* 149:27–35.

Wilson, B. A., S. R. Blanke, J. R. Murphy, A. M. Pappenheimer, Jr., and R. J. Collier. 1990. Does diphtheria toxin have nuclease activity. *Science.* 258:834–836.

Zalman, L. S., and B. J. Wisnieski. 1984. Mechanism of insertion of diphtheria toxin: peptide entry and pore size determinations. *Proceedings of the National Academy of Sciences, USA.* 81:3341–3345.

Zalman, L. S., and B. J. Wisnieski. 1985. Characterization of the insertion of *Pseudomonas* exotoxin A into membranes. *Infection and Immunity.* 50:630–635.

Zhao, J.-M., and E. London. 1986. Similarity of the conformation of diphtheria toxin at high temperature to that in the membrane-penetrating low pH state. *Proceedings of the National Academy of Sciences, USA.* 83:2002–2006.

Zhao, J.-M., and E. London. 1988. Conformation and model membrane interactions of diphtheria toxin fragment A. *Journal of Biological Chemistry.* 263:15369–15377.

Chapter 6

Reconstitution of Human Erythrocyte Band 3 into Two-dimensional Crystals

Max Dolder, Thomas Walz, Andreas Hefti, and Andreas Engel

Maurice E. Müller Institute for High-Resolution Electron Microscopy at the Biocenter, University of Basel, 4056 Basel, Switzerland

Molecular Biology and Function of Carrier Proteins © 1993 by The Rockefeller University Press

Introduction

While the functions of many membrane proteins have been elucidated in detail by biochemical and physiological assays, the structures of only a small number are currently available. This discrepancy results from difficulties in assembling three-dimensional (3D) crystals of solubilized membrane proteins and the lack of powerful methods for the structural analysis of large (>30 kD) macromolecules in a disordered solution. The progress in 3D crystallization experiments is rather slow, but fortunately a growing number of membrane proteins have been reconstituted into two-dimensional (2D) crystals in the presence of lipids (Jap et al., 1992; Kühlbrandt, 1992; Engel et al., 1992). These results suggest that the constraints for 2D crystal formation are less stringent than those for 3D crystallization. In addition, proteins are integrated in a lipid bilayer during reconstitution, which probably maintains their native structure and possibly a functional state.

Such regular layers can be imaged at high resolution using an electron microscope, the attainable resolution being limited by the crystallinity of the membrane, preparation artifacts, and electron beam damage. Recently, high-resolution topographs of biological membranes have been acquired in buffer solution by scanning force microscopes, eventually allowing structure–function relationships to be directly analyzed. Specimen preparation methods, high-resolution electron microscopes, as well as recording and image processing techniques are available to achieve atomic scale resolution. This goal requires highly ordered crystals that contain 10^4–10^6 unit cells, and sophisticated experimental techniques. However, molecular resolution (2–3 nm) revealing the tertiary structure may be obtained routinely, in spite of residual lattice disorder that is often difficult to eliminate.

Here, we illustrate the potential of this approach with trigonal and rectangular 2D crystals reconstituted from human erythrocyte band 3 and dimyristoyl phosphatidylcholine (DMPC). Assembly of band 3 crystals depends critically on the solubilization and isolation protocol and is less reproducible than 2D crystallization of proteins such as bacterial porins that in some cases form regular arrays in vivo.

Materials and Methods

Materials

Octyl-polyoxyethylene (POE) was a gift from Prof. J. Rosenbusch, Biocenter, University of Basel, Basel, Switzerland. Aminoethyl-Sepharose was synthesized according to Casey et al. (1989). Monoclonal mouse anti–human band 3 was kindly provided by Dr. H. Sigrist, Institute of Biochemistry, University of Bern, Bern, Switzerland. Goat anti–mouse IgG coupled to 10 nm colloidal gold was purchased from Sigma Immunochemicals (St. Louis, MO). All other chemicals were from either Fluka Chemie AG (Buchs, Switzerland) or Merck (Darmstadt, Germany).

Purification and Reconstitution of Band 3

Human red blood cells (0 Rh$^+$) obtained from the local blood donor service were washed and lysed, and the resulting membranes were stripped of cytoskeletal proteins according to Bennett (1983). 1 vol of vesicles (5 ml) at a protein concentration of 2–3 mg/ml in 7.5 mM sodium phosphate, pH 7.5, was extracted with 5 vol of extraction buffer (10 mM sodium phosphate, pH 7.5, 1 mM EDTA, 1 μM butylated hydroxytoluene, and 0.005% NaN_3) containing either 1% octyl-POE or 0.1% $C_{12}E_8$. Extraction was performed at room temperature with stirring for at least 1 h. The suspension was centrifuged at 100,000 g for 45 min. The supernatant (25–30 ml) was loaded to an aminoethyl-Sepharose CL-4B column (5 ml bed volume) preequilibrated with octyl-POE (1%) or $C_{12}E_8$ (0.1%) in 10 mM sodium phosphate, pH 7.5 (equilibration buffer). The column was washed with 15 ml of equilibration buffer and proteins were eluted with a linear 0–200 mM NaCl gradient. Band 3–containing fractions (as detected by SDS-PAGE) were pooled, concentrated to ~ 2 mg protein/ml using Centrisart® centrifuge tubes (Sartorius AG, Göttingen, Germany) with a cutoff of 20 kD, and dialyzed against 100 vol of 10 mM HEPES, pH 7, 1% octyl-POE (or 0.1% $C_{12}E_8$), and 0.005% NaN_3 (three changes of dialysis buffer). 200-μl aliquots were mixed with various amounts of solubilized DMPC. Lipid to protein ratios (wt/wt) varied between 0.5 and 2. The mixture was diluted to 400 μl and dialyzed against detergent-free buffer (10 mM HEPES) in a temperature-controlled dialysis apparatus (Jap et al., 1992). Different dialysis buffers were explored: NaCl concentration was varied between 0 and 300 mM, $MgCl_2$ between 0 and 100 mM, and pH between 6.5 and 7.5. In a typical dialysis experiment the sample was preincubated in the solubilized lipid for 6 h, filtered through a 0.2-μm polycarbonate filter (Nucleopore® Corporation, Pleasanton, CA), and dialyzed for two temperature cycles (a linear ramp from 22 to 37°C within 1 h, a constant interval at 37°C for 18 h, a linear ramp from 37 to 22°C within 6 h, and a constant interval at 22°C for 6 h).

Purification of Reconstituted Vesicles

A two-step sucrose gradient was prepared by overlaying a 45% (wt/wt) sucrose solution ($\rho = 1.20$) in 20 mM HEPES, pH 7, 100 mM NaCl, 5 mM $MgCl_2$, and 0.01% NaN_3 (HEPES buffer) with an equal volume (450 μl) of 13% sucrose ($\rho = 1.05$) in a centrifuge tube. Reconstituted vesicles (100 μl) were applied on top of the two-step gradient and centrifuged for 2.5 h at 100,000 g in a TLS-55 swinging bucket rotor (Beckman Instruments, Inc., Fullerton, CA). After centrifugation, two distinct bands were found: a faint band at the interface of the light and heavy sucrose solutions and a second band between the interface and the bottom of the tube. The two fractions were each carefully sucked off using a peristaltic pump, diluted 1:10 with HEPES buffer, and centrifuged for 15 min at 4°C at full speed in a Biofuge A (Heraeus Instruments, Inc., South Plainfield, NJ). The supernatants were removed and the pellets were resuspended in HEPES buffer and centrifuged again. This procedure was repeated once. The final pellets were stored in 50 μl of HEPES buffer and used for SDS-PAGE, Western blotting, and electron microscopy.

Production of Polyclonal Antibodies

Stripped vesicles (0.5 ml, 2–3 mg protein/ml) were diluted 1:10 with 7.5 mM sodium phosphate, pH 7.5, and 0.005% NaN_3 and digested with 3 μg/ml of trypsin under stirring at 4°C for 30 min. The digestion was stopped by addition of 0.5 ml PMSF (20 mg/ml in DMSO). The mixture was centrifuged (45 min, 100,000 g, 4°C) and the supernatant was purified on an aminoethyl-Sepharose column. Washing and elution of proteins was done as described for the purification of whole band 3; however, corresponding buffers were free of detergent. By this procedure the cytoplasmic domain is cut into two fragments of M_r 22,000 and 20,000, respectively. The fragments (1–2 mg/ml) were mixed with Freund's complete (primary immunization) or incomplete adjuvant (booster immunizations) and injected subcutaneously into rabbits. Immunoglobulins were purified from the antiserum by precipitation with an equal volume of saturated ammonium sulfate solution for 2 h at 4°C under gentle stirring. The mixture was centrifuged at 4°C for 15 min at 10,000 g and the pellet was resuspended in 0.1 M Na_2HPO_4, pH 7, and 0.5 M NaCl to a final concentration of 3–4 mg protein/ml. The specificity of the antibodies was tested by Western blotting.

Formation of Gold-IgG Complexes

Colloidal gold sols containing 8-nm gold particles were prepared as described by Slot and Geuze (1985). A defined amount of IgG solution as determined by the method of Roth and Binder (1978) was added to the gold sol under stirring. After 5 min a 10% polyethylene glycol (M_r 20,000) solution was added to 0.5% as a stabilizer and stirring was continued for another 5 min. The gold-IgG complexes were purified by centrifugation at 4°C for 30 min at 45,000 g. The loose sediment of complexes was resuspended in 20 ml PBS containing 0.2 mg/ml BSA. The suspension was underlaid with a layer of 50% glycerol in PBS and recentrifuged. The gold-IgG complexes were tested by immunoblotting.

Labeling Procedures

Reconstituted membranes (5 μl) were adsorbed to an electron microscope grid and incubated for 30 min with the gold-conjugated polyclonal rabbit antibody in dialysis buffer containing 1% BSA. The grid was washed with three drops of dialysis buffer (total washing time, 15 min) and with two drops of water. Finally, the grid was stained with 0.75% uranyl formate. Alternatively, reconstituted membranes were first incubated for 30 min with a monoclonal mouse antibody directed against the cytoplasmic domain of band 3. Subsequently, the grid was washed with three drops of dialysis buffer (total washing time, 15 min) and incubated with 10 nm gold-conjugated goat anti–mouse IgG for another half an hour. The grid was washed with three drops of buffer (15 min) and two drops of water, and stained with uranyl formate. Electron microscopy was done as described below.

Electron Microscopy

Reconstituted band 3 phospholipid vesicles were adsorbed to carbon-coated grids for 1 min, washed with three drops of water, and stained with 0.75% uranylformate. The grids were rendered hydrophilic by glow discharge in air or in pentylamine. Adsorbed vesicles were also quickly frozen in liquid nitrogen, dried at −70°C in a Balzers BAF 300, and unidirectionally shadowed with a 1.5-nm-thick platinum-iridium-carbon layer. The samples were observed in a Hitachi H-7000 electron microscope operated at 100 kV acceleration voltage and 50,000 magnification. Images were recorded on Kodak SO-163 plates without preirradiation at a dose of typically 2,000 electrons/nm².

Image Analysis

Suitable areas from electron micrographs were selected by optical diffraction and digitized as described elsewhere (Stauffer et al., 1992). Averages from negatively stained 2D crystals were obtained by Fourier peak filtration and correlation averaging using a reference containing two unit cells (Saxton and Baumeister, 1982).

Averages from freeze-dried, metal-shadowed lattices were calculated in a similar manner and surface reliefs were reconstructed according to Smith and Kistler (1977).

Results

Inside-out vesicles prepared from human red cells according to the protocol of Bennett (1983) contained band 3 and glycophorin as major proteins. Band 3

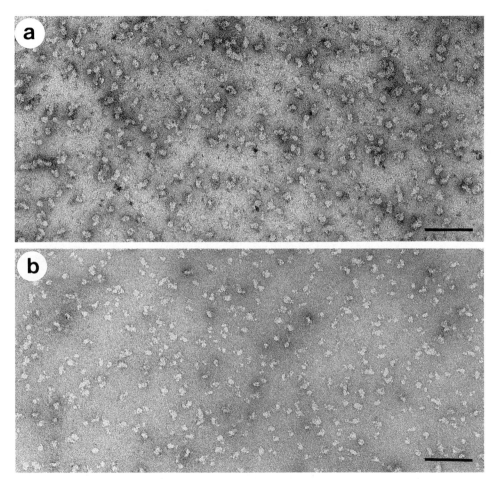

Figure 1. Electron micrograph of negatively stained solubilized and purified band 3 particles in octyl-POE (*a*) and $C_{12}E_8$ (*b*). Scale bars, 100 nm.

extracted from such vesicles by various detergents exhibited different mobility upon SDS-PAGE. With POE alkylethers containing short alkyl chains (i.e., hexyl-, octyl-, and decyl-POE), a mixture of dimers, trimers, tetramers, and higher aggregates was detected on SDS gels. With different dodecyl-POEs, monomers and dimers were observed exclusively (Dolder et al., 1993). Accordingly, negative stain electron microscopy revealed that the mean size of solubilized and purified band 3 particles

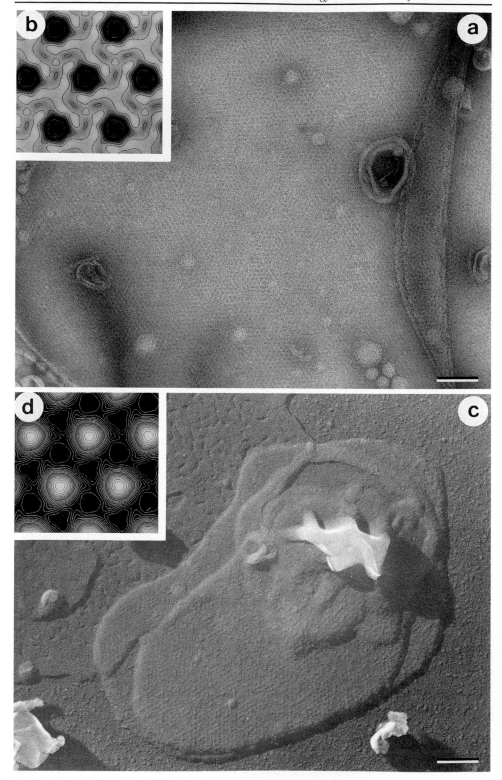

was smaller when the purification was performed with long chain than with short chain POE alkylethers (Fig. 1).

For reconstitution, purified band 3 oligomers solubilized in octyl-POE were mixed with DMPC-octyl-POE micelles in 10 mM HEPES buffer and dialyzed against various detergent-free solutions for several days. The pH, ionic strength, and lipid to protein ratio (LPR) were varied systematically to explore their influence on the crystallization process (Dolder et al., 1993). In several experiments 2D crystals were found to constitute a significant fraction of the structures adsorbed to the electron microscope grid. The best crystals assembled in 10 mM HEPES buffer, pH 7, 25–50 mM NaCl, and 20 mM $MgCl_2$, and LPR $\approx$ 1. The major lattice type was trigonal (Fig. 2) and exhibited unit cell dimensions of $a = b = 11 \pm 0.5$ nm ($n = 30$). Lattices were usually stacked in pairs or multiple pairs as expected from air-dried vesicles spread-flattened on the grid by surface tension (Fig. 2 *a*). Averaged projections (Fig. 2 *b*) showed a root mean square (RMS) deviation from threefold symmetry of typically 2–3%. The unit cell (Fig. 2 *b*) reveals one prominent stain-filled depression that is surrounded by three elongated bilobed stain-excluding domains ~7 nm in length. They extend to contact sites at threefold axes.

Most vesicles retained a curved surface after freeze-drying and metal-shadowing when carbon-coated grids were rendered hydrophilic by glow-discharge in air. In contrast, a large fraction of flat planar lattices as shown in Fig. 2 *c* was observed when the grids were glow-discharged in pentylamine, a procedure yielding positively charged films (Dubochet et al., 1971). Surface reliefs calculated from four different correlation-averaged projections of well preserved freeze-dried and metal-shadowed band 3 arrays exhibit a hexagonal pattern of peaks protruding from a smooth surface (Fig. 2 *d*). These protrusions possess an average diameter of 6 nm.

At a much lower frequency, rectangular lattices with unit cell dimensions of $a \approx b = 12 \pm 0.8$ nm ($n = 6$) were observed (Fig. 3 *a*). They were less well ordered than the trigonal lattices and appeared to have an intrinsic flexibility as documented by thc large variation of the angle between lattice vectors. The lattice disorder is also illustrated by the poor sharpness of diffraction peaks (Fig. 3 *b*, *inset*). Thus, correlation averaging was used to determine the unit cell morphology of rectangular band 3 lattices (Fig. 3 *a*, *inset*). To this end a reference was chosen (Fig. 3 *c*, *top left inset*) and its cross-correlation with a selected image area was calculated (Fig. 3 *c*). Peaks indicating the positions of reference-like motifs were fitted with a lattice by linear regression, and the displacement vectors between lattice points and motif positions were determined (Fig. 3 *c*). Their length is a measure of the lattice distortions, and can be used to reject motifs from highly distorted patches (Saxton and Baumeister, 1982). The average of the least distorted unit cells thus calculated (Fig. 3 *c*, *top right inset*) revealed a twofold symmetry (RMS deviation 3–4%), which was imposed on thc final map (Fig. 3 *a*, *inset*).

Band 3 lattices sometimes did not assemble for unknown reasons. In such cases,

Figure 2. (*a*) Electron micrograph of a negatively stained band 3 lattice reconstituted in the presence of DMPC showing a P3 lattice type. (*b*) Threefold rotationally symmetrized averaged projection exhibiting stain-filled indentations (*dark regions*) surrounded by three bilobed stain-excluding domains that merge at threefold axes (*bright regions*). (*c*) Freeze-dried, metal-shadowed reconstituted vesicle. (*d*) Surface relief reconstructed from four crystals after threefold symmetrization. Scale bars, 100 nm.

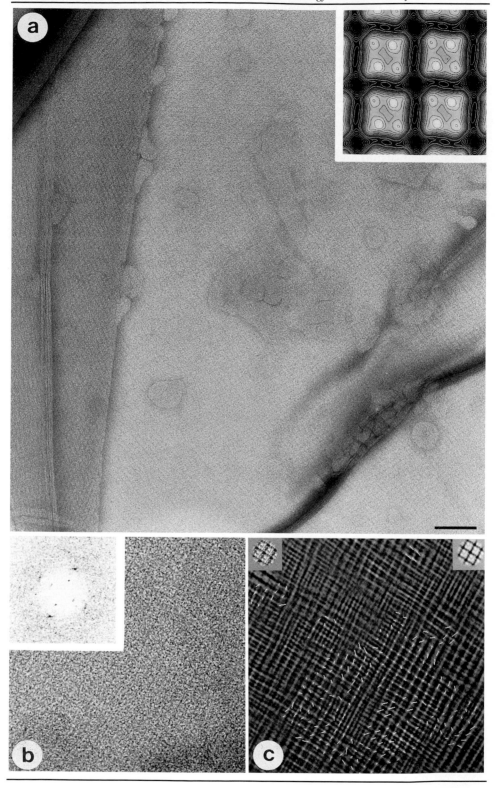

the protein precipitated and structureless (probably protein-free) vesicles formed. Even in preparations with a large fraction of vesicles exhibiting crystallinity, some aggregates were observed. Hence we sought to identify the protein in the lattices by antibody gold labeling. As shown in Fig. 4, the antibodies directed against the cytoplasmic domain of band 3 did not bind to the lattices, but labeled the precipitated protein.

In a further experiment, the precipitates and the reconstituted vesicles were separated on a sucrose step gradient. Band 3 and its degradation products produced during reconstitution were detected by SDS-PAGE and Western blotting (Fig. 5 *a*, lanes *b* and *f*). No protein band could be discerned, either in the light or in the heavy fraction from the sucrose step, that would not have been labeled by the anti–band 3 antibody (Fig. 5 *a*, lanes *c*, *d*, *g*, and *h*). As examined by electron microscopy, the heavy band contained mainly the precipitated protein and the light band the vesicles. Fig. 5 *b* shows an overview of such vesicles, while Fig. 5 *c* displays a higher magnification image, illustrating the periodically arranged band 3.

Discussion

The picture emerging from a still expanding series of reconstitution experiments with different membrane proteins (Jap et al., 1992; Engel et al., 1992) suggests that the first essential step in reconstitution of 2D membrane protein crystals is the correct integration of the protein in the lipid bilayer. During or after integration in the membrane, directional protein–protein interactions may induce an ordered conformation of the tightly packed membrane proteins that are to some extent translationally, but mainly rotationally, mobile. Depending on the LPR, different contact sites may develop, leading to different packing arrangements. In addition, the nature of both the lipid head and the hydrophobic tail may significantly influence the packing arrangement. Interactions can possibly be induced by approaches akin to those used in 3D crystallization experiments; addition of multivalent counterions or the use of precipitating agents has been reported (Jap et al., 1992; Kühlbrandt, 1992). The situation is more complex than with soluble proteins, as two further components, the lipid and the detergent, are involved. However, the constraints concern directional interactions in a plane rather than in the 3D space, hence allowing 2D crystals to grow even when irregular shapes, glycosylation, and other effects prevent a perfect crystallographic stacking of the layers.

Our finding that 2D crystals assemble from a heterogeneous mixture of band 3 oligomers rather than from the monomeric band 3 preparation suggests that the

---

Figure 3. (*a*) Rectangular band 3 lattices reconstituted in the presence of DMPC exhibit a pronounced flexibility. This is exemplified by the large local variations of the angle included by the lattice vectors. To extract the morphology of the unit cell ($a \approx b = 12$ nm) shown in the inset, correlation averaging has been used. The patch shown in *b*, yielding diffraction orders to 4 nm (*inset*), has been Fourier peak filtered. A reference selected from the filtered image (*c*, *top left inset*) cross-correlated with the unprocessed image (*b*) produced the correlation peaks displayed in *c*. Displacement vectors (drawn magnified five times) indicate the translation of unit cell with respect to a fitted lattice. The correlation average (*c*, *top right inset*) was calculated from unit cells closest to the fitted lattice. Twofold symmetrized unit cells exhibit two bilobed domains of ~8 nm length (*a*, *inset*).

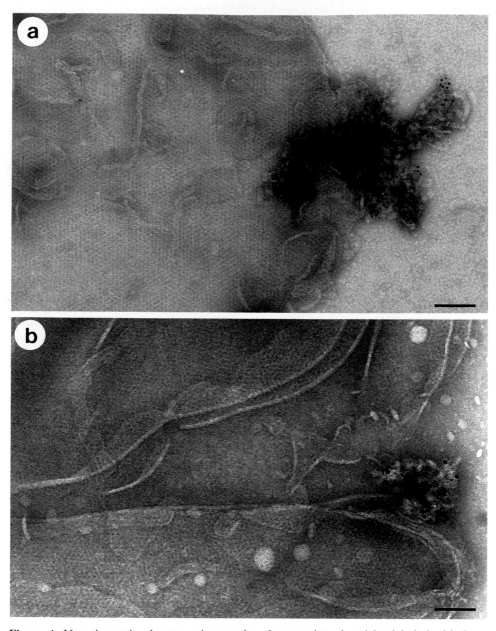

Figure 4. Negative stain electron micrographs of reconstituted vesicles labeled with 8 nm gold-conjugated polyclonal antibody (*a*) and with monoclonal antibody followed by 10 nm gold-conjugated goat anti–mouse IgG (*b*). Both primary antibodies are directed against the cytoplasmic domain of human band 3. Scale bars, 100 nm.

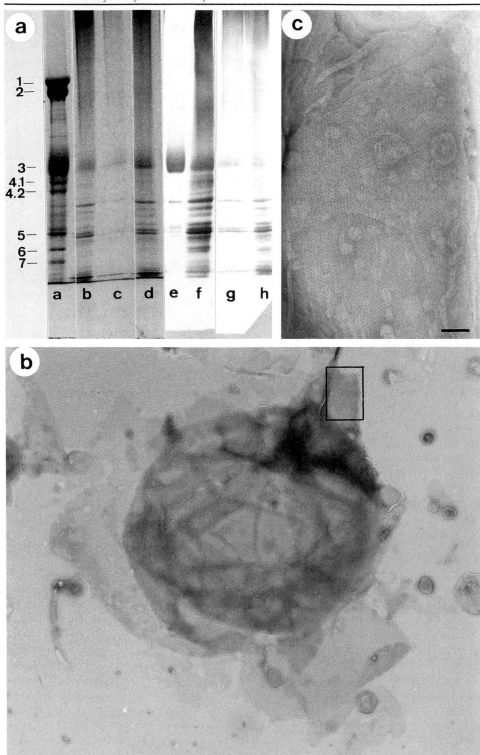

molecular symmetry of a specific oligomeric form favors packing into regular arrays. We have observed that only 16% of octyl-POE solubilized band 3 particles exhibit a size close to the trigonal unit cell dimension (Dolder et al., 1993). Accordingly, the yield of 2D crystals was poor in each reconstitution experiment. So far, no conditions have been reported in which band 3 would exist in a single defined oligomeric state. Hence, future experiments will focus on improving the yield of a single oligomeric state during solubilization and isolation of band 3.

Sucrose step gradient experiments revealed that the protein incorporated into vesicles is band 3. As these vesicles were not labeled by antibodies against the cytoplasmic domain, band 3 molecules are probably incorporated in the right-side out orientation. The flattened shape of freeze-dried vesicles adsorbed to grids that were glow-discharged in pentylamine reflects the strong interactions between the positively charged grid surface and the negatively charged sugar moiety of band 3. The latter extends into the extracellular space, supporting the assumption that reconstituted band 3 vesicles are right-side out.

Models of the transmembrane arrangement of the anion transporter have been proposed (Jay and Cantley, 1986; Lux et al., 1989). According to these models, extracellular loops are short and cannot extend far into the extracellular space. Therefore, we conclude that the protrusions seen on the surface relief (Fig. 2 d) represent the sugar moiety of band 3. We suppose that the sugar chains are located at contact sites between adjacent unit cells and that they might play a role for crystallization, which in fact requires the presence of Mg^{2+}.

The 2D crystals of band 3 reconstituted in the presence of DMPC are not sufficiently well ordered for high-resolution structural analysis. Therefore, membranes have been prepared by negative staining or freeze-drying and metal shadowing. As illustrated with the rectangular lattice type, correlation averaging (Saxton and Baumeister, 1982) has been essential in retrieving the unit cell morphology. Averaged projections of both the trigonal and rectangular lattices exhibit prominent stain-filled cavities. Bilobed stain-excluding domains adjacent to deeply stained regions are pronounced in rectangular lattices (Fig. 3 a, *inset*), but are less distinct in the trigonal crystals, where they merge at threefold axes. However, they are of similar length (7 and 8 nm, respectively) and could represent band 3 dimers. The packing unit in trigonal lattices would therefore be a hexamer, whereas in rectangular lattices it would represent a tetramer of band 3. Considering the unit cell area of trigonal crystal (105 nm^2) and the cross-section of 14 α-helices per band 3 monomer (Jay and Cantley, 1986; Lux et al., 1989), a band 3 hexamer would occupy 80% of the unit cell, leaving 20% to the lipid moiety. This value is characteristic for densely packed 2D protein–lipid crystals. If we assume that one rectangular unit cell contains one band 3 tetramer, the same calculation shows that ~60% of the surface would be taken by

Figure 5. (*a*) Coomassie blue–stained SDS-PAGE (lanes *a–d*) and immunoblot (lanes *e–h*) of erythrocyte membranes (*a, e*), reconstituted sample containing band three lattices as well as precipitated protein (*b, f*), light (*c, g*), and heavy fraction (*d, h*) after separation of crystals and aggregates on a sucrose step gradient. Immunoblotting was performed using a polyclonal antibody directed against the cytoplasmic domain of band 3. Numbering of protein bands was done according to Fairbanks et al. (1971). (*b*) Negative stain electron micrograph showing an overview of the sample collected as light fraction from the sucrose step. (*c*) High magnification image of the region outlined in *b*. Scale bars are 1 μm in *b* and 0.1 μm in *c*.

the lipid bilayer. This unusually high lipid content might well explain the pronounced flexibility of the rectangular band 3 crystals. Nevertheless, final conclusions concerning the unit cell stoichiometry require further experiments such as scanning transmission electron microscopy–mass determination, which are currently in progress.

At present we cannot provide any further specifications on the oligomeric state of band 3 in the two crystal forms. Better defined oligomeric states of solubilized band 3 in solutions from which 2D crystals are assembled are expected to yield better ordered crystals, a prerequisite for high-resolution structural analysis. Nevertheless, these first band 3 crystals described here represent an encouraging starting point to solve the structure of the anion transporter.

Acknowledgments

This work was supported by the M. E. Müller Foundation of Switzerland and grant 31-25684.88 from the Swiss National Science Foundation to A. Engel.

References

Bennett, V. 1983. Proteins involved in membrane-cytoskeleton association in human erythrocytes: spectrin, ankyrin, and band 3. *Methods in Enzymology.* 96:313–324.

Casey, J. R., D. M. Lieberman, and R. A. F. Reithmeier. 1989. Purification and characterization of band 3 protein. *Methods in Enzymology.* 173:494–512.

Dolder, M., T. Walz, A. Hefti, and A. Engel. 1993. Human erythrocyte band 3: solubilization and reconstitution into two-dimensional crystals. *Journal of Molecular Biology.* In press.

Dubochet, J., M. Ducommun, M. Zollinger, and E. Kellenberger. 1971. A new preparation method for dark-field electron microscopy of biomacromolecules. *Journal of Ultrastructure Research.* 35:147–167.

Engel, A., A. Hoenger, A. Hefti, C. Henn, R. C. Ford, J. Kistler, and M. Zulauf. 1992. Assembly of 2-D membrane protein crystals: dynamics, crystal order, and fidelity of structure analysis by electron microscopy. *Journal of Structural Biology.* In press.

Fairbanks, G., T. L. Steck, and D. F. H. Wallach. 1971. Electrophoretic analysis of the major polypeptides of the human erythrocyte membrane. *Biochemistry.* 10:2606–2617.

Jap, B. K., M. Zulauf, T. Scheybani, A. Hefti, W. Baumeister, U. Aebi, and A. Engel. 1992. 2D crystallization: from art to science. *Ultramicroscopy.* 46:45–84.

Jay, D., and L. Cantley. 1986. Structural aspects of the red cell anion exchange protein. *Annual Review of Biochemistry.* 55:511–538.

Kühlbrandt, W. 1992. Two-dimensional crystallization of membrane proteins. *Quarterly Reviews of Biophysics.* 25:1–49.

Lux, S. E., K. M. John, R. R. Kopito, and H. F. Lodish. 1989. Cloning and characterization of band 3, the human erythrocyte anion-exchange protein (AE1). *Proceedings of the National Academy of Sciences, USA.* 86:9089–9093.

Roth, J., and M. Binder. 1978. Colloidal gold, ferritin and peroxidase as markers for electron microscopic double labeling lectin techniques. *Journal of Histochemistry and Cytochemistry.* 26:163–169.

Saxton, W. O., and W. Baumeister. 1982. The correlation averaging of a regularly arranged bacterial cell envelope protein. *Journal of Microscopy.* 127:127–138.

Slot, J. W., and H. J. Geuze. 1985. A new method of preparing gold probes for multiple-labeling cytochemistry. *European Journal of Cell Biology.* 38:87–93.

Smith, P. R., and J. Kistler. 1977. Surface reliefs computed from micrographs of heavy metal-shadowed specimens. *Journal of Ultrastructure Research.* 61:124–133.

Stauffer, K. A., A. Hoenger, and A. Engel. 1992. Two-dimensional crystals of *Escherichia coli* maltoporin and their interaction with the maltose-binding protein. *Journal of Molecular Biology.* 223:1155–1165.

MDR, CFTR, and

Related Proteins

Chapter 7

Bacterial Periplasmic Permeases as Model Systems for Multidrug Resistance (MDR) and the Cystic Fibrosis Transmembrane Conductance Regulator (CFTR)

Giovanna Ferro-Luzzi Ames

Department of Molecular and Cell Biology, University of California at Berkeley, Berkeley, California 94720

Molecular Biology and Function of Carrier Proteins © 1993 by The Rockefeller University Press

Introduction

Over the last few years advances in the molecular characterization of several eukaryotic genes have yielded an exciting case of cross-fertilization between fields as different as bacterial active transport and medical molecular biology. The sequences of several genes responsible for a number of important medical problems (the P-glycoprotein of multidrug resistance [MDR] and the cystic fibrosis transmembrane conductance regulator [CFTR], among others) have become available and have been shown to bear a striking level of similarity to the sequences of several prokaryotic transporters. This finding raises the question of whether it is appropriate to extrapolate the extensive knowledge presently available for the prokaryotic systems to the eukaryotic ones (and vice versa) when trying to establish a mechanism of action for the respective gene products. In this communication I present arguments in favor of such an extrapolation. I will give a brief review of the state of the field with regard to the prokaryotic research, with special reference to the work performed in my own laboratory, and then I will raise the question of extrapolation between the prokaryotic and the eukaryotic systems.

The term "traffic ATPases" has been coined (Ames et al., 1990) to describe proteins that form a large superfamily characterized by an extensive sequence similarity that includes, but is not limited to, two motifs typical of an ATP-binding site (Walker et al., 1982) and by very similar predicted overall secondary structures. This definition specifically implies that, because of the sequence similarity, an ATPase activity is likely to be a distinguishing characteristic of all of the members of this superfamily. The term traffic was chosen as a generalized descriptive term for the superfamily for the following reasons: (*a*) it distinguishes them from the ion-motive (and other) membrane ATPases; (*b*) many of these proteins are known to be involved in translocation activities; (*c*) the translocated substrates are extremely heterogeneous, both in nature and size; and (*d*) the direction of translocation is either toward the interior or the exterior of the cell. Thus the overall picture is one of a multiplicity of thoroughfares with the same general structure, energized by the hydrolysis of ATP, and adapted to translocate a very heterogeneous set of substrates in either direction. The concept of superfamily implies the existence of a common evolutionary origin and a similar mechanism of action. Another name suggested for proteins that carry the same ATP-binding motifs is ABC (ATP-binding cassette) proteins (Hyde et al., 1990). However, the latter nomenclature has included proteins such as, for example, the F_0F_1 ATPase, which does not possess other characteristic of the superfamily. Several lists of prokaryotic (Furlong, 1987; Kerppola, 1990; Kolter and Fath, 1993) and eukaryotic (Buschman and Gros, 1993) traffic ATPases have been published.

Periplasmic Permeases

Prokaryotic periplasmic transport systems (periplasmic permeases) form a family contained within the superfamily of traffic ATPases. The family comprises many representatives that have been extensively studied (reviewed in Ames, 1986*a*; Shuman, 1987; Ames et al., 1990, 1992; Ames and Lecar, 1992). Periplasmic permeases are typically composed of a soluble receptor (the periplasmic substrate-binding protein) and a membrane-bound complex. The periplasm of gram-negative bacteria is an operationally defined space that contains proteins which can be released from the cell by osmotic shock and that is placed between the inner (cytoplasmic) and outer cell membranes (Neu and Heppel, 1965). Thus the periplasmic permease receptors are located at the periphery of the cell and incoming substrates are first exposed to them. The concentration of receptors in the periplasm is, or can be induced to be, very high (calculated to be on the order of millimolar) and the affinity for their respective substrates is usually very good, between 0.01 and 1 mM. The consequence of these two facts is that a substrate is trapped and concentrated in the periplasm, greatly exceeding the concentration of the free substrate in the external medium. This concentrative action which occurs before the actual translocation step may be an important distinguishing characteristic between prokaryotic and eukaryotic traffic ATPases, as will be discussed later. Another likely consequence of the presence of the soluble receptor is that these permeases typically concentrate substrates against a large concentration gradient when the external concentration of the free substrate is used as a comparison.

The substrate trapped in the periplasm is then translocated through the cytoplasmic membrane by the action of the membrane-bound complex. The existence and composition of this complex and the stoichiometry of its subunits have been established biochemically in two separate cases (the histidine and the maltose permeases [Davidson and Nikaido, 1991; Kerppola et al., 1991]) as including four separate polypeptides. That such composition represents the general situation is also clearly supported by the analogy afforded by inspection of the sequences of numerous other permeases: in several cases in which two of the subunits were evolutionarily fused into a single polypeptide, the sequence of the fused product still supports the notion of a four-domain structure because the two halves of the fused product clearly maintain the characteristics of the corresponding separate polypeptides. The possibility exists that the membrane-bound complex normally exists as a multimer of such a basic four-domain structure.

The Histidine Permease

The histidine permease of *Salmonella typhimurium* and *Escherichia coli* has been studied extensively in my laboratory for numerous years and is reasonably well characterized. Among other permeases, the maltose (Shuman, 1987) and the oligopeptide (Hiles et al., 1987) permeases also have been well characterized. Together with the data obtained on several other periplasmic permeases analyzed to various levels of detail, the study of the above systems has yielded an overall picture that is very consistent. Here I will describe the histidine permease, using it as a model system for the purpose of discussion and comparison.

The receptor of the histidine permease is the histidine-binding protein, HisJ; its membrane-bound complex comprises two hydrophobic integral membrane proteins, HisQ and HisM, and two copies of a hydrophilic membrane-bound, ATP-binding protein, HisP. The latter is the repository of the aforementioned sequence similarity; all the equivalent components from all traffic permeases are also referred to as the conserved components (Mimura et al., 1991). A schematic representation of the histidine permease which is applicable to periplasmic permeases in general is shown in Fig. 1.

A cycle of transport through the histidine permease starts in the periplasm with the binding of histidine by HisJ, which consequently undergoes a conformational change that allows it to be specifically recognized by the membrane-bound complex (Ames, G. F.-L., and K. Nikaido, manuscript in preparation). This interaction is postulated to initiate a series of conformational changes in the components of the complex resulting in ATP hydrolysis and translocation of the substrate (Fig. 2). The initial step, the liganding of the receptor and its conformational change, are fairly well understood, as would be expected from the fact that receptors, being soluble, are easily purified and biochemically analyzed. Because they are not an indispensable characteristic of the basic mechanism of transport (see below) and they are apparently not present in the eukaryotic systems, a brief summary will suffice.

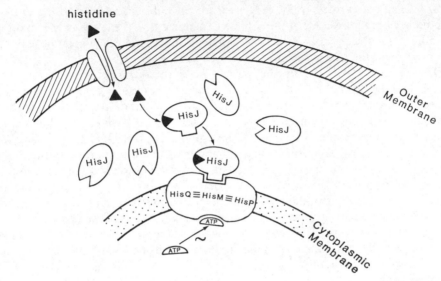

Figure 1. Schematic representation of a periplasmic permease as detailed for the histidine permease. The three membrane components (HisQ, HisM, and HisP) are represented as forming a complex within the cytoplasmic membrane. The periplasmic binding protein (HisJ) changes conformation upon binding histidine and then interacts with the membrane complex. Since the architecture of this complex has not yet been fully elucidated, the interaction is shown as occurring with the complex as a whole; a direct contact between HisJ and HisQ has been proven biochemically. The squiggle indicates the involvement of ATP in energy coupling. The histidine molecule can penetrate the outer membrane through nonspecific hydrophilic pores; larger molecules may require specific pores. (Reproduced from *FEMS Microbiology Reviews,* 1990, 75:429–446, by copyright permission of Elsevier Science Publishers B.V.)

The Receptor

Many receptors have been purified, and several have been crystallized either in the liganded or the unliganded forms and their structures resolved (reviewed in Adams and Oxender, 1989). A general picture has emerged of a two-domain structure in the shape of a kidney bean, in which the two lobes are well separated in the absence of substrate; when substrate finds its way to the specific substrate-binding site in the cleft between the two lobes, one lobe rotates relative to the other, trapping the substrate inside a deep pocket. Recently one receptor, the lysine-, arginine-, ornithine-binding protein (LAO, closely related to HisJ) has finally been purified in both the liganded and unliganded forms (Nikaido and Ames, 1992), and the resolution of both structures has given a very clear picture of the likely sequence of events that accompany the liganding process (Oh et al., 1993). The picture basically confirms the "Venus flytrap" hypothesis put forward by Quiocho and his collaborators on the basis of the structures of different receptors, none of which had been previously resolved in both the liganded and unliganded forms (Mao et al., 1982). Genetic

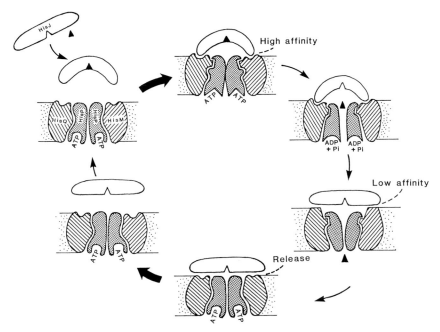

Figure 2. A speculative model for periplasmic transport. The components undergo a series of conformational changes initiated by the binding of histidine to the periplasmic binding protein HisJ. The liganded HisJ binds to the membrane complex. This interaction presumably occurs with both the hydrophobic membrane components (HisQ and HisM) and elicits a conformation change in the ATP-binding membrane protein (HisP), thus causing ATP hydrolysis. Hydrolysis of ATP leads to the opening of a pore that allows the unidirectional diffusion of the substrate to the interior. After the substrate has been released to the interior, an additional conformation change (release of ADP?) closes the pore. The unliganded binding protein, having a poorer affinity for the membrane complex, dissociates from it. The membrane complex again binds ATP, ready to start the cycle. Alternatively, the substrate might be transferred from the binding protein to specific binding site(s) on the membrane complex, from where it would then finally be transferred to the interior, the rest of the cycle remaining the same. (Reproduced from *FEMS Microbiology Reviews,* 1990, 75:429–446, by copyright permission of Elsevier Science Publishers B.V.)

analysis of the various receptor domains responsible for substrate binding and interaction with the membrane complex is underway for several receptors and, combined with x-ray studies, will undoubtedly unravel many of the secrets of the mechanism of action of these proteins.

　　Since it is at the level of the membrane-bound complex that the obvious similarity between eukaryotic and prokaryotic systems resides, it is relevant to ask whether the soluble receptor is a necessary characteristic of the prokaryotic systems, and whether it reflects a basic mechanistic difference. As discussed below, it appears that receptors may be a convenience added onto the common membrane-bound translocating machinery by prokaryotes for increased efficiency, rather than being an indispensable characteristic. Therefore, it is reasonable to focus on the molecular mechanism of the membrane translocator itself as the common puzzle to be solved.

The Membrane-bound Complex

The interaction of liganded HisJ with the membrane-bound complex is necessary to activate the translocation mechanism. It has been postulated that it is this interaction that triggers the series of events resulting in both the release of the bound molecule of substrate and the opening of a specific pore in the complex through which the substrate is translocated (Ames et al., 1990; Petronilli and Ames, 1991; Davidson et al., 1992). Clearly, in order to understand the mechanism of translocation it is necessary to obtain information concerning the physical structure of the translocating complex. In the case of the histidine permease, the two hydrophobic components, HisQ and HisM, are typical integral membrane proteins that span the membrane five times each, with their amino termini located on the external (periplasmic) side and the carboxy termini on the cytoplasmic side (Kerem et al., 1989; Kerppola et al., 1991; Kerppola and Ames, 1992). These two proteins are the smallest (25,000 D) among all known hydrophobic components of periplasmic permeases; therefore, we can take their organization, with a total of 10 membrane spanners between the two of them, as the maximum indispensable to perform the basic translocation step. It has been suggested that HisQ and HisM function as a pseudodimer since they exhibit a high level of sequence similarity (Ames, 1985). The possibility that the hydrophobic components always form a symmetrical or pseudosymmetrical dimer receives support from the finding that several permeases (such as the glutamine permease, which displays a very high level of sequence similarity with the histidine permease) comprise a single hydrophobic component; the latter, therefore, must form a dimer within the complex of two identical subunits.

Besides HisQ and HisM, the complex contains two copies of HisP, the ATP-binding, conserved component (Kerppola et al., 1991). A similar organization was determined for the membrane-bound complex of the maltose permease (Davidson and Nikaido, 1991). Thus, we visualize the basic unit of a membrane-bound complex as forming a pseudosymmetrical structure, with two hydrophobic proteins and two subunits of a hydrophilic, ATP-binding protein. The association with and the organization of HisP within the membrane is of particular interest since its sequence (similar to that of all conserved components) is definitely hydrophilic. HisP does not behave like either a typical integral or a peripheral membrane protein (Kerppola et al., 1991). Interestingly, HisP prefers to be associated with the membrane even in the absence of HisQ and HisM, but in such case it assumes an improper association, behaving like a typical peripheral membrane protein and acquiring an unusual level of sensitivity to proteases and detergents. A possible explanation for the association of HisP with the membrane when HisQ and HisM are absent may be found in its predicted tertiary structure, which has been modeled to include a large loop with some amphipathic characteristics (Mimura et al., 1991; see below). This loop might allow HisP's improper association with the membrane in the absence of the hydrophobic components and its proper insertion in their presence, in which case they would shield it from excessive hydrophobic contacts with the lipid bilayer. The notion that the conserved component normally may be (more or less deeply) embedded in the complex, and therefore in the membrane, derives support from the recent finding that HisP is accessible to proteases and to an impermeant biotinylating reagent from the exterior surface of the membrane when the complex has been assembled in the presence of HisQ and HisM, but not in their absence (Baichwal et al., 1993). No analogous evidence has yet been gathered for equivalent conserved components from other permeases. Thus, a membrane-spanning disposition as a general characteristic of conserved components awaits confirmation. Fig. 3 is a schematic rendering of a periplasmic permease which shows the conserved component with its ATP-binding sites accessible to the cytoplasm while another portion interacts with the membrane spanners of the hydrophobic components. The extent of the interaction is not to be taken literally, since no nearest neighbor analysis has yet been performed.

Because conserved components have maintained a high level of sequence similarity in the ATP-binding motifs, an attempt was made at defining the ATP-binding pocket at a three-dimensional level by computer modeling. Since HisP is a relatively small protein (25,000 D), this is a particularly useful exercise, because once the domain responsible for ATP binding has been defined, a rather small segment of the molecule remains that can be defined as responsible for the performance of its other function(s), whatever they may be. Two independent structural modeling efforts for conserved components (Hyde et al., 1990; Mimura et al., 1991) have yielded somewhat differing results. Possible reasons for the discrepancy have been discussed (Mimura et al., 1991). Our laboratory has predicted a structure which comprises a mononucleotide-binding pocket occupying about two thirds of the HisP molecule, with the remaining third forming the aforementioned large loop. The large loop has been temporarily named the helical domain because of its predicted secondary structure which includes four α-helices, although no experimental evidence is available to demonstrate the existence of such a helical structure. The helical domain is presumably involved in transport-related activities other than ATP binding and hydrolysis. One obvious possibility is its interaction with HisQ and HisM in the transmission of signals to and from the site(s) of interaction with HisJ and the ATP-binding and -hydrolyzing pocket. A particularly well-conserved region of the helical domain is a glycine-, glutamine-rich sequence (consensus: LSGGQQQ) stretch that was postulated to be a flexible arm connecting the two domains and tentatively named the linker peptide (Mimura et al., 1991; Ames et al., 1992).

Support for the predicted ATP-binding pocket was derived both from in vitro chemical modification studies using the photolabeling analogue of ATP, 8-azido ATP (Mimura et al., 1990) and by a structure–function analysis of HisP mutants (Shyamala et al., 1991). In the latter study the ability of each of the mutant proteins to function in transport and/or to bind ATP was related to the location of the mutation within both the sequence and the predicted structure. A good correlation was found between loss of ATP-binding ability (and transport) and location of the mutation within the predicted ATP-binding pocket, while transport negative mutations within the helical domain generally did not affect ATP hydrolysis. In addition, a set of interesting mutations that remove the dependence on signaling between the liganded receptor and ATP hydrolysis also supports the model because they are all located within the ATP-binding pocket, presumably having changed the binding

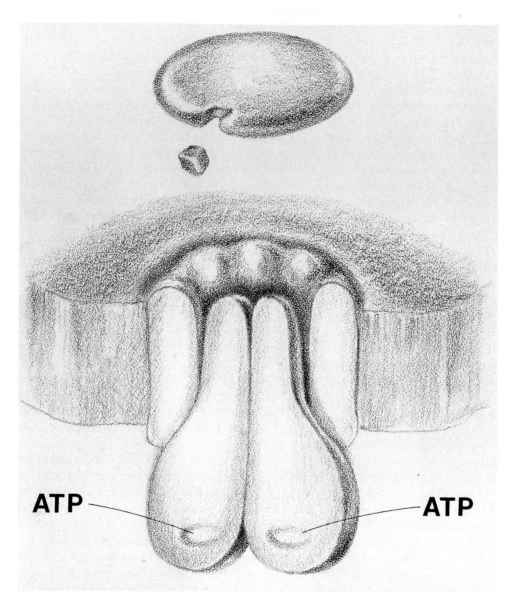

Figure 3. Three-dimensional representation of a traffic ATPase. The membrane has been sliced vertically to expose the ATP-binding domain (conserved component) extending partly into the membrane and encased by the hydrophobic domain (hydrophobic membrane components for prokaryotic members). The ATP-binding sites are located in a cytoplasmic portion of the ATP-binding domain. A substrate-binding protein is represented and is to be imagined as present and necessary only for those traffic ATPases for which such a requirement has been shown. (Reproduced from *Advances in Enzymology,* 1992, 65:1–47, by copyright permission of John Wiley & Sons, Inc.)

pocket so that it has acquired an increased facility to hydrolyze ATP (Petronilli and Ames, 1991; Shyamala et al., 1991; Speiser and Ames, 1991).

Energy Coupling

What is the function of ATP in transport? A number of in vivo and in vitro studies (Ames et al., 1989; Joshi et al., 1989; Mimmack et al., 1989; Prossnitz et al., 1989; Davidson and Nikaido, 1990) established clearly that ATP is the source of energy for transport by periplasmic permeases, resolving a decade-long controversy, in particular excluding the involvement of the proton-motive force (reviewed in Ames, 1990). Reconstitution of transport in membrane vesicles (Ames et al., 1989; Dean et al., 1989; Prossnitz et al., 1989) and in proteoliposomes (Bishop et al., 1989; Davidson and Nikaido, 1990) has finally made available the route of biochemical analysis. Using these in vitro systems it was shown clearly that ATP is hydrolyzed only concomitantly with transport. Nonhydrolyzable ATP analogues did not support transport and a photoactivatable analogue of ATP eliminated transport (Ames et al., 1989). Attempts at deriving a stoichiometry between ATP hydrolyzed and substrate molecules transported yielded values averaging about five (Bishop et al., 1989; Davidson and Nikaido, 1990). This is probably due to damage incurred by the complex during reconstitution, resulting in uncoupling between hydrolysis and transport (Bishop et al., 1989). This notion that ATP hydrolysis can be uncoupled from transport is supported by the existence of mutations located in the membrane-bound subunits that indeed do so (Petronilli and Ames, 1991; Davidson et al., 1992).

It should be noted that HisQ has also been shown to bind ATP (Shyamala, V., and G.F.-L. Ames, manuscript submitted for publication). Therefore, while it is likely that the conserved components are the units responsible for converting ATP energy into translocation work because of the homology which is well conserved throughout the entire superfamily, the possibility exists that their ATP-binding sites may instead have a regulatory function, with the hydrophobic subunit(s) performing the hydrolytic step.

Signaling Mechanism

A reasonable working model postulates that the liganded binding protein sends a signal to the membrane-bound complex which triggers conformational changes resulting in ATP hydrolysis and the opening of a pore, leading to subsequent translocation of the substrate and release of the unliganded receptor (Ames et al., 1990). In support of this model are several facts: (*a*) the receptor interacts physically with at least one of the hydrophobic components, as shown by chemical crosslinking experiments (Prossnitz et al., 1988); (*b*) the unliganded receptor interacts poorly or not at all with the membrane bound complex (Ames, G. F.-L., and K. Nikaido, unpublished data); (*c*) ATP is only hydrolyzed upon interaction of the receptor with the membrane-bound complex (Bishop et al., 1989); (*d*) mutants in the membrane-bound proteins exist that hydrolyze ATP and translocate the substrate in the absence of the receptor (Petronilli and Ames, 1991; Davidson et al., 1992). Thus ATP hydrolysis is one of the processes that must be activated after the interaction of the liganded binding protein with the membrane complex; the energy released might be used either to open the pore (or channel; see below) or close it, thus returning the system back to a state of readiness for the next cycle of transport. While the signaling mechanism is of general interest, also in the hope of using it as a model for more complex eukaryotic signaling mechanisms, here we are concerned mainly with the subsequent steps in transport.

Direction of Translocation

If the working model depicting the opening of a pore as dependent on the expenditure of ATP energy is correct, then it would be reasonable to expect that periplasmic permeases translocate irreversibly. This was indeed found to be true in reconstituted proteoliposomes (Petronilli, V., and G. F.-L. Ames, manuscript submitted for publication). Accumulated histidine did not exit from proteoliposomes, and the incorporation of ADP, inorganic phosphate, and histidine inside the proteoliposomes did not result in either the exit of histidine or the synthesis of ATP.

In an analysis related to the question of reversibility, it was shown that both the internally accumulated histidine and the hydrolytically produced ADP inhibit transport, suggesting that the internal pool of histidine (and ADP) regulates the rate of translocation. This finding, which automatically implies that there is an internal substrate-recognizing site, is particularly important because it suggests that the membrane-bound complex could bind and translocate an internal substrate molecule to the outside after an appropriate series of conformational changes, which would be the reverse of those described for a periplasmic permease. Thus a basic ancestral structure of traffic ATPases could have easily evolved either into a system able to translocate from the interior to the exterior, or vice versa, depending on each particular evolutionary history. Fig. 4 gives a schematic representation of a noncommitted traffic ATPase that could evolve in either direction (Petronilli, V., and G. F.-L. Ames, manuscript submitted for publication).

Another important consequence of this finding is that it supports the notion that the membrane-bound complex, being able to recognize the substrate, has the intrinsic ability to function in the absence of a soluble receptor. This is relevant for the question of extrapolation between periplasmic permeases and the (apparently) receptor-less eukaryotic traffic ATPases and will be discussed later. Thus elimination of the receptor from the scheme shown in Fig. 2 would only require that the substrate be present in a sufficiently high concentration to bind to the presumed membrane site.

Comparison with Eukaryotic Traffic ATPases

The general structure of eukaryotic traffic ATPases is that of a single polypeptide composed of two homologous halves, each half comprising a hydrophobic and a hydrophilic domain. This monocomponent structure is in apparent contrast to that of the multicomponent prokaryotic equivalents. However, the hydrophilic domain bears an extensive sequence similarity to the conserved components of periplasmic permeases, suggesting a common ancestry and similar mechanism of action. This obvious evolutionary relationship has been noted and commented upon repeatedly (Ames, 1986b; Roninson, 1991; Ames et al., 1992; Buschman and Gros, 1993; Gros and Buschman, this volume). No sequence similarity has been uncovered between the hydrophobic domains of the various families of traffic ATPases; however, the predicted secondary structures of all of these domains are similar, generally with a total of 10–12 membrane spanners and with a stoichiometry of two hydrophobic

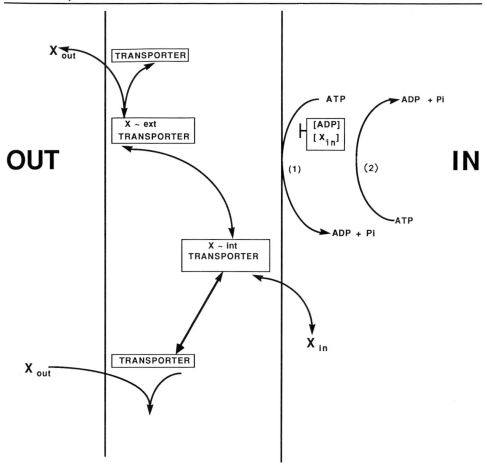

Figure 4. Reversibility of a generic traffic ATPase. The transporter is represented as becoming liganded at either the outer or inner membrane surface. In the mechanism labeled *1*, ATP hydrolysis causes a rearrangement that renders the transporter-bound substrate accessible at the inner surface, allowing its release. In the mechanism labeled *2*, ATP hydrolysis causes the reverse rearrangement, resulting in the extrusion of the substrate through an essentially identical transporter. The presence of a binding protein is not necessary for either mechanism. Accessibility of the transporter at the inner or outer membrane is hypothesized to be mediated through a series of conformational changes. (Reproduced from *International Review of Cytology,* 1992, 137a:1–35, by copyright permission of Academic Press.)

domains to two ATP-binding domains. A possible evolutionary history for traffic ATPases envisions an ancestral system composed of genes coding for a single hydrophobic domain and a single ATP-binding domain, which underwent various duplications and fusions (Ames and Lecar, 1992). The latter are a common finding when proteins from eukaryotes are compared with those from prokaryotes.

Several lists of eukaryotic traffic ATPases have been made and the list is rapidly getting longer (see this volume). Here I will concentrate on only one representative, CFTR, to be used as a model system (in part because of the availability of numerous cystic fibrosis mutations that can be viewed with a structure–function analysis in mind). CFTR has been shown to act as a chloride channel (Anderson et al., 1991*b*; Bear et al., 1992), thus providing a biochemical explanation for some of the physiological characteristics of the disease. CFTR comprises two homologous fused halves, as described above, but it also contains an extra domain, the R (i.e., regulatory) domain fused in between the two halves. Interestingly, this domain, which is phosphorylated, seems to have a regulatory function (Anderson et al., 1991*a*), which is reminiscent once more of a similar finding in a periplasmic permease: the conserved component of the maltose permease, MalK, in which the carboxyl-terminal third of the protein is involved in sensing and transmitting cellular signals rather than in translocation (Kühnau et al., 1991). A fusion involving an ancestor carrying such a regulatory region would indeed place it in between the two fused halves. The CFTR channel requires ATP in order to be opened, and evidence has been provided that ATP hydrolysis is required (Anderson et al., 1991*a*).

If the recently discovered notion that the conserved component is accessible from the external surface of the membrane, therefore spanning it to some extent, is correct (Baichwal et al., 1993), then it is possible that the nucleotide-binding domains of the eukaryotic traffic ATPases also take the same configuration. This would be particularly important in the case of CFTR, since the interpretation of the numerous mutants available would be consequently affected. For example, a large group of mutants is located in the glutamine-, glycine-rich region that has been postulated to be involved in signal transmission between the ATP-binding and -hydrolyzing pocket and the helical domain. At present this entire domain is postulated to be located on the inside of the membrane, possibly interacting with the hydrophobic domains. This possibility should be considered when interpreting the nature of the defect in CFTR mutants. Similarly, a number of mutants are located in regions assigned to the ATP-binding pocket. Apart from those that are clearly located in the sequences characteristic of ATP-binding motifs, many may be the equivalent of the uncoupled, receptor-independent mutants identified in the periplasmic permeases and as such their defect may reside in their being uncoupled and therefore constantly open, thus resulting in a continuous and improper chloride ion flux.

Analogies between Channels and Transporters

The channel activity uncovered in CFTR raises the question of whether periplasmic permeases, which have been referred to in the past as transporters or active transporters, can be used as model systems for understanding the mechanism of action of CFTR and other eukaryotic traffic ATPases. If channels were known to function in a unique fashion and had a unique biochemical structure different from that of any known transporter, the argument would be a valid one. However, it appears that there is no inherent and insurmountable discrepancy between channels and periplasmic transporters and that much of the distinction may be semantic (Ames and Lecar, 1992).

The distinction between channels and transporters has been addressed in reviews and textbooks (e.g., Harold, 1987). In brief, both channels and transporters can be viewed as proteinaceous structures that undergo conformational changes to allow the passage of substrate. An "incipient channel lined with polar groups" (Harold, 1986) can be visualized as an ancestral characteristic of both entities. Such an ancestral system could evolve in the direction of a membrane-spanning pathway that, when properly activated, allows substrate access to binding sites simultaneously from both sides of the membrane, thus forming a typical channel. In this respect, a periplasmic permease might be viewed as a channel that maintains occlusion of one of the two outlets of the channel (e.g., by way of the binding protein); alternatively, it might be viewed as having evolved a conformational change mechanism in which channel activation would result in alternating access of a binding site first to one side of the membrane and then to the other, which would be a representation closer to the classical view of a transporter. In either case, the distinction between channels and at least this kind of transporters would become blurred, residing in whether or not an asymmetry of the binding sites exposed by the respective conformational changes is created.

A comparison of several properties considered to be characteristic of either channels or transporters has been made (Ames and Lecar, 1992) which also indicates that many of the presumed distinctions may be semantic. In fact, it seems possible that in either a periplasmic transporter or a classic channel the substrate travels through a pore along which there are binding sites with varying affinities which would contribute to determining the speed of travel. Channels, the activity of which is routinely measured, may have low-affinity binding sites that allow a fast flow of the substrate, combined with relatively slow opening and closing events. These are indeed the properties that render them detectable. Channels with very high affinity binding sites may not be detectable by present technologies and may indeed be so slow as to be indistinguishable from transporters as far as the rate of transport is concerned. In this respect it is interesting that several translocating proteins, e.g., the H^+-ATPase of yeast (Ramirez et al., 1990) and the mitochondrial aspartate/ glutamate carrier (Dierks et al., 1990) have been shown to exhibit channel activity under particular circumstances.

The evidence that the CFTR channel is opened via the expenditure of ATP energy (Anderson et al., 1991*a*) eliminates the question of the mechanism of energization as being an insurmountable difference between channels and transporters and aligns the mechanism of action of CFTR very closely with that of periplasmic permeases.

The question at this point should rather be whether periplasmic permeases could be viewed as channels more or less extensively modified.

Periplasmic Permeases as Channels?

This possibility calls for some speculation since no attempts have been made at measuring channel activity in these systems. Such measurements would have to surmount rather large problems for the following reasons. Let us imagine that a periplasmic permease is indeed a channel that is opened only when the liganded protein signals its presence and ATP energy is consumed. In such a case, the receptor would be located right at the external mouth of the channel and would be presenting it with a molecule (a single one!) of substrate. The molecule might be transferred to a substrate-binding site situated somewhere within the channel, and by virtue of the fact that the channel is occluded on one side, it would be able to travel in only one direction. Alternatively, the substrate would be released from the binding protein into a small chamber where its local concentration would be quite high; it could then move down its concentration gradient toward the interior of the cell. In either case the substrate concentration, in the liganded form, would be quite high at the outer mouth of the channel because it would reflect the high concentration of the receptor to which it is liganded; thus, reaching equal substrate concentrations on the two sides of the membrane would amount to creating a concentration gradient when compared with the free external substrate concentration (see above).

The problem obviously is to measure the movement of a single molecule at the time in a situation in which the channel cannot be opened without the receptor simultaneously blocking its opening. The solution to the problem would be to mutate a periplasmic permease so that it does not require the receptor for the channel to open. Such mutants were indeed obtained in both the histidine and the maltose

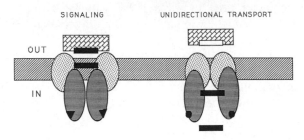

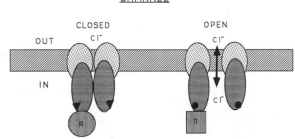

Figure 5. Are transporters and channels entirely different? Schematic representation of traffic ATPases, as they might function as either a transporter or a channel. Transporters are postulated to carry a substrate-binding site in the membrane-bound complex, and are imagined to open into a gated conformation when translocating the substrate. The channel shown here includes an R domain, known to exist for CFTR; the R domain is postulated to have a regulatory function, holding the channel in a locked conformation which is unlocked upon its phosphorylation (*solid box*), thus allowing ATP hydrolysis to open the channel. (Modified from Ames and Lecar, 1992.)

permeases (Petronilli and Ames, 1991; Davidson et al., 1992). They hydrolyze ATP continuously and in the total absence of a signal from the receptor, and they require high concentrations of substrate, as would be predicted for a substrate-binding site with poorer affinity located on the translocation complex. In other words, these may be channels that are uncoupled from an external signal and are continuously open, equilibrating the substrate across the membrane, but not concentrating it. These mutants offer a good chance of investigating whether periplasmic permeases function as channels by using standard methodologies.

Another important conclusion to be drawn from these receptor-independent mutants is that the resemblance with eukaryotic traffic ATPases, which have not been found to require a soluble receptor, becomes even more obvious because they indicate that the prokaryotic receptor's main function may be that of supplying the membrane-bound site with high concentrations of substrate, having no translocation function of its own. Teleogically speaking, the receptors might have evolved in prokaryotes to supply usually scarce substrates in concentrated form at the membrane surface, thus acting as a trapping mechanism to allow scavenging of precious nutrients. Eukaryotic cells, being normally exposed to high concentrations of substrates, may have lost the receptor in evolution because it became obsolete (Ames, 1992).

Conclusions

It seems that the numerous similarities between the prokaryotic and eukaryotic traffic ATPases would justify considering them as close relatives and extrapolating mechanistic conclusions between them, to be done judiciously and supported experimentally. A reconciliation between the properties of channels and transporters (Fig. 5) would receive considerable support if periplasmic permeases were shown to function as channels and if new information relative to the secondary and topological organizations of the eukaryotic traffic ATPases was obtained that confirmed a resemblance with those of the prokaryotic ones. The purification and biochemical characterization of the individual components will undoubtedly provide support for many of the features of various working models, and extensive structure–function studies will aid in the effort.

Acknowledgments

This paper, resulting as it does from a conference that brings together people working in such disparate areas, would not be complete without a special recognition of the role played by the National Institutes of Health, which provided support for many years to the research performed in my laboratory. Even when no immediate practical aspects were obvious, their support and faith in the rich fruits to be reaped from the support of basic research never failed. My special thanks go to the National Institute of Diabetes and Digestive and Kidney Diseases, to the American public for supporting so much of the research, and to all the present and past members of my laboratory who contributed various aspects of the knowledge accumulated over the years on periplasmic permeases.

References

Adams, M. D., and D. L. Oxender. 1989. Bacterial periplasmic binding protein tertiary structures. *Journal of Biological Chemistry*. 256:560–562.

Ames, G. F.-L. 1985. The histidine transport system of *Salmonella typhimurium*. *Current Topics in Membranes and Transport*. 23:103–119.

Ames, G. F.-L. 1986a. Bacterial periplasmic transport systems: structure, mechanism, and evolution. *Annual Review of Biochemistry*. 55:397–425.

Ames, G. F.-L. 1986b. The basis of multidrug resistance in mammalian cells: homology with bacterial transport. *Cell*. 47:323–324.

Ames, G. F.-L. 1990. Energetics of periplasmic transport systems: the histidine permease as a model system. *Research in Microbiology*. 141:341–348.

Ames, G. F.-L. 1992. Bacterial periplasmic permeases as model systems for the superfamily of traffic ATPases which includes MDR and CFTR. *International Review of Cytology.* 137a:1–35.

Ames, G. F.-L., and H. Lecar. 1992. Mechanism of action of ATP-dependent bacterial transporters and of cystic fibrosis: some thoughts about analogies between channels and transporters. *FASEB Journal.* 6:2660–2666.

Ames, G. F.-L., C. Mimura, S. Holbrook, and V. Shyamala. 1992. Traffic ATPases: a superfamily of transport proteins operating from *Escherichia coli* to humans. *Advances in Enzymology.* 65:1–47.

Ames, G. F.-L., C. Mimura, and V. Shyamala. 1990. Bacterial periplasmic permeases belong to a family of transport proteins operating from *E. coli* to humans: traffic ATPases. *FEMS Microbiology Reviews.* 75:429–446.

Ames, G. F.-L., K. Nikaido, J. Groarke, and J. Petithory. 1989. Reconstitution of periplasmic transport in inside-out membrane vesicles: energization by ATP. *Journal of Biological Chemistry.* 264:3998–4002.

Anderson, M. P., H. A. Berger, D. P. Rich, R. J. Gregory, A. E. Smith, and M. J. Welsh. 1991a. Nucleoside triphosphates are required to open the CFTR chloride channel. *Cell.* 67:775–784.

Anderson, M. P., R. J. Gregory, S. Thompson, D. W. Souza, S. Paul, R. C. Mulligan, A. E. Smith, and M. J. Welsh. 1991b. Demonstration that CFTR is a chloride channel by alteration of its anion selectivity. *Science.* 253:202–205.

Baichwal, V., D. Liu, and G. F.-L. Ames. 1993. The ATP-binding component of a prokaryotic traffic ATPase is exposed to the periplasmic (external) surface. *Proceedings of the National Academy of Sciences, USA.* 90:620–624.

Bear, C. E., C. Li, N. Kartner, R. J. Bridges, T. J. Jensen, M. Ramjeesingh, and J. R. Riordan. 1992. Purification and functional reconstitution of the cystic fibrosis transmembrane conductance regulator (CFTR). *Cell.* 68:809–818.

Bishop, L., R. J. Agbayani, S. V. Ambudkar, P. C. Maloney, and G. F.-L. Ames. 1989. Reconstitution of a bacterial periplasmic permease in proteoliposomes and demonstration of ATP hydrolysis concomitant with transport. *Proceedings of the National Academy of Sciences, USA.* 86:6953–6957.

Buschman, E., and P. Gros. 1993. The mouse multidrug resistance gene family: structural and functional analysis. *International Review of Cytology.* 137C:169–197.

Davidson, A. L., and H. Nikaido. 1990. Overproduction, solubilization, and reconstitution of the maltose transport system from *Escherichia coli. Journal of Biological Chemistry.* 265:4254–4260.

Davidson, A. L., and H. Nikaido. 1991. Purification and characterization of the membrane-associated components of the maltose transport system from *Escherichia coli. Journal of Biological Chemistry.* 266:8946–8951.

Davidson, A. L., H. A. Shuman, and H. Nikaido. 1992. Mechanism of maltose transport in *Escherichia coli*-transmembrane signaling by periplasmic binding proteins. *Proceedings of the National Academy of Sciences, USA.* 89:2360–2364.

Dean, D. A., J. D. Fikes, K. Gehring, P. J. J. Bassford, and H. Nikaido. 1989. Active transport of maltose in membrane vesicles obtained from *Escherichia coli* cells producing tethered maltose-binding protein. *Journal of Bacteriology.* 171:503–510.

Dierks, T., A. Salentin, and R. Krämer. 1990. Pore-like and carrier-like properties of the mitochondrial aspartate/glutamate carrier after modification by SH-reagents: evidence for a preformed channel as a structural requirement of carrier-mediated transport. *Biochimica et Biophysica Acta.* 1028:281–288.

Furlong, C. E. 1987. Osmotic-shock-sensitive transport systems. *In Escherichia coli* and *Salmonella typhimurium:* Cellular and Molecular Biology. F. C. Neidhardt, editor. American Society for Microbiology, Washington, DC. 768–796.

Harold, F. M. 1987. The Vital Force: A Study of Bioenergetics. W. H. Freeman and Co., New York. 577 pp.

Hiles, I. D., M. P. Gallagher, D. J. Jamieson, and C. F. Higgins. 1987. Molecular characterization of the oligopeptide permease of *Salmonella typhimurium. Journal of Molecular Biology.* 195:125–142.

Hyde, S. C., P. Emsley, M. J. Hartshorn, M. M. Mimmack, U. Gileadi, S. R. Pearce, M. P. Gallagher, D. R. Gill, R. E. Hubbard, and C. F. Higgins. 1990. Structural model of ATP-binding proteins associated with cystic fibrosis, multidrug resistance and bacterial transport. *Nature.* 346:362–365.

Joshi, A. K., S. Ahmed, and G. F.-L. Ames. 1989. Energy coupling in bacterial periplasmic transport systems. Studies in intact *Escherichia coli* cells. *Journal of Biological Chemistry.* 264:2126–2133.

Kerem, B., J. M. Rommens, J. A. Buchanan, D. Markiewicz, T. K. Cox, A. Chakravarti, M. Buchwald, and L. C. Tsui. 1989. Identification of the cystic fibrosis gene: genetic analysis. *Science.* 245:1073–1080.

Kerppola, R. E. 1990. Architecture of membrane-bound components of the histidine permease of *Salmonella typhimurium.* Ph.D. thesis. University of California, Berkeley, Berkeley, CA.

Kerppola, R. E., and G. F.-L. Ames. 1992. Topology of the hydrophobic membrane-bound components of the histidine periplasmic permease: comparison with other members of the family. *Journal of Biological Chemistry.* 267:2329–2336.

Kerppola, R. E., V. Shyamala, P. Klebba, and G. F.-L. Ames. 1991. The membrane-bound proteins of periplasmic permeases form a complex: identification of the histidine permease HisQMP complex. *Journal of Biological Chemistry.* 266:9857–9865.

Kolter, R., and M. Fath. 1993. *Microbiological Reviews.* In press.

Kühnau, S., M. Reyes, A. Sievertsen, H. A. Shuman, and W. Boos. 1991. The activities of the *Escherichia coli* MalK protein in maltose transport, regulation, and inducer exclusion can be separated by mutations. *Journal of Bacteriology.* 173:2180–2186.

Mao, B., M. P. Pear, J. A. McCammon, and F. A. Quiocho. 1982. Hinge-bending in L-arabinose binding protein. *Journal of Biological Chemistry.* 257:1131–1133.

Mimmack, M. L., M. P. Gallagher, S. R. Pearce, S. C. Hyde, I. R. Booth, and C. F. Higgins. 1989. Energy coupling to periplasmic binding protein-dependent transport systems: stoichiometry of ATP hydrolysis during transport *in vivo. Proceedings of the National Academy of Sciences, USA.* 86:8257–8260.

Mimura, C. S., A. Admon, K. A. Hurt, and G. F.-L. Ames. 1990. The nucleotide-binding site of HisP, a membrane protein of the histidine permease: identification of amino acid residues photoaffinity-labeled by 8-azido ATP. *Journal of Biological Chemistry.* 265:19535–19542.

Mimura, C. S., S. R. Holbrook, and G. F.-L. Ames. 1991. Structural model of the nucleotide-binding conserved component of periplasmic permeases. *Proceedings of the National Academy of Sciences, USA.* 88:84–88.

Neu, H. C., and L. A. Heppel. 1965. The release of enzymes from *Escherichia coli* by osmotic shock and during the formation of spheroplasts. *Journal of Biological Chemistry.* 240:3685–3692.

Nikaido, K., and G. F.-L. Ames. 1992. Purification and characterization of the periplasmic lysine-, arginine-, ornithine-binding protein (LAO) from *Salmonella typhimurium. Journal of Biological Chemistry.* 267:20706–20712.

Oh, B.-H., J. Pandit, C.-H. Kang, K. Nikaido, S. Gokcen, G. F.-L. Ames, and S.-H. Kim. 1993. 3-Dimensional structures of the periplasmic lysine-, arginine-, ornithine-binding protein with and without a ligand. *Journal of Biological Chemistry.* In press.

Petronilli, V., and G. F.-L. Ames. 1991. Binding protein-independent histidine permease mutants: uncoupling of ATP hydrolysis from transmembrane signaling. *Journal of Biological Chemistry.* 266:16293–16296.

Prossnitz, E., A. Gee, and G. F.-L. Ames. 1989. Reconstitution of the histidine periplasmic transport system in membrane vesicles: energy coupling and interaction between the binding protein and the membrane complex. *Journal of Biological Chemistry.* 264:5006–5014.

Prossnitz, E., K. Nikaido, S. J. Ulbrich, and G. F.-L. Ames. 1988. Formaldehyde and photoactivatable crosslinking of the periplasmic binding protein to a membrane component of the histidine transport system of *Salmonella typhimurium. Journal of Biological Chemistry.* 263:17917–17920.

Ramirez, J. A., H. Lecar, S. Harris, and J. Haber. 1990. Patch-clamp studies of print and regulatory mutants of yeast plasma membrane ATP-ase (PMAI). *Biophysical Journal.* 57:321a. (Abstr.)

Roninson, I. B. 1991. Molecular and Cellular Biology of MDR in Tumor Cells. I. B. Roninson, editor. Plenum Publishing Corp., New York. 400 pp.

Shuman, H. A. 1987. The genetics of active transport in bacteria. *Annual Review of Genetics.* 21:155–177.

Shyamala, V., V. Baichwal, E. Beall, and G. F.-L. Ames. 1991. Structure-function analysis of the histidine permease and comparison with cystic fibrosis mutations. *Journal of Biological Chemistry.* 266:18714–18719.

Speiser, D. M., and G. F.-L. Ames. 1991. Mutants of the histidine periplasmic permease of *Salmonella typhimurium* that allow transport in the absence of histidine-binding proteins. *Journal of Bacteriology.* 173:1444–1451.

Walker, J. E., M. Saraste, M. J. Runswick, and N. J. Gay. 1982. Distantly related sequences in the α and β-subunits of ATP synthase, myosin, kinases and other ATP-requiring enzymes and a common nucleotide binding fold. *EMBO Journal.* 1:945–951.

Chapter 8

The Multidrug Resistance Transport Protein: Identification of Functional Domains

Philippe Gros and Ellen Buschman

Department of Biochemistry, McGill University, Montreal, Quebec, Canada, H3G 1Y6

Molecular Biology and Function of Carrier Proteins © 1993 by The Rockefeller University Press

Introduction

One major limitation to the successful chemotherapeutic treatment of many types of human tumors is the emergence and outgrowth of subpopulations of drug-resistant cells. These populations of tumor cells often display cross-resistance to cytotoxic compounds that share little or no structural or functional similarity, and to which they have not been previously exposed (Moscow and Cowan, 1988). Drugs that form the multidrug resistance (MDR) spectrum include anthracyclines (such as Adriamycin), Vinca alkaloids (vincristine, vinblastine), colchicine, actinomycin D, etoposides (VP-16, VM-21), topoisomerase inhibitors (amsacrine), and many others (for example, see Gupta et al., 1988). There are only a few structural and functional similarities between these compounds: they are small, biplanar, hydrophobic molecules that have a basic nitrogen atom (hydrophobic cations) and that enter the cell by passive diffusion across the membrane lipid bilayer (Gupta et al., 1988; Zamora et al., 1988; Pearce et al., 1989; Beck, 1990).

Most of our understanding of this phenomenon has been derived from the in vitro study of MDR cell lines obtained by continuous exposure to increasing concentrations of a given cytotoxic MDR drug (stepwise selection; Biedler and Riehm, 1970; Ling and Thompson, 1974; Fojo et al., 1985; Lemontt et al., 1988). The phenotypic characteristics of independently derived MDR cell lines are remarkably similar despite the diversity of their tissue origin and in vitro selection protocols used in their production (for comprehensive reviews, see Bradley et al., 1988; Endicott and Ling, 1989; Roninson, 1991). First, MDR cell lines display a strikingly similar pattern of cross-resistance to the group of natural products that form the MDR spectrum. Second, the emergence of MDR is linked to a decreased degree of intracellular drug accumulation and to a concomitant increase in cellular drug efflux, both of which are ATP dependent (Dano, 1973; Skovsgaard, 1978). Third, although the levels of several polypeptides such as cytoplasmic calcium-binding proteins (Koch et al., 1986), enzymes of the glycosylation pathway (Peterson et al., 1979), and membrane glycoproteins (Peterson et al., 1983; Richert et al., 1985) have been found to be modulated in MDR cell lines, the most ubiquitous phenotypic marker of MDR is the overexpression of a membrane protein called P-glycoprotein or P-gp. P-gps are a heterogeneous group of membrane phosphoglycoproteins of apparent molecular weight of 170,000–200,000, first described by Juliano and Ling (1976). P-gp has since been shown to bind photoactivatable analogues of ATP (Cornwell et al., 1987*b*; Schurr et al., 1989) and cytotoxic drugs (Cornwell et al., 1986; Safa et al., 1986, 1989), and has been shown to possess ATPase activity (Hamada and Tsuruo, 1988), leading to the proposal that P-gp functions as an ATP-dependent drug efflux pump to reduce the intracellular accumulation of cytotoxic compounds of the MDR spectrum. Finally, the MDR phenotype can be reversed in drug-resistant cells overexpressing P-gp by a heterogeneous group of compounds including verapamil (channel blocker; Tsuruo et al., 1981), trifluoperazin (calmodulin inhibitor; Tsuruo et al., 1982), cyclosporin A (immunosuppressor; Slater et al., 1986), quinidine (Tsuruo et al., 1984), reserpine (Pearce et al., 1989), and several others (reviewed by Beck and Danks, 1991). Although the mechanism of MDR reversal by these compounds is not yet fully understood, it appears that some of them behave as competitive inhibitors for drug binding sites on P-gp (Cornwell et al., 1987*a*; Safa et al., 1987; Akiyama et al., 1988; Safa, 1988; Kamiwatari et al., 1989; Pearce et al., 1989; Tamai and Safa, 1990).

The mechanism by which P-gp can apparently recognize and prevent the intracellular accumulation of a large group of structurally and functionally unrelated compounds remains mysterious. In particular, P-gp domains and discrete amino acid residues implicated in substrate recognition and transport have yet to be identified. The successful isolation of molecular clones encoding several P-gp isoforms has shed considerable light on the proposed structure of this group of proteins and has provided new analytical tools for the study of its function. In addition, molecular cloning experiments have shown that P-gps are part of a large group of structurally and functionally related membrane proteins that have been highly conserved during evolution.

Cloned *mdr* Genes and Predicted P-gp Polypeptides

Molecular clones corresponding to human, mouse, and hamster P-gps have been independently obtained. These experiments have shown that P-gps are encoded by a small group of closely related genes, termed *mdr,* and full-length cDNA clones corresponding to the cellular transcripts of two human *MDR* genes, *MDR1* and *MDR2* (Chen et al., 1986*a*; Van der Bliek et al., 1988), and three rodent *mdr* genes, *mdr1, mdr2,* and *mdr3,* have been isolated and sequenced from mouse (Gros et al., 1986*b*, 1988; Devault and Gros, 1990; Hsu et al., 1990) and hamster (Ng et al., 1989). Nucleotide and predicted amino acid sequence analyses indicate that these genes encode highly homologous P-gps. The prototype P-gp is composed of 1,276 amino acids (mouse) and can be subdivided into two halves, each half encoding a large hydrophilic domain and a large hydrophobic domain. Each hydrophobic region contains six highly hydrophobic segments characteristic of membrane-associated domains, which can be arranged into three transmembrane (TM) loops. Each hydrophilic segment contains a consensus sequence for nucleotide binding (NB site) that has been described in a variety of ATP binding proteins and kinases (Higgins et al., 1986). These sequence analyses are in agreement with previous biochemical characterizations of P-gp and suggest that P-gp is a membrane glycoprotein capable of binding ATP. The proposed membrane organization of P-gp, i.e., 12 TM domains and two intracellular ATP binding sites, is in agreement with the results of epitope mapping of proteolytic fragments of P-gp identified with specific antibodies directed against synthetic P-gp peptides of known sequence, or against fusion proteins containing P-gp subfragments (Bruggemann et al., 1989; Yoshimura et al., 1989). The two halves of P-gp share up to 38% identical residues (62% overall homology), and it has been proposed that at least part of the protein evolved from the duplication of an ancestral unit. The conservation of the position of specific introns within each of the two predicted NB sites (Raymond and Gros, 1989; Chen et al., 1990) sustains this contention. Sequence homology between the two halves of individual P-gps is highest within the predicted NB domains, decreases within the predicted TM domains, and disappears in the extreme 5' end of each half, which corresponds to the amino terminus of the protein and the highly charged so-called "linker" domain, which links the two halves together. Surprisingly, P-gp segments overlapping the NB sites were found to be homologous (30%) to a group of bacterial periplasmic transport proteins such as HisP (Higgins et al., 1982), MalK (Gilson et al., 1982), OppD (Ames, 1986), and PstB (Surin et al., 1985) implicated in the high affinity TM transport of histidine, maltose, oligopeptides, and phosphates into *Escherichia coli.* An even stronger homology (50%) is detected with a second group of bacterial proteins such as HlyB (Felmlee et al., 1985), ChvA (Cangelosi et al., 1989), CyaB (Glaser et al., 1988), ArsA (Chen et al., 1986b), and LktB (Stanfield et al., 1988), implicated in the ATP-dependent export of hemolysin, toxins, and arsenate in Gram-negative bacteria. In the case of HlyB, the homology to P-gp not only overlaps the NB sites but also involves the TM domains, and both proteins share almost superimposable hydropathy profiles (Gros et al., 1988). A partial listing of bacterial proteins sharing sequence homology and predicted secondary structure with P-gp is presented in Table I. The predicted structural features of P-gp as well as its remarkable sequence homology to prokaryotic transport proteins implicated in extracellular export of specific substrates are also in agreement with the proposition that P-gp functions as an ATP-dependent efflux pump.

In rodents, sequence analyses indicate that the three genes arose from a common ancestor by two successive gene duplication events, the most recent one producing *mdr1* and *mdr3,* which are more homologous to each other than they are to *mdr2* (Devault and Gros, 1990). In humans, a single gene duplication event would have generated the two genes, *MDR1* and *MDR2* (Chin et al., 1989; Ng et al., 1989).

The *mdr* Super Gene Family

The *mdr* gene family is itself part of a larger supergene family of sequence related members whose structural and functional aspects have been preserved throughout evolution. *mdr* homologues and *mdr*-like genes from this family have been identified in a number of distantly related organisms. The amino acid sequence homology between *mdr* genes and these *mdr*-like homologues is modest (20–40%), and is clustered mostly around the predicted NB sites. However, the predicted polypeptides have very similar length and identical hydropathy profiles, and consequently share highly similar predicted secondary structures. In the yeast *Saccharomyces cerevisiae,* an *mdr* homologue called *STE-6* has been discovered (McGrath and Varshavsky, 1989) and found responsible for the transport of the mating pheromone *a* factor across the membrane of *a* cells (Kuchler et al., 1989). In the malarial parasite *Plasmodium falciparum,* the widespread emergence of resistance to chloroquine (CLQ) is a major health threat to human populations of the third world. CLQ resistance in *P. falciparum* is caused by an increased drug efflux from the parasite which can be competed by CLQ analogues, but surprisingly, also by vinblastine and verapamil, two inhibitors of mammalian P-gp (Krogstad et al., 1987; Martin et al., 1987; Bitoni et al., 1988). An *mdr* homologue called *pfmdr 1* has been cloned from *P. falciparum* and has been found amplified, overexpressed, and/or mutated in CLQ-resistant isolates of the parasite (Foote et al., 1989; Wilson et al., 1989). Likewise, in the kinetoplastidae *Leishmania tarentolae,* a group of five *mdr* homologues designated *ltpgp* has been isolated, with some members of this family found amplified and overexpressed in arsenate- and methotrexate-resistant isolates of this parasite (Ellenberger and Beverley, 1989; Ouellette et al., 1990). Also, the emergence of emetine resistance in the enteric protozoan *Entamoeba histolytica* has been found associated with the overexpression of at least three *mdr*-related mRNA species (Samuelson et al., 1990). Finally, three *mdr* homologues have been cloned and

TABLE I
Proteins That Share Sequence Homology with P-gp (*mdr* and *mdr* Homologues)

Protein	Organism	Substrate	Reference
Import (prokaryotes)			
AraG	*E. coli*	Arabinose	Scripture et al., 1987
BtuD	*E. coli*	Vitamin B12	Friedrich et al., 1986
ChlD	*E. coli*	Molybdate	Johann and Hinton, 1987
FecE	*E. coli*	Fe hydroxamate	Staudenmaier et al., 1989
FhuC	*E. coli*	Fe citrate	Coulton et al., 1987
GlnQ	*E. coli*	Glutamine	Nohno et al., 1986
HisP	*S. typhimurium*	Histidine	Higgins et al., 1982
MalK	*E. coli*	Maltose	Gilson et al., 1982
OppD	*S. typhimurium*	Peptides	Ames, 1986
OppF	*S. typhimurium*	Peptides	Ames, 1986
PstB	*E. coli*	Phosphate	Surin et al., 1985
RbsA	*E. coli*	Ribose	Ames, 1986
Export (prokaryotes)			
ArsA	*E. coli*	Arsenate	Chen et al., 1986*b*
HlyB	*E. coli*	Hemolysin	Felmlee et al., 1985
CyaB	*B. pertussis*	Hemolysin	Glaser, et al., 1988
LktB	*R. meliloti*	Cyclic sugars	Stanfield et al., 1988
ChvA	*A. tumefaciens*	Cyclic sugars	Cangelosi et al., 1989
Others (prokaryotes)			
FtsE	*E. coli*	Cell division	Higgins et al., 1986
NodI	*R. leguminosarum*	Nodulation	Higgins et al., 1986
UvrA	*E. coli*	DNA repair	Doolittle et al., 1986
Eukaryotes			
White	*D. melanogaster*	Pteridine	O'Hare et al., 1984
Brown	*D. melanogaster*	Pteridine	Dreesen et al., 1988
Mdr (49, 65)	*D. melanogaster*	?	Wu et al., 1991
Cepgp (1,2,3)	*C. elegans*	?	Borst et al., 1993
Ltpgp (A,B,C,D,E,)	*L. tarentolae*	Multiple drugs	Ouellette et al., 1990
Pfmdr (1,2)	*P. falciparum*	Chloroquine	Foote et al., 1989
Ehpgp (1,2,3)	*E. histolytica*	Emetine	Samuelson et al., 1990
STE-6	*S. cerevisiae*	Pheromone	McGrath and Varshavosky, 1989
MpbX	*M. polymorpha*	?	Ohyama et al., 1986
Mammals			
MDR (1,2)	*Human*	Multiple drugs	Chen et al., 1986*a*
mdr (1,2,3)	*Mouse*	Multiple drugs	Gros et al., 1986*b*, 1988
CFTR	*Human*	Halides	Riordan et al., 1989
PMP70	*Human*	?	Kamijo et al., 1990
HAM (1,2)	*Mouse*	Peptides	Monaco et al., 1990
mtp (1,2)	*Rat*	Peptides	Deverson et al., 1990 Powis et al., 1991
RING (4,11)	*Human*	Peptides	Trowsdale et al., 1990; Spies et al., 1990

sequenced from the nematode *Caenorhabditis elegans* (Borst et al., 1993), and two *mdr* homologues have been cloned from the fly *Drosophila melanogaster* (Wu et al., 1991). The function of these *mdr* homologues has not been elucidated in these organisms, although their overexpression appears to confer resistance to emetine and colchicine, respectively. Finally, a chloroplast gene of as yet unknown function but encoding a protein sharing sequence homology with the bacterial MalK and HisP proteins has been identified in *Marchantia polymorpha* (Ohyama et al., 1986).

In humans, at least three *mdr*-like genes have recently been cloned and shown to participate in key physiological events. Cystic fibrosis (CF) is an inherited disorder characterized by a defect in chloride transport, and for which a candidate gene designated *CFTR* (for cystic fibrosis transmembrane conductance regulator) has been isolated by reverse genetics (Riordan et al., 1989). The *CFTR* gene is mutated in CF patients, with vast majority of mutations found within the predicted NB sites of *CFTR* (Kerem et al., 1989; Cutting et al., 1990). Direct transfection experiments of wild-type or mutant *CFTR* cDNAs in cells otherwise not expressing this gene very strongly suggest that *CFTR* is indeed the membrane-bound chloride channel that is defective in CF patients (Drumm et al., 1990; Rich et al., 1990; Anderson et al., 1991). The identification in CF patients of several inactivating, naturally occuring mutations in the *CFTR* gene has pointed out key functional residues in the protein and facilitated its structure function analysis in in vitro transfection systems (Cheng et al., 1991; Rich et al., 1991; Tabcharani et al., 1991). A pair of closely related *mdr* homologues have been discovered within the major histocompatibility complex (MHC) of both humans and rodents. The chromosomal region carrying these two genes is either deleted or rearranged in cell lines that fail to properly present antigens to their surface in association with class I MHC antigens. The genes designated *HAM* (Monaco et al., 1990), *mtp* (Deverson et al., 1990; Powis et al., 1991), *RING* (Trowsdale et al., 1990), and *PSF* (Spies et al., 1990), encode for proteins exactly half the size of P-gp, and containing only six TM domains and one NB site. It has been proposed that these two proteins work in concert to transport short soluble peptides from proteolyzed antigens across the reticulum endoplasmic membrane, and eventually to the cell surface in association with class I MHC antigens for presentation to other cells of the immune system (Spies and De Mars, 1991). Finally, a protein sharing sequence homology with P-gp has been identified in the membrane of peroxysomes (Kamijo et al., 1990). This protein, termed PMP70, has an apparent molecular weight of 70,000, and shows a structure similar to the *HAM* gene products and half the size of P-gp. The function of this protein is still uncertain, but it has been proposed to be part of the import machinery of peroxysomes in general, and possibly an acyl CoA carrier in particular (Kamijo et al., 1990). Taken together, these observations indicate that the *mdr* supergene family is composed of an increasing number of genes that encode proteins in which the basic structural unit is conserved (at least at the predicted secondary structure level), and that carry out membrane-associated transport of unrelated substrates.

Functional Analysis of *mdr* Genes in Transfected Cells

The functional role of *mdr* genes in the establishment of MDR has been addressed in cell clones transfected and stably expressing individual members of the mouse (Gros et al., 1986*a*) and human (Ueda et al., 1987) *mdr* family. These studies have also provided a starting point and general model system to carry out functional analysis of wild-type, chimeric, and mutant P-gp. In our laboratory, full-length cDNAs for individual *mdr* genes were cloned into the mammalian expression vector pMT2 and introduced in drug-sensitive cells either directly or after cotransfection with a dominant selectable marker such as the Tn5 (*neo*) gene (Southern and Berg, 1982). In this assay, the mouse *mdr1* cDNA can directly confer resistance to Adriamycin, colchicine, vinblastine, actinomycin D, and gramicidin D, but not gramicidin S or bleomycin, two non-MDR drugs (Devault and Gros, 1990). Drug-resistant *mdr1* transfectants express a 180–200-kD membrane glycoprotein that is highly phosphorylated and capable of combining photoactivatable analogues of ATP and known P-gp substrates (Schurr et al., 1989). The binding of ATP is specific and restricted to two discrete tryptic peptides detected by two-dimensional electrophoresis and chromatographic analysis (Schurr et al., 1989). The emergence of MDR in these *mdr1* transfectants is linked to a decreased intracellular drug accumulation and an increased drug efflux, both of which are dependent on the presence of intact intracellular nucleotide triphosphate pools and can be inhibited by combined treatment with deoxyglucose and rotenone (Hammond et al., 1989). By contrast, transfection and overexpression of the mouse *mdr2* cDNA fails to confer an MDR phenotype (Gros et al., 1988). Clones stably transfected with *mdr2* overexpress a 160-kD glycoprotein that is enriched in the membrane fraction (Buschman, E., and P. Gros, unpublished observations). However, *mdr2* overexpression fails to confer any detectable increase in resistance of LR73 cells not only to classical MDR drugs, but also to non-MDR drugs such as 5-fluorouracil, *cis*-platinum, cyclophosphamide, or chlorambucil (Gros, P., unpublished observations). Finally, a full-length cDNA for the *mdr3* gene has also been shown to be capable of conferring MDR to LR73 cells (Devault and Gros, 1990). Interestingly, the apparent MDR phenotype conferred by *mdr3* appears to be qualitatively distinct from that encoded by the other biologically active *mdr* gene, *mdr1*. In independent cell clones expressing similar amounts of both polypeptides, it appeared that although both cDNAs could confer similar degrees of resistance to vinblastine, *mdr1* conferred levels of colchicine and Adriamycin resistance 10-fold greater than those conveyed by *mdr3*. In the case of actinomycin D the situation was reversed, as *mdr3* conveyed higher levels of resistance to this drug than did *mdr1* (Devault and Gros, 1990). Taken together, these results clearly indicate strong functional differences (*mdr1/3* vs. *mdr2*) between individual members of the mouse *mdr* gene family, despite considerable sequence homologies. Similar findings have been obtained from the independent analysis of the human *MDR* gene family. Although the *MDR1* gene can confer MDR (Ueda et al., 1987), transfection and overexpression of the protein encoded by the *MDR2* gene fails to alter the drug-sensitive phenotype of recipient BRO cells (Schinkel et al., 1991).

Analysis of Chimeric Genes Obtained by Exchanging Homologous Domains of *mdr1* and *mdr2*

We have taken advantage of the strong sequence homology and striking functional differences detected in transfection experiments between *mdr1* and *mdr2* to identify the domains of *mdr1* that are essential for MDR and that may be functionally distinct in *mdr2*. We have constructed chimeric cDNA molecules in which discrete domains of *mdr2* have been introduced into the homologous region of *mdr1* and analyzed these chimeras for their capacity to transfer MDR (Buschman and Gros, 1991). The two predicted nucleotide-binding (NB) sites of *mdr2* share almost complete sequence identity with those of *mdr1*. Both *mdr2* NB sites were found to be functional, since either could independently complement the biological activity of *mdr1*. Thus, it appears that the NB domains of *mdr2* are functional and it is likely that these sites are involved in the common mechanism of action of the two proteins, probably the energy coupling component of the transport system. Likewise, a chimeric molecule in which the highly sequence-divergent linker domain of *mdr2* had been introduced at the homologous position of *mdr1* could also confer resistance to Adriamycin and colchicine. It is interesting to note that this linker domain contains in *mdr1* the only two consensus sites for cyclic AMP-dependent protein kinase A found in *mdr1*. Since these sites are not conserved in *mdr2*, it is unlikely that they play a key mechanistic or regulatory role in the drug efflux function of *mdr1*. However, the replacement of either the amino or carboxy terminus membrane-associated domains (TM domain, intra- and extracellular loops) of *mdr1* by the homologous segments of *mdr2* resulted in inactive chimeras. The replacement of as few as two TM domains from either the amino (TM5-6) or the carboxy (TM7-8) half of *mdr1* by the homologous *mdr2* region was sufficient to destroy the activity of *mdr1*. These results suggest that the TM regions of *mdr1* are essential for its capacity to confer MDR and are functionally distinct in *mdr2*. The data also suggest that the functional differences detected between the two proteins are not encoded by a single TM segment but rather involve protein domains present in both homologous halves of each protein. Finally, the observation that *mdr2* differs from *mdr1* and *mdr3* by only a few nonconserved residues within the TM5-6 and TM7-8 intervals suggests that functional differences detected between *mdr1* and *mdr2* in these regions may be encoded by a very small number of amino acid residues. Taken together, these findings are compatible with the proposal that *mdr1* and *mdr2* encode for membrane-associated transport proteins that act by the same mechanism on perhaps distinct sets of substrates. It is interesting to speculate that NB domains represent functional sites common to both transport systems and that TM domains may be involved in specific substrate recognition and transport.

Functional Analysis of Chimeric Proteins Constructed by Exchanging Homologous Domains of Two P-gps Conferring Distinct Drug Resistance Profiles

P-gps encoded by the mouse *mdr1* and *mdr3* (Phe^{939}, *mdr3F*) genes confer distinct drug resistance profiles. While the *mdr1* and *mdr3F* clones confer comparable levels of vinblastine resistance, *mdr3F* confers actinomycin D resistance levels two times greater than *mdr1*, while *mdr1* confers colchicine resistance levels seven times greater than *mdr3F*. We wished to identify in chimeric proteins discrete protein domains responsible for the distinct drug resistance profiles of *mdr1* and *mdr3F*. Homologous protein domains were exchanged in hybrid cDNA clones, and the specific drug resistance profiles of chimeric proteins were determined in stably transfected cell clones expressing comparable amounts of wild-type or chimeric P-gps (Dhir and Gros, 1992). Immunoblotting experiments showed that all chimeras could be found expressed in membrane enriched fractions of transfected cell clones, and all could convey cellular drug resistance at levels above the background of nontransfected, drug-sensitive LR73 cells. For vinblastine, all chimeric constructs were found to convey similar levels of resistance. For colchicine and actinomycin D, the levels of resistance confered by the various chimeras were heterogenous, being either similar to the parental *mdr1* or *mdr3F* clones, or in many cases intermediate between the two. The preferential colchicine and actinomycin D resistance phenotypes of *mdr1* and *mdr3F,* respectively, did not segregate in chimeric proteins with any specific protein segment. Taken together, our results suggest that the preferential drug resistance phenotypes encoded by the *mdr1* and *mdr3F* clones implicate multiple protein determinants and complex interactions between the two homologous halves of the respective P-gps.

Mutational Analysis of the Predicted NB Domains of *mdr1*

The capacity of *mdr1* and *mdr3* to prevent intracellular accumulation of cytotoxic drugs in stably transfected cells is strictly ATP dependent. Analysis of the predicted amino acid sequence of P-gps encoded by these two genes reveals the presence of two putative NB folds, originally compiled by Walker et al. (1982) for a number of ATP binding proteins and ATPases, which are highly conserved in *mdr1* and *mdr3,* but also in *mdr2* and other members of the *mdr* supergene family such as *STE-6, pfmdr,* and *CFTR.* Each predicted NB site is formed by two consensus motifs, a G-(X)4-G-K-(T)-(X)6-I/N segment at positions 419 and 1061 (known as motif A), and a hydrophobic pocket (known as motif B) of sequence R/K-(X)3-G-(X)3-L-(hydrophobic)4-D, located 100 residues downstream of motif A at positions 542 and 1186. The A motif is believed to form a flexible loop between a beta strand and an alpha helix. Through conformational changes induced by ATP binding, this loop would control access to the NB domain (Fry et al., 1986). The conserved lysine residue is thought to interact directly with one of the phosphate groups, and would be essential for ATPase activity. On the other hand, the B motif is thought to be close to the glycine-rich flexible loop, and to form a homing pocket for the adenine moiety of the ATP molecule, possibly conferring the nucleotide specificity to the binding site. The aspartate residue is believed to interact directly with the magnesium atom of the nucleotide triphosphate complex (Hyde et al., 1990; Mimura et al., 1991). We have introduced discrete amino acid substitutions within the core consensus sequence for nucleotide binding, GXGKST (motif A). Mutants bearing the sequence GXAKST or GXGRST at either of the two NB sites of *mdr1,* and a double mutant harboring the sequence GXGRST at both NB sites were generated (Azzaria et al., 1989). The integrity of the two NB sites was essential for the biological acitivty of *mdr1,* since all five mutants were, unlike the wild-type *mdr1,* unable to confer drug resistance to drug-sensitive cells in transfection experiments. Conversely, a lysine to arginine substitution outside the core consensus sequence (position 1100) had no effect on the activity of *mdr1.* The loss of activity detected in cell clones stably expressing individual *mdr1* mutants was paralleled by a failure to reduce the intracellular accumulation of radiolabeled vinblastine. However, the ability to combine the photoactivatable analogue of ATP, 8-azido ATP, was retained in the five inactive *mdr1* mutants, suggesting that a step subsequent to ATP binding, possibly ATP hydrolysis, was impaired in the *mdr1* mutants (Azzaria et al., 1989). We have observed that the mutant *mdr1* carrying a lysine to arginine substitution in both NB sites has indeed lost its ATPase activity (Sharom, F., and P. Gros, unpublished observations). Taken together, these results indicate that both NB sites are essential for the overall expression of MDR by *mdr1*. Moreover, these two sites do not appear to function independently of each other since mutating either site completely abolishes the biological activity of *mdr1*. This suggests that the dual nature of the P-gp encoded by *mdr1* is pivotal for drug efflux, not merely by duplicating the biological activity of a functional monomer, but rather suggesting cooperative interactions between the two halves of the molecule. These cooperative interactions could involve either concerted or sequential ATP binding and hydrolysis at both sites. Concerted hydrolysis of two ATP molecules may be required to produce a conformational change responsible for efflux; alternatively, sequential mechanisms could involve ATP hydrolysis at one site to effect efflux and at a second site to regain the original conformation.

Analysis of *mdr1* and *mdr3* Mutants in Predicted TM Domain 11

The drug resistance profiles of transfected cells expressing a cloned cDNA for the *mdr3* gene isolated from a pre–B cell cDNA library (Devault and Gros, 1990) differ qualitatively and quantitatively from that of MDR J7-V3-1 macrophages overexpressing the endogenous *mdr3* (designated *mdr1a* in that study) gene (Yang et al., 1990). A comparison of the amino acid sequences predicted for *mdr3* (Devault and Gros, 1990) and *mdr1a* (Hsu et al., 1990) clones identifies a single amino acid difference between the two clones, located within the predicted TM11 of the protein (a Ser to Phe substitution at position 939 of *mdr3*). The Ser residue at this position within TM11 is conserved in all human and rodent P-gps sequenced to date. Direct DNA sequencing from control RNA specimens from drug-sensitive and -resistant cells in that region of *mdr3* reveals that the Phe^{939} in our clone results from a polymerase error during cDNA construction (Gros et al., 1991). Surprisingly, this mutation maps at the homologous position of TM11 of two nonconservative amino acid substitutions present in the mutant 7G8 allele of the *pfmdr1* gene detected in a large proportion of chloroquine-resistant isolates of *P. falciparum* (Foote et al., 1990). To directly test the functional role of this Ser to Phe substitution on *mdr* function, site-directed mutagenesis was used to convert the Phe^{939} to Ser^{939} in *mdr3* and to replace the wild-type Ser^{941} with Phe^{941} at the homologous position of *mdr1*. The biological activity of wild-type and mutant *mdr1* and *mdr3* clones was tested and compared in transfected cell clones expressing similar amounts of each protein (Gros et al., 1991). In *mdr3,* the Phe^{939} (*mdr3F*) to Ser^{939} (*mdr3S*) substitution caused a considerable and general increase in drug resistance levels conferred by this protein. Interestingly, the increase in resistance was not identical for all drugs tested and was small for vinblastine (threefold) and large for Adriamycin and colchicine (15- and 30-fold, respectively). These results suggest that the mutation at position 939 alters the overall activity of *mdr3* but possibly also its substrate specificity. To test this possibility further a Ser (*mdr1S*) to Phe (*mdr1F*) replacement was introduced at the homologous position (941) in TM11 of *mdr1,* and the activity of each protein was tested in transfected clones. The Ser to Phe substitution at position 941 of *mdr1* resulted in a general decrease in the activity of the protein. As noted earlier for *mdr3,* the modulating effect of this mutation was very different for the four drugs tested: it was strongest for colchicine and adriamycin (33- and 15-fold decreases in levels of resistance) and was smallest for vinblastine (twofold reduction). In fact, *mdr1F* encodes a unique P-gp that can clearly confer resistance to vinblastine but has lost the capacity to confer resistance to Adriamycin and colchicine. Labeling experiments with photoactivatable P-gp ligands iodoarylazidoprazozin and azidopine indicated a strong reduction in binding of these photoactivatable probes to the mutant P-gps (*mdr1F, mdr3F*) as compared with their wild-type counterparts (*mdr1S, mdr3S*). Taken together, these findings suggest that the TM11 mutations studied affect the ability of P-gp to reduce intracellular drug accumulation by reducing drug binding (Kajiji, S., and P. Gros, unpublished observations). The combined analysis of these *mdr1* and *mdr3* mutants identifies Ser$^{939/941}$ in TM11 as a critical residue for the overall activity and substrate specificity of the efflux pumps encoded by the two genes. In addition, the finding that mutations at the homologous position of TM11 in proteins encoded by distantly related members of the *mdr* supergene family (in this case *pfmdr1*) suggests that this specific protein domain underlies a common structural and functional determinant of transport in the drug efflux by *mdr* proteins and that of chloroquine by *pfmdr1*. Besides their high degree of hydrophobicity, no sequence homology is detected between TM11 of *mdr* and *pfmdr1*. However, both segments are capable of forming amphiphilic helices and the mutations in both proteins map at the interface of the hydrophobic and hydrophilic segments of this domain.

Functional Complementation of the Yeast *STE6* Mutation by Mouse *mdr3*

The yeast *S. cerevisiae STE6* gene encodes a membrane transport protein (Kuchler et al., 1989; McGrath and Varshavsky, 1989) which is responsible for the extracellular secretion of the **a** mating pheromone, a farnesylated dodecapeptide required for mating (Anderegg et al., 1988; Michaelis and Herskowitz, 1988). The yeast *STE6* gene product and mammalian P-gps share 57% homology, including conservative amino acid substitutions, and have similar predicted secondary structure and proposed membrane topology (Kuchler et al., 1989; McGrath and Varshavsky, 1989). To determine if the structural similarities between the yeast and the mammalian transporters translate into a functional homology, we asked if the mouse *mdr3* gene could complement the function of the yeast *STE6* in a yeast *ste6* mutant strain (Raymond et al., 1992). Yeast *MATa* cells carrying a *ste6* deletion produce no extracellular **a** factor and therefore are defective in mating. Expression of the full-length cDNA for the mouse *mdr3* gene in a yeast *ste6* deletion strain restored their ability to export **a** factor, as measured by a growth arrest assay, and their ability to mate to alpha cells, as measured by their capacity to form diploids. In these cells, the *mdr3* gene product was expressed as a 140-kD unglycosylated polypeptide found enriched in the membrane fraction of these yeast cells. The introduction and overexpression of the mutant *mdr3* carrying a single amino acid substitution within TM11 (Ser^{939} to Phe^{939}) abolished its ability to complement the *Ste6* mutation in the same experimental protocol. These results indicate that the mammalian P-gp can transport the **a** peptide in yeast cells and suggest that functions of P-gp in normal tissues may include transport of endogenous peptides. Our findings also suggest that the mechanism for peptide transport by P-gp in yeast cells is similar to that of MDR drugs, since mutations known to affect drug transport also affect peptide transport in the heterologous yeast system. Finally, the functional expression of mammalian P-gp in a simple organism such as yeast should greatly facilitate the structure–function analysis of P-gp and perhaps allow the identification of intramolecular protein interactions through the selection of second site revertants from known mutations such as the Phe[939].

mdr Gene Expression in Normal Tissues

Although the precise function of *mdr* genes in normal physiological events remains unknown, the organ and cellular distribution of *mdr* mRNA transcripts and polypeptides appears to be tightly regulated in a tissue- and cell-specific fashion. In addition, subcellular distribution of the discrete P-gp isoforms also appears to be narrowly restricted to the plasma membrane, in particular in polarized epithelia. In normal mouse tissues, *mdr1* is expressed at high levels in the pregnant uterus and adrenals, while lower levels are found in the kidneys and placenta (Arceci et al., 1988; Croop et al., 1989). *mdr2* is expressed almost exclusively in the liver, and *mdr3* is expressed in the intestine and detected at lower levels in the heart and brain (Croop et al., 1989). Immunohistochemical localization of P-gp in epithelial cells with a polarized secretory function has shown that it is expressed on the apical surface of cells apposed to a lumen: it is found on the canalicular face of hepatocytes, the apical surface of biliary ductules, the apical surface of columnar epithelial cells of the intestine, the brush border of renal proximal tubules, and the pancreatic ductule cells (Thiebaut et al., 1987; Bradley et al., 1990; Georges et al., 1990; Buschman et al., 1992). P-gp is also expressed on endothelial cells of the blood–brain barrier (Cordon-Cardo et al., 1989; Thiebaut et al., 1989). In the adrenal glands, P-gp is diffusely distributed on the cell surface in the zona glomerulosa and fasciculata of the cortex, and is expressed at lower levels on the cell surface of all medullary cells (Sugarawa et al., 1988). P-gp (*mdr1*) is expressed at very high levels on the luminal surface of glandular epithelial cells of the endometrium late during pregnancy (Arceci et al., 1988). The role of P-gp in this tissue is unknown, but its expression is tightly regulated by estrogen and progesterone (Arceci et al., 1990). Finally, a recent study has detected high levels of P-gp (*MDR1*) expression in the hematopoietic system, in particular in a population of proposed pluripotent stem cells of the bone marrow (Chaudhary and Roninson, 1991). Overall, the precise substrates of P-gp in normal tissues remain to be identified. It has been proposed that it could act as a normal detoxifying mechanism that would protect the normal environment against toxic insults by xenobiotics, and that would simply become overexpressed in drug-resistant tumor cells. The high levels of expression and specific subcellular localization of P-gp in the intestine, kidney, endothelial cells of the blood–brain barrier, and, in particular, pluripotent precursor stem cells of the bone marrow would certainly argue in favor of this proposal. Others have proposed that chemotherapeutic drugs really represent fraudulent substrates for P-gp which would, under normal physiological conditions, transport separate classes of yet to be identified substrates. P-gp expression detected in adrenals and pregnant uterus, together with the observation that P-gp can directly bind progesterone (Qian and Beck, 1990) and transport a peptide (Raymond et al., 1992), would favor the latter proposal.

General Discussion and Conclusion

The combined biochemical and genetic analyses of distantly related members of the *mdr* supergene family have allowed the identification of structural and functional aspects common to these transport systems. A simple working hypothesis is that proteins encoded by the *mdr* and *mdr*-like gene family transport different types of substrates according to the same mechanism, and that evolutionary pressure has acted to diversify the substrate specificity of these systems while preserving their common mechanistic basis. The predicted NB sites probably represent the protein domains responsible for the common mechanism of action, while the membrane-associated domains are implicated in substrate recognition and hence specificity.

A large body of experimental evidence supports the contention that NB sites represent the major determinants in the proposed mechanism of action common to these proteins. First, these segments are most highly conserved at the primary amino acid sequence level among *mdr* and *mdr*-like genes (Foote et al., 1989; Kuchler et al., 1989; Riordan et al., 1989; Hyde et al., 1990). This high degree of evolutionary conservation indicates that very few amino acid replacements are tolerated in these regions to maintain function. Second, our studies with *mdr1*/*mdr2* chimeras identify those domains as functionally interchangeable and capable of functioning in the context of different membrane-associated domains (Buschman and Gros, 1991). This suggests that the NB domains can carry out the same functional steps in perhaps distantly related proteins. Third, in vitro mutational analysis of the mouse *mdr1* gene clearly demonstrates that even highly conservative amino acid substitutions cannot be tolerated at either of these sites and completely abrogate the activity of this protein (Azzaria et al., 1989). A similar mutational analysis of the yeast *STE6* gene also shows that discrete mutations at these sites abolish normal **a** pheromone transport (Berkower and Michaelis, 1991). Finally, the vast majority of naturally occurring mutations in the *CFTR* gene are within the predicted NB domains of the proteins (Kerem et al., 1989; Riordan et al., 1989; Cutting et al., 1990), again consistent with the notion that these sites are key functional determinants of this protein superfamily.

Biochemical and genetic evidence also suggests that the membrane-associated domains of these proteins (transmembrane segment, extra- and intracellular loops) participate in substrate interactions and may confer substrate specificity to the transport system. In the case of P-gps, the hydrophobic nature of the substrates, their ability to partition in the lipid phase of the membrane, and their mode of cell entry by passive diffusion intuitively suggest that P-gp segments embedded in the membrane are the primary sites of substrate interaction. This proposition is sustained by two sets of biochemical data. First, mapping of the binding site(s) of photoactivatable P-gp ligands by tryptic digestion of photolabeled P-gp, followed by epitope mapping with specific antibodies, identified the membrane-associated segments as primary sites of photolabeling (Bruggemann et al., 1989; Yoshimura et al., 1989). Moreover, a more refined analysis identified the TM11 and TM12 region as the primary site of binding of two photoactivatable analogues of known P-gp ligands (Greenberger et al., 1991); Second, P-gp can be directly labeled by daunomycin by energy transfer from a photoactivatable probe, 5-[^{125}I]iodonaphthalene-1-azide (Raviv et al., 1990). The study of chimeric proteins constructed between two parental P-gps showing distinct substrate specificities suggests that both complete sets of TM domains are required to recreate the full phenotype of the respective parent (Dhir and Gros, 1992). Indirect evidence from the yeast *mdr* homologue *STE-6* also suggests a role of TM domains in substrate recognition: *STE-6* shares with *STE-3,* the receptor for the **a** mating pheromone expressed on alpha cells, short segments of sequence homology near TM7 and TM12, possibly implicated in **a** factor binding (Kuchler et al., 1989). A number of mutations have been found to have a modulating effect on the substrate specificity of proteins encoded by the *mdr* or *mdr*-like genes. Invariably, these mutations have been found within the membrane-associated segments of the respective proteins: a single Val to Gly substitution near TM3 (pst 185) in the human *MDR1* gene strongly modulates the degree of colchicine resistance conferred by this gene in transfected cells (Choi et al., 1988). In the hamster *pgp1,* a Gly^{338}-Ala^{339} to Ala^{338}-Pro^{339} replacement within TM6 was associated with continuous selection and increased resistance to actinomycin D (Devine et al., 1992). A single Ser to Phe substitution in TM11 of *mdr1* and *mdr3* modulates the overall activity but also substrate specificity of the two mouse P-gps, and in the case of *mdr1* uncouples colchicine and anthracycline resistance from vinblastine resistance (Gros et al., 1991). In the *P. falciparum mdr* homologue *pfmdr1,* the emergence of chloroquine resistance (increased efflux) coincides with the presence of two mutant alleles at the *pfmdr1* locus, either a single amino acid substitution near TM1 (Asn^{86} to Tyr^{86}) or a Ser^{1024}/Asn^{1042} to Cys^{1034}/Asp^{1042} replacement within TM11 (Foote et al., 1990). In the proposed transporters associated with antigen processing in lymphocytes (*TAP1, TAP2;* Bodmer et al., 1992), differential transport/loading of distinct populations of peptides into the endoplasmic reticulum for association with rat class I MHC molecules is associated with dramatic sequence variation in two *mtp2* alleles (*cima, cimb*). Of the 25 amino acid sequence variations detected between the two alleles, most map in the membrane-associated portion of the protein, with clusters of two or three residues adjacent to TM domains (Powis et al., 1992). Finally, mutating charged residues within TM1, TM6, and TM10 of the CFTR protein has been found to modulate the halide specificity of the channel (Anderson et al., 1991).

The parallel study of these various P-gps and P-gp-like proteins in their respective assay systems has allowed us to postulate some common mechanistic aspects of transport. It appears likely that the different substrates are recognized in association with the membrane lipid bilayer or interact directly with the membrane-associated segments of these proteins. It also appears that ATP binding and hydrolysis occurs at both predicted NB sites, in either a concerted or sequential fashion. This ATP hydrolysis could then transduce a signal to membrane-associated domains to mediate transport of the various substrates. High resolution crystal structures have been obtained for NB domains containing Walker's A and B motifs in adenylate kinase and other soluble ATP binding proteins and ATPases. The most important difference between the NB site of adenylate kinase and *mdr*-like proteins is the presence of an extra alpha-helical loop, which has been proposed (Hyde et al., 1990) to be responsible for signal transduction to TM domains. It is interesting to note that the major CF mutation, del F508, maps within this loop (Riordan et al., 1989). Finally, it has recently been observed that deletion of the R domain in CFTR suppresses the inactivating effect of a mutation in NB2 but not NB1, providing direct evidence that the R domain interacts specifically with NB2 (Rich et al., 1991). The intramolecular interactions between NB sites and other segments of these proteins, and the details of signal transduction from these sites, will be most easily deciphered in lower eukaryotes such as yeast (*STE-6* gene), where this type of suppressor mutation is most easily generated and analyzed.

References

Akiyama, S. I., M. M. Cornwell, M. Kuwano, I. Pastan, and M. M. Gottesman. 1988. Most drugs that reverse multidrug resistance also inhibit photoaffinity labeling of P-glycoprotein by a vinblastine analog. *Molecular Pharmacology.* 33:144–147.

Ames, G. F.-L. 1986. Bacterial periplasmic transport systems: structure, mechanism and evolution. *Annual Review of Biochemistry.* 55:397–425.

Anderegg, R. J., R. Betz, S. A. Carr, J. W. Crabb, and W. Duntze. 1988. Structure of *Saccharomyces cerevisae* mating hormone a-factor: identification of S-farnesyl cysteine as a structural component. *Journal of Biological Chemistry.* 263:18236–18240.

Anderson, M. P., R. J. Gregory, S. Thompson, D. W. Souza, P. Sucharita, R. C. Mulligan, A. E. Smith, and M. J. Welsh. 1991. Demonstration that CFTR is a chloride channel by alteration of its anion selectivity. *Science.* 253:202–205.

Arceci, R. J., F. Baas, R. Raponi, S. B. Horwitz, D. E. Housman, and J. Croop. 1990. Multidrug resistance gene expression is controlled by steroid hormones in the secretory epithelium of the uterus. *Molecular Reproduction and Development.* 25:101–109.

Arceci, R. J., J. Croop, S. B. Horwitz, and D. E. Housman. 1988. The gene encoding multidrug resistance is induced and expressed at high levels during pregnancy in the secretory epithelium of the uterus. *Proceedings of the National Academy of Sciences, USA.* 85:4350–4354.

Azzaria, M., E. Schurr, and P. Gros. 1989. Discrete mutations introduced in the predicted nucleotide-binding sites of the *mdr1* gene abolish its ability to confer multidrug resistance. *Molecular and Cellular Biology.* 9:5289–5297.

Beck, W. T. 1990. Multidrug resistance and its circumvention. *European Journal of Cancer.* 26:513–515.

Beck, W. T., and M. K. Danks. 1991. Characteristics of multidrug resistance in human tumor cells. *In* Molecular and Cellular Biology of Multidrug Resistance in Tumor Cells. I. B. Roninson, editor. Plenum Publishing Corp., New York. 3–55.

Berkower, C., and S. Michaelis. 1991. Mutational analysis of the yeast a-factor transporter STE6 a member of the ATP binding cassette (ABC) protein superfamily. *EMBO Journal.* 10:3777–3785.

Biedler, J. L., and H. Riehm. 1970. Cellular resistance to actinomycin D in Chinese hamster cells in vitro: cross-resistance, radioautographic, and cytogenetic studies. *Cancer Research.* 30:1174–1184.

Bitoni, A. J., A. Sjoerdsma, P. P. McCann, D. E. Kyle, A. M. J. Oduola, A. N. Rossan, W. K. Milhous, and D. E. Davidson. 1988. Reversal of chloroquine resistance in malaria parasite *P. falciparum* by desipramine. *Science.* 242:1301–1303.

Bodmer, J. G., S. G. E. Marsh, E. D. Albert, W. F. Bodmer, B. Dupont, H. A. Erlich, B. Mach, W. R. Mayr, P. Parham, T. Sasazuki, G. M. T. Schreuder, J. L. Strominger, A. Svejgaard, and P. I. Terasaki. 1992. Nomenclature for factors of the HLA system. *Tissue Antigens.* 39:161–173.

Borst, P., F. Baas, C. R. Lincke, M. Ouellete, A. H. Schinkel, and J. J. M. Smit. 1993. Proceedings of the Bristol-Meyers Squibb symposium on cancer research: drug resistance as a biochemical target in cancer chemotherapy. Vol. 14. T. Tsuruo, M. Ogawa, and S. K. Carter, editors. Tokyo, Japan. In press.

Bradley, G., E. Georges, and V. Ling. 1990. Sex dependent and independent expression of the P-glycoprotein isoforms in Chinese hamster. *Journal of Cellular Physiology.* 145:398–408.

Bradley, G., P. F. Juranka, and V. Ling. 1988. Mechanism of multidrug resistance. *Biochimica et Biophysica Acta.* 948:87–128.

Bruggemann, E. P., U. A. Germann, M. M. Gottesman, and I. Pastan. 1989. Two different regions of phosphoglycoprotein are photoaffinity-labeled by azidopine. *Journal of Biological Chemistry.* 264:15483–15488.

Buschman, E., R. J. Arceci, J. M. Croop, M. Che, I. M. Arias, D. E. Housman, and P. Gros. 1992. Mouse *mdr2* encodes P-glycoprotein expressed in the bile canalicular membrane as determined by isoform specific antibodies. *Journal of Biological Chemistry.* 267:18093–18099.

Buschman, E., and P. Gros. 1991. Functional analysis of chimeric genes obtained by exchanging homologous domains of the mouse *mdr1* and *mdr2* genes. *Molecular and Cellular Biology.* 11:595–603.

Cangelosi, G. A., G. Martinetti, J. A. Leigh, C. C. Lee, C. Theines, and E. W. Nester. 1989. Role of *Agrobacterium tumefaciens* ChvA protein in export of β-1,2-glucan. *Journal of Bacteriology.* 171:1609–1615.

Chaudhary, P. M., and I. B. Roninson. 1991. Expression and activity of P-glycoprotein, a multidrug efflux pump, in human hematopoietic stem cells. *Cell.* 66:85–94.

Chen, C. J., J. E. Chin, K. Ueda, D. P. Clark, I. Pastan, M. M. Gottesman, and I. B. Roninson. 1986*a*. Internal duplication and homology with bacterial transport proteins in the *mdr1* (P-glycoprotein) gene from multidrug resistant human cells. *Cell.* 47:381–389.

Chen, C. J., D. Clark, K. Ueda, I. Pastan, M. M. Gottesman, and I. B. Roninson. 1990. Genomic organization of the human multidrug resistance (MDR1) gene and origin of P-glycproteins. *Journal of Biological Chemistry.* 265:506–514.

Chen, C. M., T. K. Misra, S. Siver, and B. P. Rosen. 1986*b*. Nucleotide sequence of the structural genes for an anion pump: the plasmid-encoded arsenical resistance operon. *Journal of Biological Chemistry.* 261:15030–15038.

Cheng, S. H., D. P. Rich, J. Marshall, R. J. Gregory, M. J. Welsh, and A. E. Smith. 1991. Defective intracellular transport and processing is the molecular basis for most cystic fibrosis. *Cell.* 66:1027–1036.

Chin, J. E., R. Soffir, K. E. Noonan, K. Choi, and I. B. Roninson. 1989. Structure and expression of the human MDR (P-glycoprotein) gene family. *Molecular and Cellular Biology.* 9:3808–3820.

Choi, K., C. J. Chen, M. Kriegler, and I. B. Roninson. 1988. An altered pattern of cross-resistance in multidrug resistance human cells results from spontaneous mutations in the *MDR1* (P-glycoprotein) gene. *Cell.* 53:519–529.

Cordon-Cardo, C., J. P. O'Brien, D. Casals, L. Rittman-Grauer, J. L. Biedler, M. R. Melamed, and J. R. Bertino. 1989. Multidrug-resistance gene (P-glycoprotein) is expressed by endothelial cells at the blood-brain barrier sites. *Proceedings of the National Academy of Sciences, USA.* 86:695–698.

Cornwell, M. M., I. Pastan, and M. M. Gottesman. 1987*a*. Certain calcium channel blockers bind specifically to multidrug resistant human KB carcinoma membrane vesicles and inhibit drug binding to P-glycoprotein. *Journal of Biological Chemistry.* 262:2166–2170.

Cornwell, M. M., A. R. Safa, R. L. Felsted, M. M. Gottesman, and I. Pastan. 1986. Membrane vesicles from multidrug resistant cancer cells contain a specific 150-170 kDa protein detected by photoaffinity labelling. *Proceedings of the National Academy of Sciences, USA.* 83:3847–3850.

Cornwell, M. M., T. Tsuruo, M. M. Gottesman, and I. Pastan. 1987*b.* ATP-binding properties of P-glycoprotein from multidrug resistant KB cells. *FASEB Journal.* 1:51–54.

Coulton, J. W., P. Mason, and D. D. Allatt. 1987. *FhuC* and *FhuD* genes for Iron (III)-ferrichrome transport into *Escherichia coli* K-12. *Journal of Bacteriology.* 169:3844–3849.

Croop, J. M., M. Raymond, D. Haber, A. Devault, R. J. Arceci, P. Gros, and D. E. Housman. 1989. The three mouse multidrug-resistance genes are expressed in a tissue specific manner in normal mouse tissues. *Molecular and Cellular Biology.* 9:1346–1350.

Cutting, G. R., L. M. Kash, B. J. Rosenstein, J. Zielenski, L. C. Tsui, S. E. Antonarakis, and H. Kazazian. 1990. A cluster of cystic fibrosis mutations in the first nucleotide-binding fold of the cystic fibrosis conductance regulator protein. *Nature.* 346:366–369.

Dano, K. 1973. Active outward transport of daunomycin in resistant Ehrlich ascites tumor cells. *Biochimica et Biophysica Acta.* 323:466–483.

Devault, A., and P. Gros. 1990. Two members of the mouse *mdr* gene family confer multidrug resistance with overlapping but distinct drug specificities. *Molecular and Cellular Biology.* 10:1652–1663.

Deverson, E. V., I. R. Gow, J. Coadwell, J. J. Monaco, G. W. Butcher, and J. C. Howard. 1990. Major histocompatibility complex class II region encoding proteins related to the multidrug resistance family of transmembrane transporters. *Nature.* 348:738–741.

Devine, S. E., V. Ling, and P. W. Melera. 1992. Amino acid substitutions in the sixth transmembrane domain of P-glycoprotein alter multidrug resistance. *Proceedings of the National Academy of Sciences, USA.* 89:4564–4568.

Dhir, R., and P. Gros. 1992. Functional analysis of chimeric proteins constructed by exchanging homologous domains of two P-glycoproteins conferring distinct multidrug resistance profiles. *Biochemistry.* 31:6103–6110.

Doolittle, R. F., M. S. Johnson, I. Husain, B. Van Houten, D. C. Thomas, and A. Sancar. 1986. Domainal evolution of a prokaryotic DNA repair protein and relationship to active transport proteins. *Nature.* 323:451–453.

Dreesen, T. D., D. H. Johnson, and S. Henikoff. 1988. The brown protein of *Drosophila melanogaster* is similar to the white protein and to components of active transport complexes. *Molecular and Cellular Biology.* 8:5206–5215.

Drumm, M. L., H. A. Pope, W. H. Cliff, J. M. Rommens, S. A. Marvin, L. C. Tsui, F. Collins, R. A. Frizzel, and J. M. Wilson. 1990. Correction of the cystic fibrosis defect in vitro by retrovirus-mediated gene transfer. *Cell.* 62:1227–1233.

Ellenberger, T. E., and S. M. Beverley. 1989. Multiple drug resistance and conservative amplification of the H region in *Leishmania major. Journal of Biological Chemistry.* 264:15094–15103.

Endicott, J. A., and V. Ling. 1989. The biochemistry of P-glycoprotein mediated multidrug resistance. *Annual Review of Biochemistry.* 58:137–171.

Felmlee, T., S. Pellett, and R. A. Welch. 1985. Nucleotide sequence of a chromosomal hemolysin. *Journal of Bacteriology.* 163:94–105.

Fojo, A., S. Akiyama, M. M. Gottesman, and I. Pastan. 1985. Reduced drug accumulation in multiple drug-resistant human KB carcinoma cell lines. *Cancer Research.* 45:3002–3007.

Foote, S. J., D. E. Kyle, R. K. Martin, A. M. J. Oduola, K. Forsyth, D. J. Kemp, and A. F. Cowman. 1990. Several alleles of the multidrug resistance gene are closely linked to chloroquine resistance in *Plasmodium falciparum*. *Nature*. 345:255–258.

Foote, S. J., J. K. Thompson, A. F. Cowman, and D. J. Kemp. 1989. Amplification of the multidrug resistance gene in some chloroquine-resistant isolates of *P. falciparum*. *Cell*. 57:921–930.

Friedrich, M. J., L. C. Deveaux, and R. J. Kadner. 1986. Nucleotide sequence of the *btuCED* genes involved in vitamin B-12 transport in *Escherichia coli* and homology with components of periplasmic-binding-protein-dependent transport systems. *Journal of Bacteriology*. 167:928–934.

Fry, D. C., S. A. Kuby, and A. S. Mildvan. 1986. ATP-binding site of adenylate kinase: mechanistic implications of its homology with ras-encoded p21, F1-ATPase and other nucleotide-binding proteins. *Proceedings of the National Academy of Sciences, USA*. 83:907–911.

Georges, E., G. Bradley, J. Gariepy, and V. Ling. 1990. Detection of P-glycoprotein isoforms by gene-specific monoclonal antibodies. *Proceedings of the National Academy of Sciences, USA*. 87:152–156.

Gilson, E., C. F. Higgins, M. Hofnung, G. F. L. Ames, and H. Nikaido. 1982. Extensive homology between membrane associated components of histidine and maltose transport systems of *Salmonella typhimurium* and *Escherichia coli*. *Journal of Biological Chemistry*. 257:9915–9918.

Glaser, P., H. Sakamoto, J. Bellalou, A. Ullmann, and A. Danchin. 1988. Secretion of cyclolysin, the calmodulin-sensitive adenylate cyclase-heamolysin bifunctional protein of *Bordettella pertussis*. *EMBO Journal*. 7:3997–4004.

Greenberger, L. M., C. J. Lisanti, J. T. Silva, and S. B. Horwitz. 1991. Domain mapping of the photoaffinity drug binding sites in P-glycoprotein encoded by mouse *mdr1b*. *Journal of Biological Chemistry*. 266:20744–20751.

Gros, P., Y. Ben Neriah, J. M. Croop, and D. E. Housman. 1986*a*. Isolation and expression of a complementary DNA that confers multidrug resistance. *Nature*. 323:728–731.

Gros, P., J. M. Croop, and D. E. Housman. 1986*b*. Mammalian multidrug resistance gene: complete cDNA sequence indicates strong homology to bacterial transport proteins. *Cell*. 47:371–380.

Gros, P., R. Dhir, J. M. Croop, and F. Talbot. 1991. A single amino acid substitution strongly modulates the activity and substrate specificity of the mouse *mdr1* and *mdr3* drug efflux pumps. *Proceedings of the National Academy of Sciences, USA*. 88:7289–7293.

Gros, P., M. Raymond, J. Bell, and D. E. Housman. 1988. Cloning and characterization of a second member of the mouse *mdr* gene family. *Molecular and Cellular Biology*. 8:2770–2778.

Gupta, R. S., W. Murray, and R. Gupta. 1988. Cross resistance pattern towards anticancer drugs of a human carcinoma multidrug-resistant cell line. *British Journal of Cancer*. 58:441–447.

Hamada, H., and T. Tsuruo. 1988. Purification of the 170- to 180-kilodalton membrane glycoprotein associated with multidrug resistance. *Journal of Biological Chemistry*. 263:1454–1458.

Hammond, J., R. M. Johnstone, and P. Gros. 1989. Enhanced efflux of [³H] vinblastine from Chinese hamster ovary cells transfected with a full length complementary DNA clone for the *mdr1* gene. *Cancer Research*. 49:3867–3871.

Higgins, C. F., P. D. Haag, K. Nikaido, F. Ardeshir, G. Garcia, and G. F. L. Ames. 1982. Complete nucleotide sequence and identification of membrane components of the histidine transport operon of *Salmonella typhimurium. Nature.* 298:723–727.

Higgins, C. F., I. D. Hyles, G. P. C. Salmond, D. R. Gill, J. A. Downie, I. J. Evans, I. B. Holland, L. Gray, S. D. Buckel, A. W. Bell, and M. A. Hermondson. 1986. A family of related ATP-binding subunits coupled to many distinct biological processes in bacteria. *Nature.* 323:448–450.

Hsu, S. I. H., D. Cohen, L. S. Kirschner, L. Lothstein, M. Hartstein, and S. B. Horwitz. 1990. Structural analysis of the mouse *mdr1a* (P-glycoprotein) promoter reveals the basis for differential transcript heterogeneity in multidrug resistant J774.2 cells. *Molecular and Cellular Biology.* 10:3596–3606.

Hyde, S. C., P. Emsley, M. J. Hartshorn, M. M. Mimmack, U. Gileadi, S. R. Pearce, M. P. Gallagher, D. R. Gill, R. E. Hubbard, and C. F. Higgins. 1990. Structural model for ATP-binding proteins associated with cystic fibrosis, multidrug resistance and bacterial transport. *Nature.* 346:362–365.

Johann, S., and S. M. Hinton. 1987. Cloning and nucleotide sequence of the chlD locus. *Journal of Bacteriology.* 169:1911–1916.

Juliano, R. L., and V. Ling. 1976. A surface glycoprotein modulating drug permeability in Chinese hamster ovary cell mutants. *Biochimica et Biophysica Acta.* 455:152–162.

Kamijo, K., S. Taketani, S. Yokota, T. Osumi, and T. Hashimoto. 1990. The 70 kDa peroxisomal membrane protein is a member of the MDR (P-glycoprotein) related ATP-binding protein superfamily. *Journal of Biological Chemistry.* 265:4534–4540.

Kamiwatari, M., Y. Nagata, H. Kikuchi, A. Yoshimura, T. Sumizawa, N. Shudo, R. Sakoda, K. Seto, and S. I. Akiyama. 1989. Correlation between reversing of multidrug resistance and inhibiting of 3H-azidopine photolabeling of P-glycoprotein by newly-synthesized dihydropyridine analogues in a human cell line. *Cancer Reserach.* 49:3190–3195.

Kerem, B. S., J. M. Rommens, J. A. Buchanan, D. Markiewicz, T. Cox, A. Chakravarti, M. Buchwald, and L. C. Tsui. 1989. Identification of the cystic fibrosis gene: genetic analysis. *Science.* 245:1073–1080.

Koch, G., M. Smith, P. Twentyman, and K. Wright. 1986. Identification of a novel calcium-binding protein (CP22) in multidrug-resistant murine and hamster cells. *FEBS Letters.* 195:275–279.

Krogstad, D. J., I. Y. Gluzman, D. E. Kyle, A. M. Oduola J., S. K. Martin, W. K. Milhous, and P. H. Schlesinger. 1987. *Science.* 238:1283–1285.

Kuchler, K., R. E. Sterne, and J. Thorner. 1989. *Saccharomyces cerevisiae* STE6 gene product: a novel pathway for protein export in eukaryotic cells. *EMBO Journal.* 8:3973–3984.

Lemontt, J. F., M. Azzaria, and P. Gros. 1988. Increased *MDR* gene expression and decreased drug accumulation in multidrug-resistant human melanoma cells. *Cancer Research.* 48:6348–6353.

Ling, V., and L. H. Thompson. 1974. Reduced permeability in CHO cells as a mechanism of resistance to colchicine. *Journal of Cellular Physiololgy.* 83:103–116.

Martin, S. K., A. M. J. Oduola, and W. K. Milhous. 1987. Reversal of chloroquine resistance in *Plasmodium falciparum* by verapamil. *Science.* 235:899–901.

McGrath, J. P., and A. Varshavsky. 1989. The yeast STE6 gene encodes a homologue of the mammalian multidrug resistance P-glycoprotein. *Nature.* 340:400–404.

Michaelis, S., and I. Herskowitz. 1988. The a-factor pheromone of *Saccharomyces cerevisiae* is essential for mating. *Molecular and Cellular Biology.* 8:1309–1318.

Mimura, C. S., S. R. Holbrook, and G. F. L. Ames. 1991. Structural model of the nucleotide-binding conserved component of periplasmic permeases. *Proceedings of the National Academy of Sciences, USA.* 88:84–88.

Monaco, J. J., S. Cho, and M. Attaya. 1990. Transport protein genes in the murine MHC: possible implications for antigen processing. *Science.* 250:1723–1726.

Moscow, J. A., and K. H. Cowan. 1988. Multidrug resistance. *Journal of the National Cancer Institute.* 80:14–20.

Ng, W. F., F. Sarangi, R. L. Zastawny, L. Veinot-Drebot, and V. Ling. 1989. Identification of members of the P-glycoprotein multigene family. *Molecular and Cellular Biology.* 9:1224–1232.

Nohno, T., T. Saito, and J. S. Hong. 1986. Cloning and complete nucleotide sequence of the *Escherichia coli* glutamine permease operon (glnHPQ). *Molecular and General Genetics.* 205:260–269.

O'Hare, K., C. Murphy, R. Levis, and G. M. Rubin. 1984. DNA sequence of the *white* locus of *Drosophila melanogaster*. *Journal of Molecular Biology.* 180:437–455.

Ohyama, K., H. Fukuzawa, T. Kohchi, H. Shirai, T. Sano, S. Sano, K. Umesono, Y. Shiki, M. Takeuchi, Z. Chang, S. I. Aoto, H. Inokuchi, and H. Ozeki. 1986. Chloroplast gene organization deduced from complete sequence of liverwort *Marchantia polymorpha* chloroplast DNA. *Nature.* 322:572–574.

Ouellette, M., F. Fase-Fowler, and P. Borst. 1990. The amplified H circle of methotrexate resistant *Leishmania tarentolae* contains a novel P-glycoprotein gene. *EMBO Journal.* 9:1027–1033.

Pearce, H. L., A. R. Safa, N. J. Bach, M. A. Winter, M. C. Cirtain, and W. T. Beck. 1989. Essential features of the P-glycoprotein pharmacophore as defined by a series of reserpine analogs that modulate multidrug resistance. *Proceedings of the National Academy of Sciences, USA.* 86:5128–5132.

Peterson, R. H., W. J. Beutler, and J. L. Biedler. 1979. Ganglioside composition of malignant and actinomycin D resistant nonmalignant Chinese hamster cells. *Biochemical Pharmacology.* 28:579–582.

Peterson, R. H., M. B. Meyers, B. A. Spengler, and J. L. Biedler. 1983. Alteration of plasma membrane glycopeptides and gangliosides of Chinese hamster cells accompanying development of resistance to daunorubicin and vincristine. *Cancer Research.* 43:222–228.

Powis, S. J., E. V. Deverson, W. J. Coadwell, A. Ciruela, N. S. Huskisson, H. Smith, G. W. Butcher, and J. C. Howard. 1992. Effect of polymorphism of an MHC-linked transporter on the peptides assembled in a class I molecule. *Nature.* 357:211–215.

Powis, S. J., A. R. M. Townsend, E. V. Deverson, J. Bastin, G. W. Butcher, and J. C. Howard. 1991. Restoration of antigen presentation to the mutant cell line RMA-S by an MHC-linked transporter. *Nature.* 354:528–531.

Qian, X. D., and W. T. Beck. 1990. Progesterone photoaffinity labels P-glycoprotein in multidrug resistant human leukemic lymphoblasts. *Journal of Biological Chemistry.* 265:18753–18756.

Raviv, Y., H. B. Pollard, E. P. Bruggemann, I. Pastan, and M. M. Gottesman. 1990. Photosensitized labeling of a functional multidrug transporter in living drug-resistant tumor cells. *Journal of Biological Chemistry.* 265:3975–3980.

Raymond, M., and P. Gros. 1989. Mammalian multidrug resistance gene: correlation of genome organization with structural domains and duplication of an ancestral gene. *Proceedings of the National Academy of Sciences, USA.* 86:6488–6492.

Raymond, M., P. Gros, M. Whiteway, and D. Y. Thomas. 1992. Functional complementation of yeast *ste6* by a mammalian multidrug resistance gene. *Science.* 256:232–234.

Rich, D. P., M. P. Anderson, R. J. Gregory, S. H. Cheng, S. Paul, D. M. Jefferson, J. D. McCann, K. W. Klinger, A. E. Smith, and M. J. Welch. 1990. Expression of cystic fibrosis transmembrane conductance regulator corrects defective chloride channel regulation in cystic fibrosis airway epithelial cells. *Nature.* 347:358–363.

Rich, D. P., R. J. Gregory, M. P. Anderson, P. Manavalan, A. E. Smith, and M. J. Welsh. 1991. Effect of deleting the R domain on CFTR-generated chloride channels. *Science.* 253:205–207.

Richert, N., S. Akiyama, D. Shen, M. M. Gottesman, and I. Pastan. 1985. Multiply drug resistant human KB carcinoma cells have decreased amounts of a 75 kDa and a 72 kDa glycoprotein. *Proceedings of the National Academy of Sciences, USA.* 82:2330–2333.

Riordan, J. R., J. M. Rommens, B.-S. Kerem, N. Alow, R. Rozmahel, Z. Grzelczak, J. Zielenski, S. Lok, N. Plavsic, J.-L. Chou, M. L. Drumm, M. C. Ianuzzi, F. S. Collins, and L.-C. Tsui. 1989. Identification of the cystic fibrosis gene: cloning and characterization of complementary DNA. *Science.* 245:1066–1073.

Roninson, I. B. 1991. Structure and evolution of P-glycoproteins. *In* Molecular and Cellular Biology of Multidrug Resistance in Tumor Cells. I. B. Roninson, editor. Plenum Publishing Corp., New York. 189–209.

Safa, A. R. 1988. Photoaffinity labeling of the multidrug resistance related P-glycoprotein with photoactive analogs of verapamil. *Proceedings of the National Academy of Sciences, USA.* 85:7187–7191.

Safa, A. R., C. J. Glover, M. B. Meyers, J. L. Biedler, and R. L. Felsted. 1986. Vinblastine photoaffinity labeling of high molecular weight surface membrane glycoprotein specific for multidrug resistance cells. *Journal of Biological Chemistry.* 261:6137–6140.

Safa, A. R., C. J. Glover, J. L. Sewell, M. B. Meyers, J. L. Biedler, and R. L. Felsted. 1987. Identification of the multidrug resistance-related membrane glycoprotein as an acceptor for calcium channel blockers. *Journal of Biological Chemistry.* 262:7884–7888.

Safa, A., N. D. Metha, and M. Agresti. 1989. Photoaffinity labelling of P-glycoprotein is multidrug resistant cells with photoactive analogs of colchicine. *Biochemical and Biophysical Research Communications.* 162:1402–1408.

Samuelson, J., P. Ayala, E. Oroczo, and D. Wirth. 1990. Emetine-resistant mutants of *Entamoeba histolytica* overexpress mRNAs for multidrug resistance. *Molecular and Biochemical Parasitology.* 38:281–290.

Schinkel, A. H., M. E. M. Roelofs, and P. Borst. 1991. Characterization of the human *MDR3* P-glycoprotein and its recognition by P-glycoprotein specific monoclonal antibodies. *Cancer Research.* 51:2628–2635.

Schurr, E., M. Raymond, J. Bell, and P. Gros. 1989. Characterization of the multidrug resistance protein expressed in cell clones stably transfected with the mouse *mdr1* cDNA. *Cancer Research.* 49:2729–2734.

Scripture, J. B., C. Voelker, S. Miller, R. T. O'Donnell, L. Polgar, J. Rade, B. F. Hrazdovsky, and R. W. Hogg. 1987. High-affinity L-arabinose transport operon: nucleotide sequence and analysis of gene products. *Journal of Molecular Biology.* 197:37–46.

Skovsgaard, T. 1978. Mechanism of cross-resistance between vincristine and daunorubicin in Erlich ascites tumor cells. *Cancer Research.* 39:4722–4727.

Slater, L. M., P. Sweet, M. Stupecky, and S. Gupta. 1986. Cyclosporin A reverses vincristine and daunorubicin resistance in acute lymphatic leukemia in vitro. *Journal of Clinical Investigation.* 77:1405–1408.

Southern, P. J., and P. Berg. 1982. Transformation of mammalian cells to antibiotic resistance with a bacterial gene under control of the SV40 early region promoter. *Journal of Molecular and Applied Genetics.* 1:327–341.

Spies, T., M. Bresnahan, S. Bahram, D. Arnold, G. Blanck, E. Mellins, D. Pious, and R. De Mars. 1990. A gene in the human major histocompatibility complex class II region controlling the class I antigen presentation pathway. *Nature.* 348:744–747.

Spies, T., and R. De Mars. 1991. Restored expression of major histocompatibility class I molecules by gene transfer of a putative peptide transporter. *Nature.* 351:323–324.

Stanfield, S. W., L. Ielpi, D. O'Brochta, D. R. Helinski, and G. S. Ditta. 1988. The *ndvA* gene product of *Rhizobium meliloti* is required for β-(1U2) glucan production and has homology to the ATP-binding export protein HlyB. *Journal of Bacteriology.* 170:3523–3530.

Staudenmaier, H., B. Van Hove, Z. Yaraghi, and V. Braun. 1989. Nucleotide sequences of the *fecBCDE* genes and locations of the proteins suggests a periplasmic binding protein dependent transport mechanism for iron (III) dicitrate in *Escherichia coli. Journal of Bacteriology.* 171:2626–2633.

Sugarawa, I., M. Nakahama, H. Hamado, T. Tsuruo, and S. Mori. 1988. Apparent stronger expression in the human adrenal cortex than in the human adrenal medulla of Mr 170,000-180,000 P-glycoprotein. *Cancer Research.* 48:4611–4614.

Surin, B. P., H. Rosenberg, and G. B. Cox. 1985. Phosphate-specific transport system of *Escherichia coli:* nucleotide sequence and gene-polypeptide relationships. *Journal of Bacteriology.* 161:189–198.

Tabcharani, J. A., X. B. Chang, J. R. Riordan, and J. W. Hanrahan. 1991. Phosphorylation-regulated Cl channel in CHO cells stably expressing the cystic fibrosis gene. *Nature.* 352:628–631.

Tamai, I., and A. R. Safa. 1990. Competitive interaction of cyclosporins with the Vinca alkaloid binding site of P-glycoprotein in multidrug resistant cells. *Journal of Biological Chemistry.* 265:16509–16513.

Thiebaut, F., T. Tsuruo, H. Hamada, M. M. Gottesman, I. Pastan, and M. C. Willingham. 1987. Cellular localization of the multidrug-resistance gene product P-glycoprotein in normal human tissues. *Proceedings of the National Academy of Sciences, USA.* 84:7735–7738.

Thiebaut, F., T. Tsuruo, H. Hamada, M. M. Gottesman, I. Pastan, and M. C. Willingham. 1989. Immunohistochemical localization in normal tissues of different epitopes in the multidrug transport protein P170: evidence for localization in brain capillaries and cross-reactivity of one antibody with a muscle protein. *Journal of Histochemistry and Cytochemistry.* 37:159–164.

Trowsdale, J., I. Hanson, I. Mockbridge, S. Beck, A. Townsend, and A. Kelly. 1990. Sequences encoded in the class II region of the major histocompatibility complex related to the 'ABC' superfamily of transporters. *Nature.* 348:741–743.

Tsuruo, T., H. Iida, Y. Kitatani, K. Yokota, S. Tsukagoshi, and Y. Sakurai. 1984. Effects of quinidine and related compounds on cytotoxicity and cellular accumulation of vincristine and Adriamycin in drug resistant tumor cells. *Cancer Research.* 44:4303–4307.

Tsuruo, T., H. Iida, S. Tsukagoshi, and Y. Sakurai. 1981. Overcoming of vincristine resistance in P388 leukemia in vivo and in vitro through enhanced cytotoxicity of vincristine and vinblastine by verapamil. *Cancer Research.* 41:1967–1972.

Tsuruo, T., H. Iida, S. Tsukagoshi, and Y. Sakurai. 1982. Increased accumulation of vincristine and Adriamycin in drug resistant tumor cells following incubation with calcium antagonists and calmodulin inhibitors. *Cancer Research.* 42:4730–4733.

Ueda, K., C. Cardarelli, M. M. Gottesman, and I. Pastan. 1987. Expression of a full-length cDNA for the human *MDR1* gene confers resistance to colchicine, doxorubicin, and vinblastine. *Proceedings of the National Academy of Sciences, USA.* 84:3004–3008.

Van der Bliek, A. M., P. M. Kooiman, C. Schneider, and P. Borst. 1988. Sequence of MDR3 cDNA encoding a human P-glycoprotein. *Gene.* 71:401–411.

Walker, J. E., M. Sraste, M. J. Runswick, and N. J. Gay. 1982. Distantly related sequences in the alpha- and beta-subunits of ATP synthase, myosin, kinases and other ATP-requiring enzymes and a common nucleotide-binding fold. *EMBO Journal.* 1:945–951.

Wilson, C. M., A. E. Serrano, A. Wasley, M. P. Bogenschutz, A. H. Shankar, and D. F. Wirth. 1989. Amplification of a gene related to mammalian *mdr* genes in drug-resistant *Plasmodium falciparum. Science.* 244:1184–1186.

Wu, C. T., M. Budding, M. S. Griffin, and J. M. Croop. 1991. Isolation and characterization of Drosophila multidrug resistance gene homologs. *Molecular and Cellular Biology.* 11:3940–3948.

Yang, C.-P. H., D. Cohen, L. M. Greenberger, S. I.-H. Hsu, and S. B. Horwitz. 1990. Differential transport properties of two *mdr* gene products are distinguished by progesterone. *Journal of Biological Chemistry.* 265:10281–10288.

Yoshimura, A., Y. Kuwazuru, T. Sumizawa, M. Ichikawa, S. I. Ikeda, T. Uda, and S. I. Akiyama. 1989. Cytoplasmic orientation and two domain structure of the multidrug transporter, P-glycoprotein, demonstrated with sequence specific antibodies. *Journal of Biological Chemistry.* 264:16282–16291.

Zamora, J. M., H. L. Pearce, and W. T. Beck. 1988. Physico-chemical properties shared by compounds that modulate multidrug resistance in human leukemic cells. *Molecular Pharmacology.* 33:454–462.

Chapter 9

Regulation of the Cystic Fibrosis Transmembrane Conductance Regulator Chloride Channel by MgATP

Michael J. Welsh and Matthew P. Anderson

*Departments of Internal Medicine and Physiology and Biophysics,
Howard Hughes Medical Institute, University of Iowa College of
Medicine, Iowa City, Iowa 52242*

Introduction

Cystic fibrosis (CF) is a common lethal genetic disease (Boat et al., 1989) caused by mutations in the gene encoding the cystic fibrosis transmembrane conductance regulator (CFTR) (Riordan et al., 1989). Recent studies have demonstrated that CFTR is a regulated Cl⁻ channel (for review see Welsh et al., 1992). That observation explains, at least in part, the best characterized defect in the tissues affected by CF: there is a loss of cAMP-regulated transepithelial Cl⁻ transport (Boat et al., 1989; Quinton, 1990).

Sequence analysis of CFTR and comparison of CFTR with a family of proteins named the traffic ATPases or *A*TP *b*inding *c*assette (ABC) transporters allowed the prediction that CFTR contains five domains (Riordan et al., 1989; Ames et al., 1990; Hyde et al., 1990): two membrane-spanning domains, each composed of six transmembrane segments; an R domain, which contains several consensus phosphorylation sequences; and two nucleotide-binding domains (NBDs), which were predicted to interact with ATP. The traffic ATPase/ABC transporter family includes periplasmic permeases of prokaryotes, such as the histidine and maltose transport systems; STE6, involved in secretion of a mating factor in yeast; and P-glycoprotein, responsible for multiple drug resistance. All these family members share a similar general topology with the exception of the R domain, which is a feature unique to CFTR.

The membrane-spanning domains contain hydrophobic residues that are predicted to cross the lipid bilayer as α-helices. Because mutation of specific residues in the first and the sixth membrane-spanning sequences altered the anion selectivity of the CFTR Cl⁻ channel (Anderson et al., 1991*b*), it has been speculated that the putative membrane-spanning α-helices may contribute to the conduction pore.

The R domain appears to regulate the Cl⁻ channel; in the unphosphorylated state, the R domain inhibits the channel and when it is phosphorylated, inhibition is relieved and the channel opens. This conclusion is supported by several observations. First, addition of the catalytic subunit of cAMP-dependent protein kinase (PKA) to the cytosolic surface of excised cell-free patches of membrane activates the CFTR Cl⁻ channel (Berger et al., 1991; Tabcharani et al., 1991). Second, CFTR is phosphorylated by PKA (Gregory et al., 1990; Cheng et al., 1991). Multiple serines in the R domain are substrates for PKA in vitro and are phosphorylated in vivo when cellular levels of cAMP increase (Cheng et al., 1991). Third, mutation of multiple serines to alanines prevents phosphorylation-dependent activation of CFTR Cl⁻ channels (Cheng et al., 1991). Finally, expression of CFTR in which most of the R domain has been deleted generates Cl⁻ channels that are open even without an increase in intracellular cAMP (Rich et al., 1991).

The NBDs, however, have remained more puzzling. The NBDs contain the conserved amino acid sequence that brought CFTR into the traffic ATPase/ABC transporter family (Riordan et al., 1989; Ames et al., 1990; Hyde et al., 1990). In the traffic ATPase/ABC transporter family the NBDs appear to be the site of ATP hydrolysis; in some members of the family, the energy released during ATP hydrolysis is used to actively transport substrate across the cell membrane. But because CFTR forms a Cl⁻ channel, it was not clear why it would contain a domain that might hydrolyze ATP. (It remains possible, however, that CFTR might have functions in addition to that of a Cl⁻ channel.) Adding to the mystery, the NBDs are the site of many naturally occurring CF mutations (for reviews see Tsui and Buchwald, 1991; McIntosh and Cutting, 1992).

ATP Is Required for CFTR Cl⁻ Channel Activity

Because CFTR functions as a Cl⁻ channel rather than a Cl⁻ pump, we tested the hypothesis that ATP might regulate the CFTR Cl⁻ channel (Anderson et al., 1991*a*). To test this hypothesis, we used cells expressing recombinant wild-type CFTR. We obtained inside-out patches of membrane from the cell and examined the effect of ATP.

Fig. 1 shows macroscopic currents through a large number of low conductance Cl⁻ channels. Addition of ATP (1 mM) alone to the cytosolic surface of excised patches had no effect, but the combination of ATP plus the catalytic subunit of PKA activated Cl⁻ channels. When we eliminated the PKA but kept the ATP on the cytosolic surface, the channels remained open. However, once the ATP was removed, current returned to basal values. Readdition of ATP alone activated the channels. These results indicate that ATP regulates the channel, but only when it has first been phosphorylated by PKA.

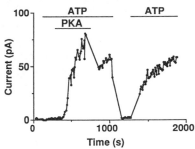

Figure 1. ATP regulates the CFTR Cl⁻ channel. ATP (1 mM, Mg salt) and the catalytic subunit of PKA (75 nM) were present during the times indicated. Data are results from a single experiment. Current is the sum of currents through numerous CFTR Cl⁻ channels in an excised, inside-out patch. The pipette solution contained 49 mM Cl⁻ and the bath solution contained 147 mM Cl⁻. Voltage was −40 mV. Currents were inward, but we represent Cl⁻ flowing out across the patch of membrane as positive current for the convenience of the reader. (Reproduced from *Cell,* 1991, 67:775–784, by copyright permission of Cell Press.)

In a series of studies we showed that PKA-dependent phosphorylation is relatively irreversible in these excised, cell-free patches. We also showed that the effect of ATP does not occur through reversible phosphorylation of the CFTR Cl⁻ channel. That conclusion was supported by a series of studies using kinase and phosphatase inhibitors. In addition, ATP regulated CFTR lacking a major part of the R domain (CFTRΔR). CFTRΔR channel had biophysical properties identical to those of wild-type CFTR, but they did not require PKA. Addition of ATP alone was sufficient to reversibly activate the channels.

These results indicate that ATP can regulate the CFTR Cl⁻ channel independent of phosphorylation and independent of the R domain.

Increasing MgATP Concentrations Increased Channel Activity

The studies described above showed that ATP regulates CFTR. However, they did not determine what effect ATP has on channel gating or whether ATP acts at more than one site. To begin to address these issues, we examined the effect of MgATP concentration on the probability that the phosphorylated channels were in the open state (P_o) (Fig. 2, *A* and *B*; Anderson and Welsh, 1992). As the concentration of MgATP increased, P_o increased. An Eadie-Hofstee plot of the data (Fig. 2 *B*) generated a curved line; this type of curve is most consistent with negative kinetic cooperativity. Kinetic cooperativity cannot be explained by a heterogeneous population of channels because the same curvature was observed with a single channel (dotted line). Most models used to explain negative kinetic cooperativity include two or more substrate effector binding sites. Therefore, we speculated that MgATP may interact with two different sites in the CFTR: NBD1 and NBD2.

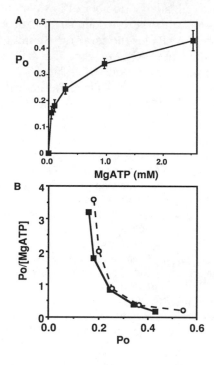

Figure 2. Effect of MgATP on channel activity. MgATP concentration was varied as indicated and current was measured as indicated in the legend of Fig. 1. Data are means ± SEM from six patches. The P_o of a single CFTR Cl⁻ channel (*broken line*) is shown for comparison. (Reproduced from *Science*, 1992, 257:1701–1704, by copyright permission of the American Association for the Advancement of Science.)

ATP Interacts with Both NBD1 and NBD2

To test the hypothesis that ATP interacts directly with CFTR and the NBDs, and to determine which of the two nucleotide binding domains are involved, we mutated, individually, amino acids in highly conserved regions of the two NBDs (Anderson and Welsh, 1992). Each NBD contains a Walker A motif and a Walker B motif (Walker et al., 1982) which are conserved in the traffic ATPase/ABC transporter family, and many other proteins that hydrolyze ATP. The Walker A motif is GXXGXGK. In the Walker A motif, Lys is thought to interact with either the α or γ phosphate of ATP. Mutation of the Walker A Lys in NBD1 to Ala (K464A)[1] (Fig. 3 *A*) or that in NBD2 to Met (K1250M) (Fig. 3 *B*) resulted in an altered relation between MgATP concentration and channel activity. In both CFTR variants in which

---

[1] Mutations were named to include the amino acid residue number preceded by the wild-type amino acid and followed by the amino acid to which the residue is changed, with the use of the single letter amino acid code. Thus, K464A means that lysine residue 464 was changed to alanine.

the Walker A Lys was mutated, MgATP was less potent at stimulating channel activity. When the conserved Asp in Walker B was mutated to Asn (D1370N), the potency of MgATP was also reduced.

The Walker A Lys is followed by two hydroxyl amino acids in most members of the ATPase/ABC transporter family. In NBD1 of CFTR, the sequence is Lys-Thr-Ser. When we switched the order of the two hydroxyls to Lys-Ser-Thr (CFTR-T465S; S466T), we found that the MgATP dose–response curve was shifted to the left. This suggests that MgATP is more potent at stimulating CFTR-T465S; S466T than wild-type CFTR.

These results indicate that ATP interacts directly with CFTR, and more specifically that it interacts with both NBDs. The data also suggest that an effect of MgATP on both NBDs is required for maximal channel activity.

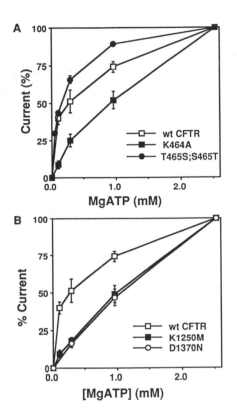

Figure 3. Effect of site-directed mutations on the function of CFTR. The figures show the effect of NBD1 (*A*) and NBD2 (*B*) mutations on MgATP-dependent stimulation of CFTR Cl⁻ channels. The current observed in each excised patch was normalized to the current measured with 2.53 mM MgATP. (Reproduced from *Science,* 1992, 257:1701–1704, by copyright permission of the American Association for the Advancement of Science.)

ATP Hydrolysis May Be Required to Activate CFTR Cl⁻ Channels

In some members of the traffic ATPase/ABC transporter family, the NBDs hydrolyze ATP. Therefore, we speculated that the NBDs may hydrolyze ATP to open CFTR Cl⁻ channels. We began to test this hypothesis by testing the effect of nonhydrolyzable ATP analogues. We found that nonhydrolyzable ATP analogues such as adenosine 5′-(β,γ-methylene) triphosphate (AMP-PCP) and adenosine 5′-(β,γ-imino) triphosphate (AMP-PNP) failed to activate CFTR Cl⁻ channels (Anderson et al., 1991*b*). Likewise, other nucleotides such as ADP and cAMP also failed to activate the Cl⁻ channels. Mg^{2+}, a cofactor required in ATP hydrolysis reactions, was also required for nucleotide regulation. These results suggest that ATP hydrolysis is required for opening of CFTR Cl⁻ channels.

The requirement for nucleoside triphosphates was not, however, highly specific. At a concentration of 1 mM, the nucleotide specificity sequence was ATP > AMP-CPP > GTP > ITP $\cong$ UTP > CTP. This broad nucleotide specificity contrasts with the high specificity for ATP observed for a number of kinases.

ATPγS is an especially useful ATP analogue because of its slow rate of hydrolysis. It can serve as a substrate for protein kinases because phosphorylation is cumulative and the transferred thiophosphate is not readily removed by protein phosphatases. But this slowly hydrolyzed ATP analogue does not substitute for ATP in many proteins that directly couple the rate of ATP hydrolysis to protein function. Fig. 4 shows that, in contrast to ATP plus PKA, ATPγS plus PKA failed to activate CFTR Cl⁻ channels. However, the channel appeared to be phosphorylated by ATPγS because after PKA and ATPγS were removed, addition of ATP increased current. When ATP was removed and ATPγS was once again added, it failed to substitute for ATP. These results indicate that ATPγS substitutes for ATP in PKA-dependent phosphorylation, but cannot substitute for ATP in nucleotide regulation of the channel.

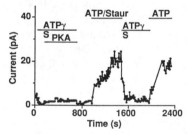

Figure 4. Effect of ATPγS on CFTR Cl⁻ channel activity. Data points are current at −40 mV from excised inside-out patches containing large numbers of channels. 1 mM ATP, 1 mM ATPγS, 75 nM PKA, and 5 μM staurosporine (to inhibit any PKA associated with the membrane patch) were added during the times indicated. Data are representative examples from one patch. (Reproduced from *Cell,* 1991, 67:775–784, by copyright permission of Cell Press.)

ADP Competitively Inhibits CFTR Cl⁻ Channels through the NBDs

ADP is a product of ATP hydrolysis and is another abundant intracellular nucleotide whose concentration is influenced by the metabolic status of a cell. Therefore, we tested the hypothesis that ADP would interact with the NBDs, thereby inhibiting CFTR Cl⁻ channel function (Anderson and Welsh, 1992). As indicated above, ADP alone did not stimulate channel activity. In the presence of ATP, increasing concentrations of ADP progressively inhibited current (Fig. 5). An inverse plot of the data suggested that ADP is a competitive antagonist. When we tested the relative inhibitory potency of nucleoside diphosphates, we found that the rank order potency was ADP > GDP $\cong$ IDP > UDP > CDP > ADPβS. The potency sequence is the same as that for nucleoside triphosphate stimulation of channel activity.

The similar nucleotide specificity and the competitive antagonism suggested that ATP and ADP may interact with the same site. It seemed most likely that that site would be the NBD. However, it was possible that ADP might act at either one or both NBDs. To determine at which NBD ADP acted, we used CFTR containing the site-directed mutations described above. We found that ADP inhibited CFTR containing mutations in NBD1. However, inhibition was abolished or significantly reduced in CFTR variants containing mutations in the conserved Walker motifs of NBD2. These results suggest that ADP inhibits CFTR by competing with ATP and that competition occurs at NBD2. Thus, the divergence in amino acid sequences of the two NBDs may be paralleled by a divergence in their function. Because ADP is a product of ATP hydrolysis, one might predict that ADP would be a competitive antagonist at the site of hydrolysis. Thus, we speculate that if CFTR hydrolyzes ATP, such hydrolysis may occur at NBD2. In this regard, CFTR may be similar to some enzymes that contain both ATP catalytic sites and ATP allosteric or regulatory sites. Alternatively, ADP may simply compete for binding of ATP to NBD2, but not cause channel opening. It will be interesting to learn whether the NBDs of other members

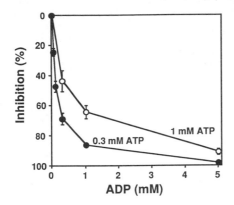

Figure 5. Inhibition of CFTR Cl⁻ channels by ADP. The effect of ADP concentration on the current in the presence of either 0.3 mM (*filled symbols*) or 1 mM MgATP (*open symbols*) is shown. Data are normalized to current in the absence of ADP. Data are from four to eight membrane patches at each point. (Reproduced from *Science,* 1992, 257:1701–1704, by copyright permission of the American Association for the Advancement of Science.)

of the traffic ATPase/ABC transporter family, particularly those presumed to have two identical NBDs, interact with ADP or have divergent functions.

Summary

These results begin to indicate that nucleoside triphosphates directly regulate CFTR Cl⁻ channels by interacting with the NBDs. Thus, they may begin to explain why some CF-associated mutations in the NBDs may block Cl⁻ channel function in the epithelia of CF patients.

These results also suggest that the intracellular ATP/ADP ratio may be more important than the absolute concentration of ATP in regulating CFTR. Thus, changes in the metabolic state of the cell that alter the ATP-ADP ratio may regulate CFTR Cl⁻ channel activity in vivo. These observations suggest that CFTR might be regulated in the physiologic range of nucleotides. Such a mechanism of regulation could provide a mechanism for coupling the metabolic status of the cell and the activity of the Na-K ATPase with the rate of transepithelial Cl⁻ secretion as regulated by apical membrane CFTR Cl⁻ channels.

References

Ames, G. F., C. S. Mimura, and V. Shyamala. 1990. Bacterial periplasmic permeases belong to a family of transport proteins operating from *Escherichia coli* to human: traffic ATPases. *FEMS Microbiology Reviews.* 75:429–446.

Anderson, M. P., H. A. Berger, D. P. Rich, R. J. Gregory, A. E. Smith, and M. J. Welsh. 1991*a*. Nucleoside triphosphates are required to open the CFTR chloride channel. *Cell.* 67:775–784.

Anderson, M. P., R. J. Gregory, S. Thompson, D. W. Souza, S. Paul, R. C. Mulligan, A. E. Smith, and M. J. Welsh. 1991*b*. Demonstration that CFTR is a chloride channel by alteration of its anion selectivity. *Science.* 253:202–205.

Anderson, M. P., and M. J. Welsh. 1992. Regulation by ATP and ADP of CFTR chloride channels that contain mutant nucleotide-binding domains. *Science.* 257:1701–1704.

Berger, H. A., M. P. Anderson, R. J. Gregory, S. Thompson, P. W. Howard, R. A. Maurer, R. Mulligan, A. E. Smith, and M. J. Welsh. 1991. Identification and regulation of the cystic fibrosis transmembrane conductance regulator-generated chloride channel. *Journal of Clinical Investigation.* 88:1422–1431.

Boat, T. F., M. J. Welsh, and A. L. Beaudet. 1989. Cystic fibrosis. *In* The Metabolic Basis of Inherited Disease. C. R. Scriver, A. L. Beaudet, W. S. Sly, and D. Valle, editors. McGraw-Hill, Inc., New York. 2649–2680.

Cheng, S. H., D. P. Rich, J. Marshall, R. J. Gregory, M. J. Welsh, and A. E. Smith. 1991. Phosphorylation of the R domain by cAMP-dependent protein kinase regulates the CFTR chloride channel. *Cell.* 66:1027–1036.

Gregory, R. J., S. H. Cheng, D. P. Rich, J. Marshall, S. Paul, K. Hehir, L. Ostedgaard, K. W. Klinger, M. J. Welsh, and A. E. Smith. 1990. Expression and characterization of the cystic fibrosis transmembrane conductance regulator. *Nature.* 347:382–386.

Hyde, S. C., P. Emsley, M. J. Hartshorn, M. M. Mimmack, U. Gileadi, S. R. Pearce, M. P. Gallagher, D. R. Gill, R. E. Hubbard, and C. F. Higgins. 1990. Structural model of ATP-binding proteins associated with cystic fibrosis, multidrug resistance and bacterial transport. *Nature.* 346:362–365.

McIntosh, I., and G. R. Cutting. 1992. Cystic fibrosis transmembrane conductance regulator and etiology and pathogenesis of cystic fibrosis. *FASEB Journal.* 6:2775–2782.

Quinton, P. M. 1990. Cystic fibrosis: a disease in electrolyte transport. *FASEB Journal.* 4:2709–2717.

Rich, D. P., R. J. Gregory, M. P. Anderson, P. Manavalan, A. E. Smith, and M. J. Welsh. 1991. Effect of deleting the R domain on CFTR-generated chloride channels. *Science.* 253:205–207.

Riordan, J. R., J. M. Rommens, B. Kerem, N. Alon, R. Rozmahel, Z. Grzelczak, J. Zielenski, S. Lok, N. Plavsic, J. L. Chou, M. L. Drumm, M. C. Iannuzzi, F. S. Collins, and L.-C. Tsui. 1989. Identification of the cystic fibrosis gene: cloning and characterization of complementary DNA. *Science.* 245:1066–1073.

Tabcharani, J. A., X.-B. Chang, J. R. Riordan, and J. W. Hanrahan. 1991. Phosphorylation-regulated Cl⁻ channel in CHO cells stably expressing the cystic fibrosis gene. *Nature.* 352:628–631.

Tsui, L.-C., and M. Buchwald. 1991. Biochemical and molecular genetics of cystic fibrosis. *Advances in Human Genetics.* 20:153–266.

Walker, J. E., M. Saraste, M. J. Runswick, and N. J. Gay. 1982. Distantly related sequences in the α- and β-subunits of ATP synthase, myosin, kinases and other ATP-requiring enzymes and a common nucleotide binding fold. *EMBO Journal.* 1:945–951.

Welsh, M. J., M. P. Anderson, D. P. Rich, H. A. Berger, G. M. Denning, L. S. Ostedgaard, D. N. Sheppard, S. H. Cheng, R. J. Gregory, and A. E. Smith. 1992. Cystic fibrosis transmembrane conductance regulator: a chloride channel with novel regulation. *Neuron.* 8:821–829.

Chapter 10

Effects of Nucleotide Binding Fold Mutations on STE6, a Yeast ABC Protein

Carol Berkower and Susan Michaelis

*Department of Cell Biology and Anatomy, Johns Hopkins University
School of Medicine, Baltimore, Maryland 21205*

Molecular Biology and Function of Carrier Proteins © 1993 by The Rockefeller University Press

Introduction

Recently, a great deal of attention has been focused on a class of membrane-bound proteins that utilize ATP to convey substrates into and out of cells and between cellular compartments. This class of proteins is designated the ATP binding cassette (ABC) family (Higgins et al., 1986; Hyde et al., 1990; Higgins, 1992), and includes such diverse members as the mammalian multidrug resistance protein (MDR; Chen et al., 1986; Gros et al., 1986; Endicott and Ling, 1989; Gottesman and Germann, 1993), the transporter associated with antigen processing (TAP; Parham, 1992), the bacterial hemolysin transporter (HlyB; Felmlee et al., 1985; Gerlach et al., 1986; Mackman et al., 1986), and a variety of bacterial sugar, amino acid, and peptide permeases (Ames, 1986). Most ABC proteins are thought to carry out energy-dependent substrate transport, though the types of substrates differ greatly from one family member to another (Higgins et al., 1988; Kuchler and Thorner, 1990). The cystic fibrosis transmembrane conductance regulator (CFTR) is also an ABC protein, although it is distinctive in that it is not known to display transport activity, but functions instead as a chloride ion channel (Riordan et al., 1989; Bear et al., 1992; Collins, 1992). Nonetheless, as with other ABC proteins, ATP utilization appears to be critical for CFTR activity (Anderson et al., 1991; Drumm et al., 1991; Gregory et al., 1991; Anderson and Welsh, 1992).

This paper focuses on the STE6 protein (pronounced "sterile 6"), an ABC family member in the yeast *Saccharomyces cerevisiae* (Kuchler et al., 1989; McGrath and Varshavsky, 1989). STE6 bears a close resemblance to human MDR1; these two proteins exhibit an overall conservation of structure and 40% identity between their ATP binding domains. STE6 and CFTR are also homologous, particularly in their ATP binding domains. The STE6 protein mediates export of the yeast pheromone, **a**-factor, a 12-residue secreted peptide that is isoprenylated and carboxyl-methylated (Anderegg et al., 1988). **a**-Factor is not exported by the classical secretory pathway (reviewed in Michaelis, 1992). Like the substrates of MDR, **a**-factor is lipophilic. Extracellular **a**-factor plays an obligatory step in initiating the yeast mating program (Michaelis and Herskowitz, 1988). A mutant defective for STE6 function is unable to export **a**-factor, and as a result is unable to mate (hence the designation *ste*rile) (Rine, 1979; Wilson and Herskowitz, 1987; Michaelis, 1992). Since yeast is amenable to both genetic and biochemical analysis, and since the native substrate of STE6 is known, STE6 provides a unique opportunity for examination of an ABC protein at the molecular level.

Comparison of ABC family members from diverse species suggests a modular organization for these proteins. STE6, MDR, and CFTR are composed of two homologous halves, each encoding six predicted membrane spans and an ATP nucleotide binding fold (NBF) domain (Kuchler and Thorner, 1990). Certain ABC proteins, such as HlyB and the TAP transporters, appear to be half-molecule versions of STE6, with only a single set of membrane spans and a single NBF domain. These presumably associate into homodimers (HlyB) or heterodimers (TAP) to form the functional transporter (Spies et al., 1992). Several bacterial permeases are further subdivided, encoding their membrane spans and NBF domains on separate polypeptides (Hyde et al., 1990; Mimura et al., 1991). By examining STE6 mutants and half-molecules, we expect to determine how the various regions of STE6 contribute to its function and assembly.

The predicted structure of STE6 is shown in Fig. 1. (The two NBFs will be referred to here as NBF1 and NBF2, designating their relative positions in the protein.) Each NBF includes two subregions, the Walker A and B motifs, which contain highly conserved residues implicated in ATP binding and hydrolysis (Walker et al., 1982). The Walker A motif, also known as the P-loop for its apparent proximity to the ATP triphosphate (Saraste et al., 1990), has the consensus sequence GX_4GKS/T, while the Walker B motif consensus is $RX_{6-8}hyd_4D$. A signature sequence, LSGGQ, precedes the Walker B motif (Riordan et al., 1989; Cutting et al., 1990). Although this sequence is highly conserved among ABC family members, its function remains unclear. In STE6 and other ABC family members, the Walker A and B motifs tend to be separated by ~100 residues. An additional Center region, occurring midway between the A and B sites (Fig. 1), is loosely conserved within the ABC family (Riordan et al., 1989; Berkower and Michaelis, 1991).

Certain highly conserved residues in the NBF domains are of universal importance for ABC proteins, since mutations in these residues disrupt function in all family members tested. For instance, in lysine and glycine residues of the P loop of CFTR, both naturally occurring cystic fibrosis (CF) mutations and artificially con-

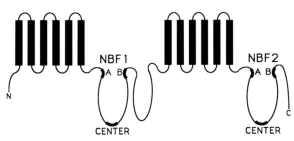

Figure 1. Model for the structure of STE6 in the membrane. The disposition of STE6 shown here is predicted by hydropathy analysis of the deduced amino acid sequence (Kuchler et al., 1989; McGrath and Varshavsky, 1989). Predicted transmembrane domains are shown (rectangles), as are the two NBF domains. Thickened areas represent the conserved Walker A and B regions and Center region of each NBF (described in Berkower and Michaelis, 1991).

structed mutations disrupt activity (Anderson et al., 1991; Drumm et al., 1991; Gregory et al., 1991; Anderson and Welsh, 1992). Corresponding mutations in MDR destroy its capacity to confer drug resistance (Azzaria et al., 1989). Our studies have demonstrated that STE6 function may be impaired by mutations in these and other conserved residues in either NBF (Berkower and Michaelis, 1991; Table II). Since we found that single point mutations in either NBF of STE6 disrupt function, it is clear that both halves of STE6 are important for function. Thus, it is likely that two distinct ATP-requiring steps are involved in STE6-mediated **a**-factor transport. In addition, we have severed STE6 into its two homologous halves and shown that, although separated half-molecules do not function, function can be restored when the two halves are coexpressed within the same cell (Berkower and Michaelis, 1991).

Our previous work involved mutating one NBF of STE6 at a time, while leaving the other NBF intact. Surprisingly, none of our single-residue substitutions or deletions completely destroyed STE6 function, leaving open the possibility that there might be an ATP-independent component to STE6-mediated **a**-factor export. We address this issue here by constructing double mutants, in which analogous residues in both NBFs of STE6 are altered simultaneously. We also begin to address the question of whether STE6 can form dimers or higher-order structures. To do this, we examine the ability of full-length STE6 molecules to form functional associations with half-molecules or other full-length molecules.

Materials and Methods

Yeast Strains, Media, and Growth Conditions

Construction of the Δ*ste6* deletion strain, SM1646, was described previously (Berkower and Michaelis, 1991). SM1646 is an **a** strain in which 87% of the *STE6* coding region has been replaced with the *URA3* marker. Plasmid-bearing strains were constructed by transformation of SM1646 and growth on selective dropout medium. The α strain used for testing **a**-factor halos is SM1086 (α *sst2-1, rme, his6, met1, can1, cyh2*) (Michaelis and Herskowitz, 1988), originally designated RC757 (Chan and Otte, 1982). The α strain used for all mating assays is SM1068 (α *lys1*) (Michaelis and Herskowitz, 1988).

Complete medium (YEPD), SD drop-out medium, and SD minimal medium were prepared as previously described (Michaelis and Herskowitz, 1988). Where necessary, SD medium was supplemented with histidine (20 μg/ml), tryptophan (20 μg/ml), or leucine (100 μg/ml). All strains were grown at 30°C, except where indicated.

Plasmid Constructions

The STE6 C-Half is encoded by plasmid pSM434 (containing *CEN6* and *TRP1*), described previously (Berkower and Michaelis, 1991). Unless otherwise noted, full-length wild-type and mutant versions of *STE6* are carried in the vector pRS315-1 (*CEN6 LEU2*) (Berkower and Michaelis, 1991). To obtain the G509D mutant on a plasmid marked by the *TRP1* gene, we subcloned DNA encoding G509D into the vector pRS314 (*CEN6 TRP1*) (Sikorsky and Hieter, 1989), which yielded pSM531.

Construction of Coexpression Strains

To generate the G1193D + C-Half cotransformant (shown in Fig. 3), we simultaneously transformed SM1646 with the plasmids pSM403, containing the full-length *STE6* mutant G1193D in pRS351-1 (*CEN6 LEU2*), and pSM434, containing the C-Half (*CEN6 TRP1*). Transformants were selected on SD minimal medium lacking tryptophan and leucine. Three independent transformants were evaluated by the patch mating assay and gave identical results. One of these strains was chosen for further studies and is designated SM1878.

To construct the strain coexpressing two full-length *STE6* mutants, listed in Table III, an existing plasmid-bearing strain was retransformed with a new plasmid. As previously described (Berkower and Michaelis, 1991), SM1646 was transformed with a plasmid containing the *STE6* NBF2 mutant G1193D on pRS315-1 (*CEN6 LEU2*), to generate the strain SM1781. We then transformed SM1781 with pSM531 by the method of Elble (1992). The resultant strain is SM1990 (G509D *TRP1* + G1193D *LEU2*).

STE6 Mutagenesis

Construction of single point mutations in the *STE6* gene was described previously (Berkower and Michaelis, 1991). Double mutants were obtained in pSM322 (*STE6 CEN6 LEU2*) either by site-directed mutagenesis with two mutagenic oligonucleotide primers (described in Berkower and Michaelis, 1991) by the method of Kunkel et al. (1987), or by ligation of DNA restriction fragments from single mutants. Mutant plasmids were transformed into the yeast strain SM1646. Plasmids were rescued from yeast and resequenced to confirm the presence of both mutations.

Metabolic Labeling and Immunoprecipitation of STE6

For immunoprecipitation of STE6, cells were grown to log phase (OD_{600} 0.4–0.8) in SD medium containing required supplements. 5 OD_{600} units of cells were resuspended in 0.5 ml fresh medium together with 300 µCi Tran-^{35}S-label (ICN Biomedicals, Inc., Costa Mesa, CA) and incubated at 30°C. Labeling was terminated after 15 min by addition of cells to ice-cold 2× azide stop mix (40 mM cysteine, 40 mM methionine, 20 mM NaN_3, 500 µg/ml BSA). Cells were washed twice in water and resuspended in 50 µl immunoprecipitation buffer (Hrycyna et al., 1991) with 0.25 g zirconium beads (0.5 mm diameter; Biospec Products, Inc., Bartlesville, OK). Cells were lysed by vigorous vortexing at 4°C in twelve 30-s intervals, interspersed with 30-s incubations on ice. The lysate was drawn off in 1.5 ml immunoprecipitation buffer, and unbroken cells were pelleted in a 5-min spin at 500 *g*. Insoluble material was removed from the supernatant by a 1-min spin at 13,000 *g*. The cleared lysate was incubated overnight at 4°C with 7.5 µl anti-STE6 antiserum C12-JH210 (Berkower and Michaelis, 1991). Antigen–antibody complexes were collected on protein A–Sepharose CL-4B beads (Pharmacia LKB Biotechnology Inc., Piscataway, NJ), gently agitated at 4°C for 90 min, and washed (Hrycyna et al., 1991). After the final wash, bound complexes were released from the beads by the addition of 30 µl 2× Laemmli sample buffer (20% glycerol, 10% β-mercaptoethanol, 4.3% SDS, 0.125 M Tris-HCl, pH 6.8, and 0.2% bromophenol blue). Samples were incubated at 37°C for 20 min before electrophoresis; boiling causes STE6 to run aberrantly near the top edge of the gel. Proteins were separated on a 10% SDS-PAGE gel and analyzed by phosphorimager (Molecular Dynamics, Inc., Sunnyvale, CA).

Metabolic Labeling and Immunoprecipitation of a-Factor

Log phase cells (3.3 OD_{600} units) were labeled under steady-state conditions for 1 h in SD-LEU dropout medium with 200 µCi [35S]cysteine (New England Nuclear, Boston, MA). Cell-associated and extracellular fractions were prepared and immunoprecipitated with **a**-factor antiserum 9-137 as described in Hrycyna et al. (1991). Immunoprecipitates were electrophoresed in a 16% SDS-PAGE gel (0.5 OD_{600} equivalent per lane), followed by fluorography and autoradiography.

Physiological Assays

Quantitative mating assays were performed by the plate mating procedure, essentially as previously described (Michaelis and Herskowitz, 1988). Mutants to be tested were grown to saturation in SD medium containing required supplements and serially diluted in 10-fold increments into YEPD. A 0.1-ml aliquot of each dilution was spread on an SD plate together with ~10^7 cells of the α mating tester, SM1068, in 0.1 ml YEPD. Plating the cells in a small amount of YEPD provides essential nutrients which allow haploids to grow into microcolonies which can contact one another, thereby increasing their opportunity to mate. All mutant **a** strains tested are His⁻ and the α mating tester is Lys⁻. Diploids resulting from mating are prototrophic and can therefore form colonies on SD. After 3 d, diploids were counted and normalized to the total number of **a** cells plated (~10–30% of wild-type **a** cells participate in mating). The frequency of diploid formation for each mutant as compared with wild type is designated mating efficiency and is expressed as a percentage of wild type. Results represent an average of two to three separate trials. Quantitative mating assays are normally performed at 30°C.

Patch halo and mating assays were carried out essentially as previously described (Wilson and Herskowitz, 1984; Michaelis and Herskowitz, 1988). **a** strains to be tested were spread in patches on YEPD (complete) or SD dropout plates and incubated at 30°C for 24 h. For the halo assay, these patches were replica-plated onto YEPD plates spread with the α halo tester SM1086 and incubated at 30°C for 24 h. The *sst2* mutation in SM1086 renders it supersensitive to growth inhibition by **a**-factor. For a strain producing **a**-factor, the patch is surrounded by a clear zone (or halo) where growth of the lawn has been inhibited. The width of the clear zone provides a qualitative measure of **a**-factor activity.

For the patch mating assay, the same original patches were replica-plated onto an SD minimal agar plate spread with a lawn (in YEPD) of the α mating tester, SM1068, and incubated at 37°C for 2–3 d. The patch mating assay shown in Fig. 4 was done by directly streaking **a** cells onto the α lawn. Only prototrophic diploids formed by mating can grow on minimal medium. Mating occurs less efficiently at 37°C than at 30°C for all strains examined, including the wild type. Since the wild-type mating patch is confluent at both temperatures, incubation at 37°C improves visualization of debilitated mutants relative to the wild-type strain. Absolute mating efficiencies vary with the thickness of the α lawn, the amount of **a** cells transferred, the condition of the **a** cells (i.e., growth phase before plating), and the temperature of the incubation; hence, results observed for particular patches can be accurately compared only with other patches derived from similarly treated original colonies on the same tester plate.

Results

Measuring STE6 Function

Mating in yeast involves the fusion of two haploid cells of opposite mating type, designated **a** and α (reviewed in Fields, 1990 and Kurjan, 1992). Each cell type secretes a distinct mating pheromone, **a**-factor or α-factor, respectively, that interacts with a receptor on the opposite cell to activate a pheromone response pathway. **a**-Factor stimulation of α cells leads to growth arrest; other morphological changes follow, culminating in mating. Loss of *ste6* function results in the inability of **a** cells to export **a**-factor, and consequently, an inability to mate (Wilson and Herskowitz, 1987; Kuchler et al., 1989; McGrath and Varshavsky, 1989; Berkower and Michaelis, 1991). To evaluate the effects of mutations on STE6 function, we have utilized three different assays. These include immunoprecipitation of **a**-factor from the culture fluid of **a** cells, mating of **a** cells, and the **a**-factor halo assay.

Immunoprecipitation of a-factor. The most direct way to measure STE6 function is to determine the amount of **a**-factor exported from **a** cells. This can be accomplished by metabolically labeling cellular proteins, separating intracellular from extracellular fractions, and immunoprecipitating **a**-factor from these fractions (Kuchler et al., 1989; Berkower and Michaelis, 1991; Hrycyna et al., 1991). On an

SDS-PAGE gel, intracellular **a**-factor runs as several bands, reflecting its processing from a precursor (P) to the mature (M), bioactive form (Fig. 2*A*, lane *1*). Mature **a**-factor is exported into the extracellular fraction (Fig. 2*A*, lane *2*). A mutant strain from which the *STE6* gene has been deleted, Δ*ste6*, exhibits a wild-type level of intracellular **a**-factor, but lacks extracellular **a**-factor (Fig. 2*A*, lanes *3* and *4*), indicating a specific export defect, rather than altered synthesis or processing of the

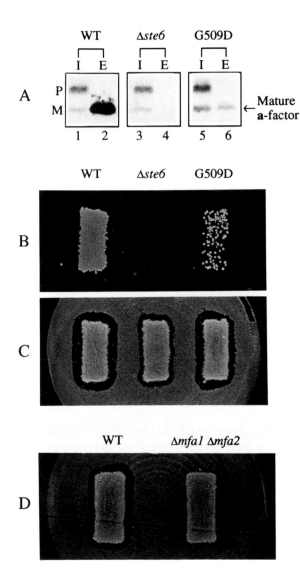

Figure 2. Three assays for STE6 function. (*A*) Immunoprecipitation of **a**-factor from wild-type and *ste6* mutant strains. Cells were radiolabeled with [^{35}S]cysteine. Intracellular (*I*) and extracellular (*E*) fractions were prepared, immunoprecipitated with **a**-factor antiserum, and analyzed by SDS-PAGE, as described in Materials and Methods. **a**-Factor precursor (*P*) and mature (*M*) forms are labeled. All strains used have a chromosomal deletion of the *STE6* gene, and harbor a pRS351-1 plasmid derivative containing *STE6* (WT), no insert (Δ*ste6*), or *STE6* with an NBF1 point mutation (G509D). (*B*) Patch mating assays were performed as described in Materials and Methods. (*C* and *D*) Patch halo assays were performed as described in Materials and Methods; the strain labeled Δ*mfa1* Δ*mfa2* is deleted for both **a**-factor structural genes (Michaelis and Herskowitz, 1988).

a-factor precursor. In contrast to the complete absence of **a**-factor export in a Δ*ste6* mutant, a strain containing a point mutation, G509D, in NBF1 of STE6 exports a low but detectable amount of **a**-factor (Fig. 2*A*, lane *6*). Densitometric analysis of the extracellular **a**-factor band from this mutant reveals ~5% of the wild-type level of extracellular **a**-factor (Table I). Lower levels become hard to discriminate by this method, so sensitivity is diminished for mutants severely defective in **a**-factor export (i.e., those that export <5% the wild-type level of **a**-factor).

Mating. Another way to measure STE6 function is by the quantitative mating assay. This assay is based on the ability of two auxotrophic haploids, deficient in different biosynthetic steps, to mate and form a prototrophic diploid (Michaelis and Herskowitz, 1988). When mating is carried out on minimal plates, only diploids will grow into visible colonies. Performed quantitatively, as described in Materials and Methods, the mating assay is exquisitely sensitive, permitting detection of mating at levels as low as 0.0001% of wild type. We have shown elsewhere that the ability of a *ste6* mutant to export **a**-factor correlates reliably with its ability to mate (Berkower and Michaelis, 1991). The mating assay is particularly useful for discriminating among *ste6* mutants that have very low levels of activity, where immunoprecipitation loses sensitivity. It should be noted that mating does not effectively discriminate among mutants that are marginally defective (>50% wild-type **a**-factor export). In these cases, it is necessary to look at **a**-factor export directly by immunoprecipitation.

TABLE I
Quantitative Immunoprecipitation, Mating, and Halo Assays for STE6 Function

STE6 mutation*	a-Factor IP[‡]	Mating[§]	Halo[‖]
		%	
WT	100	100	100
Δste6	<1	<0.0001	1
G509D	5	0.5	2

CEN plasmids bearing wild-type or mutant versions of the *STE6* gene were expressed in a Δ*ste6* mutant strain.
[‡]Mature, metabolically labeled a-factor, immunoprecipitated from the extracellular medium, was quantitated by densitometry and normalized to wild type (=100%).
[§]Quantitative mating assays were carried out at 30°C and normalized to wild type.
[‖]The halo dilution assay was performed as described in Berkower and Michaelis (1991) and normalized to wild type.

A qualitative version of the mating assay, patch mating (Fig. 2 *B*), allows for rapid and accurate visual comparison between strains. As can be seen by comparison of Fig. 2 *B* with Table I, patch mating reflects the difference in mating efficiency between wild-type (100%), Δ*ste6* (<0.001%), and G509D (0.5%) mutant strains determined quantitatively. Although the G509D mutant in Fig. 2 *B* mates at only 0.5% of wild-type levels, the patch test reveals that it is clearly a better mater than the Δ*ste6* strain.

a-Factor halo production. Another way to measure STE6 function is by the halo assay (Fig. 2 *C* and Table I). After **a**-factor is exported into the medium, it binds to a receptor on α cells, causing cells to arrest growth at the G1 phase of the cell cycle (Fields, 1990; Kurjan, 1992). While normal α cells recover, certain mutants, designated *sst2* (supersensitive), respond to extremely low levels of pheromone and remain permanently arrested (Kurjan, 1992). Growth arrest may be visualized as a clear zone of growth inhibition, or halo, surrounding a patch of **a** cells on an otherwise confluent lawn of α cells (Michaelis and Herskowitz, 1988; Berkower and

Michaelis, 1991). The a-factor titer of a strain can be measured by collecting the culture fluid from cells, spotting a dilution series onto an α *sst2* lawn, and determining the highest dilution that prevents growth of the underlying α cells (described in Materials and Methods). A low titer of exported a-factor reflects a low level of STE6 function. Precision of the halo assay for measuring STE6 function is limited by a high background of a-factor that escapes into the culture fluid of a cells independently of STE6, as evidenced by the culture supernatant from the Δ*ste6* deletion strain, which exhibits 1% wild-type levels of a-factor (Berkower and Michaelis, 1991). Thus, *ste6* mutants producing as much as 1% wild-type levels of a-factor are indistinguishable from Δ*ste6* by halo assay, even though they clearly perform better than Δ*ste6* as measured by the quantitative mating assay and immunoprecipitation of extracellular a-factor (Berkower and Michaelis, 1991).

Fig. 2 *C* shows a striking example of the production of an a-halo by a Δ*ste6* mutant. This patch halo assay compares wild-type, Δ*ste6* , and G509D mutant strains, replica-plated directly onto an α lawn. The residual halo around the Δ*ste6* deletion strain is quite dramatic, reflecting the STE6-independent release of a-factor. The diameter of this residual halo can vary, depending on plating conditions. The halo is due to a-factor activity and not to some other secreted toxic compound, since a strain lacking the a-factor structural genes, *MFA1* and *MFA2,* produces no residual halo (Fig. 2 *D*). Possible explanations for the residual a-factor halo of the Δ*ste6* strain are given in the discussion.

In summary, when comparing the three assays for a-factor export, quantitative mating has proven the most versatile and sensitive assay of STE6 function. Though we have evaluated our mutants by all three methods with consistent results, this paper presents data in terms of mating efficiency.

Double ATP Binding Domain Mutants

Walker A motif (GX$_4$GKS/T). STE6 contains two ATP binding folds, termed NBF1 and NBF2 (Fig. 1). In our previous studies, we altered conserved residues individually in either NBF of STE6 and saw a resultant decrease in function, indicating that a-factor transport involves ATP utilization (Berkower and Michaelis, 1991). However, it was notable that no single mutation caused a complete loss of a-factor export or mating (Table II). For instance, the most severe mutation had 0.3% residual mating activity, as compared with a Δ*ste6* null mutant, in which activity is <0.0001%. This residual function could be interpreted to mean that STE6 is capable of mediating a-factor transport, albeit at low levels, without using ATP. Alternatively, residual activity could reflect a residual ability of the mutated NBF to utilize ATP.

To examine this issue further, we made double mutants containing mutations in both NBFs of STE6. If ATP utilization is required for STE6-mediated a-factor transport, then we would expect that simultaneous alteration of both NBFs would completely abolish function. The double mutants were analyzed by mating (Table II) and by immunoprecipitation of exported a-factor (data not shown). The two single Walker A site mutations, G392V in NBF1 and G1087V in NBF2, reduce mating efficiency to 0.8% and 0.3%, respectively. When these mutations are combined in the same STE6 molecule, mating is obliterated (<0.0001%; Table II). This result strongly suggests that ATP utilization is, in fact, essential for STE6 activity. Interestingly, another double mutant, K398R K1093R, retains significant activity (1% of the wild-type level; Table II), suggesting that the NBFs thus mutated retain enough function to promote a low, but detectable, level of mating.

Another pair of mutations that cause severe effects individually are the LSGGQ site mutations near the Walker B site, G509D in NBF1 and G1193D in NBF2. Combining these two mutations completely abolishes STE6 activity (Table II), once again suggesting that ATP utilization is essential for function.

Center region. We had previously shown that mutations in conserved residues of the STE6 Center region have no effect on function. It seemed likely, therefore, that these residues were inessential for transport activity, or that perhaps the two Center regions performed some redundant function. This question is addressed here by examining double Center region mutants (Table II). L455 of STE6 is in a position analogous to that of F508 of CFTR, and Y1150 occupies the corresponding position in NBF2 of STE6 (Berkower and Michaelis, 1991). A deletion of F508 from CFTR

TABLE II
Mating Efficiencies of STE6 Single and Double Mutants

Location	Mutation(s)*	Mating efficiency (%)[‡]		
		NBF1	NBF2	DOUBLE
A region	G392V*, G1087V	0.8	0.3	<0.0001
	K398R, K1093R	1	15	1
	K398A, K1093A	25	26	5
Center	Q440N, Q1135N	87	88	64
	ΔF445, ΔF1140	95	102	3
	ΔL455*, ΔY1150	105	85	87
B region	G509D*, G1193D*	0.5	6	<0.0001
Δ*ste6*		—	—	<0.0001

*Plasmids bearing the indicated mutations were evaluated in a Δ*ste6* mutant strain. Starred mutations are analogous to CF mutations of CFTR (Riordan et al., 1989; Cuppens et al., 1990; Cutting et al., 1990; Beaudet et al., 1991). The impact of each mutation on STE6 function was measured by quantitative mating assays carried out at 30°C and is normalized to wild type (<100%).
[‡]Results for single mutants are taken from Berkower and Michaelis (1991).

produces the most common defective cystic fibrosis allele (Kerem et al., 1989). In STE6, the L455 Y1150 double mutant has nearly wild-type mating activity (87%), as is the case for each single mutant. In contrast, a double mutant harboring deletions of residues F445 and F1140 is notably debilitated for STE6 function (3%). F445 and F1140 are in an interesting location in STE6, since they lie near F508 in an alignment with CFTR and are highly conserved throughout the ABC family; the phenotype of the ΔF double mutant is analyzed further in the Discussion.

Several Center region mutants had wild-type function (Table II); these were tested for a temperature-sensitive loss of function by performing the mating assay at 37°C. Though mating efficiency is decreased even for the wild-type strain at 37°C, none of these mutants were observed to have altered mating relative to wild type at the higher temperature.

To examine whether any of our mutations caused a reduction in protein level (for instance, due to misfolding and premature degradation), we metabolically labeled proteins from the double mutants, immunoprecipitated with antibodies to STE6, and analyzed STE6 protein levels by SDS-PAGE (Fig. 3). Full-length STE6 is present in all strains harboring double mutants. The amount of STE6 does not differ significantly from wild type in any double mutants tested. (The twofold variation in levels of some mutant proteins is not consistently observed in other trials.) By pulse-chase analysis, heightened metabolic instability is not observed for any of the mutants (data not shown). It therefore seems likely that the loss of function observed for several STE6 double mutants is the direct result of a faulty transporter, rather than of defective processing or instability of the STE6 protein.

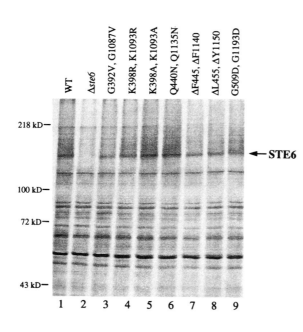

Figure 3. Immunoprecipitation of STE6 protein from wild-type and *ste6* double mutants. Strains were radiolabeled with Tran-^{35}S-label, cell extracts were prepared, and immunoprecipitation using anti-STE6 antiserum was carried out as described in Materials and Methods. Proteins were separated by SDS-PAGE and visualized by phosphorimager. The location of STE6 is shown. The density of each STE6 band was quantified and normalized to background bands in the same lane. Densities of the *STE6* mutant bands, expressed as a percentage of wild-type (100%, lane *1*) are: lane *2*, no material detected above background; lane *3*, 48%; lane *4*, 112%; lane *5*, 175%; lane *6*, 193%; lane *7*, 88%; lane *8*, 84%; lane *9*, 106%.

Complementation of Single NBF Mutants by Half-Molecules

We showed elsewhere that a single half-molecule of STE6, coding for the NH$_2$-terminal six membrane spans and ATP binding site (N-Half), could neither transport **a**-factor nor promote mating (Berkower and Michaelis, 1991). The same was seen to be true for the second half-molecule (C-Half). However, when the two half-molecules were coexpressed within the same cell, they restored STE6 function to nearly wild-type levels, suggesting that the two halves of STE6 were capable of association in the absence of any covalent bonds. We have extended this analysis to determine whether a coexpressed half-molecule could restore function to a full-length STE6 protein that harbors an NBF point mutation. We transformed SM1646 cells containing the full-length STE6 point mutant, G$_{1193}$D, with DNA encoding the "wild-type" C-Half molecule. While the $G_{1193}D$ mutation alone exhibits only 6% function (Table II), coexpression with the C-Half molecule restores mating efficiency to 100%. This dramatic restoration of function is shown by a patch mating assay in Fig. 4.

    To ensure that functional rescue was not due to recombination between homologous regions of *STE6* to regenerate a wild-type gene, the strain expressing both plasmids was cured of each one separately. The high level of mating disappeared when either the $G_{1193}D$ or C-Half plasmid was cured, indicating that the observed rescue was not due to recombination. As expected, rescue is dependent on coexpression of the appropriate half-molecule; coexpression of the N-Half molecule does not rescue $G_{1193}D$ (data not shown).

Lack of Cross-Complementation between Full-Length NBF Mutants

Given the ability of a half-molecule to rescue function of a full-length STE6 mutant, it seemed reasonable to test two full-length molecules, each mutated in a different

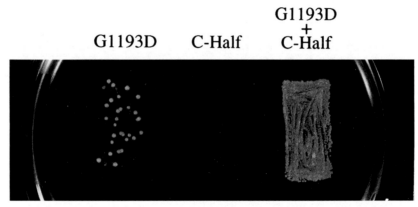

Figure 4. Functional rescue of *STE6* G1193D mutant by coexpression of C-Half. The patch mating assay was performed as described in Materials and Methods. Strains shown contain plasmids pSM403 (G1193D), pSM434 (C-Half), or pSM403 + pSM434 (G1193D + C-Half).

NBF, for cross-complementation of one another's defects. To test for cross-complementation, we coexpressed full-length STE6 bearing a single mutation in NBF1, G509D, with full-length STE6 bearing the analogous mutation in NBF2, G1193D. STE6 function was measured by the quantitative mating assay (Table III). No cross-complementation was observed for this pair of mutants (in contrast to the striking half-molecule complementation of G1193D described above). Rather, the effect of coexpression of two full-length mutants was only to raise function to levels near the sum of the two solo mutant activities.

Discussion

Assays for STE6 Function

Three assays are commonly used for evaluating the function of the STE6 protein: immunoprecipitation of extracellular **a**-factor, quantitative mating, and the **a**-factor halo assay. Of these, the quantitative mating assay is most powerful, due to its sensitivity at very low levels of STE6 function. Our observations here and elsewhere (Berkower and Michaelis, 1991) demonstrate the strong correlation between the amount of **a**-factor exported and the capacity of cells to mate. The mating assay also has the advantage of discriminating among a wide range of activities (Fig. 1 and Table I) and providing the highest degree of precision of any method. The ability of the mating assay to distinguish between very low levels of STE6 function (i.e., <5%) and total loss of function (0.0001%) permits detection of activities so low that they might otherwise go unnoticed. For instance, Raymond et al. (1992) showed that expression of mouse MDR3 in a Δ*ste6* mutant restores mating to 1% of wild-type levels, leading to the conclusion that **a**-factor is a substrate for MDR3. While 1% is a low level of residual function, it represents a significant increase over background, which is <0.0001%.

Strains unable to produce **a**-factor due to deletion of the **a**-factor structural genes make no halo on an α*sst2* lawn and cannot mate. In contrast, the Δ*ste6* deletion mutant cannot mate but produces a sizable residual halo. How can **a**-factor escape from a strain lacking the *STE6* protein? One possibility is the existence of an

TABLE III
Effect of Coexpression of Two STE6 Mutants on Mating Efficiency

G509D* (Trp$^+$)	G1193D (Leu$^+$)	Coexpressed (Trp$^+$Leu$^+$)
4.8%‡	10%	17%

*Plasmids bearing the indicated mutations were evaluated in a Δ*ste6* deletion strain. The impact of each mutation on STE6 function was measured by quantitative mating assays carried out at 30°C and is normalized to wild type (=100%). Results for single mutants represent cured versions of the coexpressing strains, which have lost one of the plasmids.
‡The unusually high mating efficiency of G509D in this experiment, when compared with the results in Tables I and II, may be due to an elevated copy number of the *TRP* plasmid (pRS314) relative to the *LEU* plasmid (pRS315-1). All strains listed in Tables I and II contain the *Leu* plasmid, pRS315-1.

alternative transporter. Recently, genes for several new ABC family members have been discovered in yeast (Wang et al., 1991; Dean, M., personal communication); it will be interesting to determine whether the residual halo of a Δ*ste6* mutant is dependent upon any of these genes. Another possible cause of the Δ*ste6* residual halo is natural death and lysis of some cells in culture, which releases their intracellular **a**-factor. While lysed cells would not be expected to mate, the **a**-factor they release might be expected to promote mating of the surrounding cells, which we do not observe. However, it has previously been shown that exogenous **a**-factor does not enhance mating of a Δ*mfa1* Δ*mfa2* strain (Michaelis and Herskowitz, 1988). Why cells must actively secrete **a**-factor in order to mate is not yet understood, but it has been suggested that the gradient of **a**-factor surrounding an **a** cell is important for correct orientation of nearby α cells before mating (Jackson and Hartwell, 1990).

Double NBF Mutants

Our previous work involved mutating one NBF of STE6 at a time, while leaving the other NBF intact. Surprisingly, none of our single-residue substitutions or deletions completely destroyed STE6 function, leaving open the possibility that there might be an ATP-independent component to STE6-mediated **a**-factor export. However, the effects of double NBF mutants reported here, two of which completely destroy STE6 function (G392V G1087V and G509D G1193D; Table II), clearly support an absolute requirement for ATP in the export of **a**-factor by STE6. Other ABC family members have been shown to require ATP for transport function (Horio et al., 1988; Bishop et al., 1989; Mimmack et al., 1989). Further characterization of ATP utilization in STE6 mutants awaits ATP analogue binding studies (see below), and ultimately, the in vitro reconstitution of **a**-factor transport.

Double Center Region Mutants

One double mutant of particular interest is ΔF445 ΔF1140. Although these Phe residues occur in the poorly conserved Center regions of NBF1 and NBF2 of STE6, they are strikingly conserved among ABC family members and can be aligned with Phe residues in NBF Center regions of mammalian, *Plasmodium, Drosophila,* and bacterial homologues, and also in NBF2 of CFTR (Riordan et al., 1989; Berkower and Michaelis, 1991). The local homology algorithm of Smith and Waterman (Devereux et al., 1984) aligns F508 of CFTR with L455 of STE6 (Berkower and Michaelis, 1991). However, shifting the center region of CFTR's NBF1 upstream by 10 residues relative to the other ABC members places F508 in line with these other conserved Phe's without greatly diminishing the quality of the alignment. Thus, it is possible that F445 and F1140 represent the closest analogues in STE6 to F508 of CFTR.

Deletion of F445 or F1140 individually from STE6 does not affect STE6 function, while the double mutant is drastically down in mating (30-fold; Table II). Thus, the combination of two minor perturbances may be more destructive than their individual effects multiplied together. The distinctly different effects of single and double ΔF mutants could, perhaps, be due to a synergistic effect on function or folding. Thomas et al. (1992) have shown that deletion of F508 from a synthetic peptide encoding a portion of NBF1 of CFTR causes the overall structure to become disordered. Denning et al. (1992) have detected a temperature-sensitive defect in the ΔF508 CFTR mutant protein, apparently due to misfolding at the restrictive temperature, which results in improper targeting and degradation (Cheng et al., 1990). Though the STE6 double ΔF mutant does not have grossly altered protein levels or a specific temperature-sensitive defect, its behavior may nonetheless reflect a damaging perturbance of the Center region. It will be interesting to see whether the double ΔF mutant has altered ATP binding properties. We have demonstrated that the radioactive ATP analogue, α-[^{32}P]8-azido ATP, binds to wild-type STE6 (Berkower, C., and S. Michaelis, unpublished results). It will be interesting to determine whether the double ΔF mutation affects binding to this analogue.

Rescue of STE6 Point Mutants by Half-Molecules

We show here that coexpression of the C-Half molecule restores wild-type mating efficiency to the full-length STE6 mutant, G1193D (Fig. 4). Though this observation does not imply a model for the structure of a STE6 transporter, at its simplest level it suggests the schematic pictured in Fig. 5. A lone half-molecule could interact with one half of the full-length molecule and, in the process, displace the half similar to itself. A related, more complex schematic could be drawn for the possible case in which the essential **a**-factor transporter consists of two or more full-length STE6 molecules.

Nothing is known about the capacity of ABC proteins to form higher order structures. One of the best-studied transporters, the *Escherichia coli* lactose permease, appears to function as a monomer (Kaback et al., 1990). The evident lack of cross-complementation we observe here between full-length *ste6* mutants (Table III) suggests that the normal transporter is not formed by interaction of the N-Half of one STE6 monomer with the C-Half of another. We are now testing other mutant pairs. To determine whether full-length STE6 monomers interact with one another in the membrane will require biochemical cross-linking experiments, which we have begun to do. Assessing the functional importance of such an interaction would involve purification of STE6 and the development of an in vitro assay for STE6-mediated **a**-factor transport.

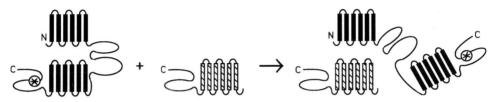

Figure 5. Schematic representation of the association of C-Half with a full-length STE6 molecule, mutated in NBF2. Full-length STE6 (*filled*) and a C-Half molecule (*striped*) are shown. The asterisk indicates a mutation in full-length STE6. A functional transporter is one in which N-Half and C-Half are aligned.

Acknowledgments

We thank Amy Kistler for helpful comments on the manuscript and Thomas Urquhart for assistance in preparing figures.

This work was supported by a grant from the Cystic Fibrosis Foundation (to S. Michaelis) and a March of Dimes predoctoral fellowship (to C. Berkower).

References

Ames, G. F.-L. 1986. Bacterial periplasmic transport systems: structure, mechanism, and evolution. *Annual Review of Biochemistry.* 55:397–425.

Anderegg, R. J., R. Betz, S. A. Carr, J. W. Crabb, and W. Duntze. 1988. Structure of Saccharomyces cerevisiae mating hormone a-factor. Identification of S-farnesyl cysteine as a structural component. *Journal of Biological Chemistry.* 263:18236–18240.

Anderson, M. P., H. A. Berger, D. P. Rich, R. J. Gregory, A. E. Smith, and M. J. Welsh. 1991. Nucleoside triphosphates are required to open the CFTR chloride channel. *Cell.* 67:775–784.

Anderson, M. P., and M. J. Welsh. 1992. Regulation by ATP and ADP of CFTR chloride channels that contain mutant nucleotide-binding domains. *Science.* 257:1701–1704.

Azzaria, M., E. Schurr, and P. Gros. 1989. Discrete mutations introduced in the predicted nucleotide-binding sites of the mdr1 gene abolish its ability to confer multidrug resistance. *Molecular and Cellular Biology.* 9:5289–5297.

Bear, C. E., C. Li, N. Kartner, R. J. Bridges, T. J. Jensen, M. Ramjeesingh, and J. R. Riordan.

1992. Purification and functional reconstitution of the cystic fibrosis transmembrane conductance regulator (CFTR). *Cell.* 68:809–818.

Beaudet, A. L., G. L. Feldman, K. Kobayashi, W. K. Lemna, S. D. Fernbach, M. R. Knowles, R. C. Boucher, and W. E. O'Brien. 1991. Mutation analysis for cystic fibrosis in a North American population. *In* The Identification of the CF (Cystic Fibrosis) Gene: Recent Progress and New Research Strategies. L.-C. Tsui, G. Romeo, R. Greger, and S. Gorini, editors. Plenum Publishing Corp., New York. *Advances in Experimental Medicine and Biology.* 290:53–54.

Berkower, C., and S. Michaelis. 1991. Mutational analysis of the yeast **a**-factor transporter STE6, a member of the ATP binding cassette (ABC) protein superfamily. *EMBO Journal.* 10:3777–3785.

Bishop, L., R. Agbayani, Jr., S. V. Ambudkar, P. C. Maloney, and G. F.-L. Ames. 1989. Reconstitution of a bacterial periplasmic permease in proteoliposomes and demonstration of ATP hydrolysis concomitant with transport. *Proceedings of the National Academy of Sciences, USA.* 86:6953–6957.

Chan, R. K., and C. A. Otte. 1982. Isolation and genetic analysis of Saccharomyces cerevisiae mutants supersensitive to G1 arrest by **a**-factor and alpha-factor pheromones. *Molecular and Cellular Biology.* 2:11–20.

Chen, C.-J., J. E. Chin, K. Ueda, D. P. Clark, I. Pastan, M. M. Gottesman, and I. B. Roninson. 1986. Internal duplication and homology with bacterial transport proteins in the mdr1 (P-glycoprotein) gene from multidrug-resistant human cells. *Cell.* 47:381–389.

Cheng, S. H., R. J. Gregory, J. Marshall, S. Paul, D. W. Souza, G. A. White, C. R. O'Riordan, and A. E. Smith. 1990. Defective intracellular transport and processing of CFTR is the molecular basis of most cystic fibrosis. *Cell.* 63:827–834.

Collins, F. 1992. Cystic fibrosis: molecular biology and therapeutic implications. *Science.* 256:774–779.

Cuppens, H., P. Marynen, C. De Boeck, and J. J. Cassiman. 1990. Study of the G542X and G458V mutations in a sample of Belgian CF patients. *Pediatric Pulmonology.* 5 (Suppl.):203.

Cutting, G. R., L. M. Kasch, B. J. Rosenstein, J. Zielenski, L.-C. Tsui, S. E. Antonarakis, and H. H. Kazazian, Jr. 1990. A cluster of cystic fibrosis mutations in the first nucleotide-binding fold of the cystic fibrosis conductance regulatory protein. *Nature.* 346:366–369.

Denning, G. M., M. P. Anderson, J. F. Amara, J. Marshall, A. E. Smith, and M. J. Welsh. 1992. Processing of mutant cystic fibrosis transmembrane conductance regulator is temperature-sensitive. *Nature.* 358:761–764.

Devereux, J., P. Haeberli, and O. Smithies. 1984. A comprehensive set of sequence analysis programs for the VAX. *Nucleic Acids Research.* 12:387–395.

Drumm, M. L., D. J. Wilkinson, L. S. Smit, R. T. Worrell, T. V. Strong, R. A. Frizzell, D. C. Dawson, and F. S. Collins. 1991. Chloride conductance expressed by ΔF508 and other mutant CFTRs in Xenopus oocytes. *Science.* 254:1797–1799.

Elble, R. 1992. A simple and efficient procedure for transformation of yeasts. *BioTechniques.* 13:18–20.

Endicott, J. A., and V. Ling. 1989. The biochemistry of P-glycoprotein-mediated multidrug resistance. *Annual Review of Biochemistry.* 58:137–171.

Felmlee, T., S. Pellett, and R. Welch. 1985. Nucleotide sequence of an Escherichia coli chromosomal hemolysin. *Journal of Bacteriology.* 163:94–105.

Fields, S. 1990. Pheromone response in yeast. *Trends in Biochemical Science.* 15:270–273.

Gerlach, J. H., J. A. Endicott, P. F. Juranka, G. Henderson, F. Sarangi, K. L. Deuchers, and V. Ling. 1986. Homology between P-glycoprotein and a bacterial haemolysin transport protein suggests a model for multidrug resistance. *Nature.* 324:485–489.

Gottesman, M., and U. Germann. 1993. P-glycoproteins: mediators of multidrug resistance. *Seminars in Cell Biology.* In press.

Gregory, R. J., D. P. Rich, S. H. Cheng, D. W. Souza, S. Paul, P. Manavalan, M. P. Anderson, M. J. Welsh, and A. E. Smith. 1991. Maturation of cystic fibrosis transmembrane conductance regulator variants bearing mutations in putative nucleotide-binding domains 1 and 2. *Molecular and Cellular Biology.* 11:3886–3893.

Gros, P., J. Croop, and D. Housman. 1986. Mammalian multidrug resistance gene: complete cDNA sequence indicates strong homology to bacterial transport proteins. *Cell.* 47:371–380.

Higgins, C. F. 1992. ABC transporters: from microorganisms to man. *Annual Review of Cell Biology.* 8:67–113.

Higgins, C. F., M. P. Gallagher, M. L. Mimmack, and S. R. Pierce. 1988. A family of closely related ATP-binding subunits from prokaryotic and eukaryotic cells. *Bioessays.* 8:111–116.

Higgins, C. F., I. D. Hiles, G. P. C. Salmond, D. R. Gill, J. A. Downie, I. J. Evans, I. B. Holland, L. Gray, S. D. Buckel, A. W. Bell, and M. A. Hermodson. 1986. A family of related ATP-binding subunits coupled to many distinct biological processes in bacteria. *Nature.* 323:448–450.

Horio, M., M. M. Gottesman, and I. Pastan. 1988. ATP-dependent transport of vinblastine in vesicles from human multidrug-resistant cells. *Proceedings of the National Academy of Sciences, USA.* 85:3580–3584.

Hrycyna, C. A., S. K. Sapperstein, S. Clarke, and S. Michaelis. 1991. The Saccharomyces cerevisiae STE14 gene encodes a methyltransferase that mediates C-terminal methylation of a-factor and RAS proteins. *EMBO Journal.* 10:1699–1709.

Hyde, S. C., P. Emsley, M. J. Hartshorn, M. M. Mimmack, U. Gileadi, S. R. Pearce, M. P. Gallagher, D. R. Gill, R. E. Hubbard, and C. F. Higgins. 1990. Structural model of ATP-binding proteins associated with cystic fibrosis, multidrug resistance and bacterial transport. *Nature.* 346:362–365.

Jackson, C. L., and L. H. Hartwell. 1990. Courtship in Saccharomyces cerevisiae: an early cell-cell interaction during mating. *Molecular and Cellular Biology.* 10:2202–2213.

Kaback, H. R., E. Bibi, and P. D. Roepe. 1990. Beta-galactoside transport in E. coli: a functional dissection of lac permease. *Trends in Biochemical Science.* 15:309–314.

Kerem, B.-S., J. M. Rommens, J. A. Buchanan, D. Markiewicz, T. K. Cox, A. Chakravarti, M. Buchwald, and L.-C. Tsui. 1989. Identification of the cystic fibrosis gene: genetic analysis. *Science.* 245:1073–1080.

Kuchler, K., R. E. Sterne, and J. Thorner. 1989. Saccharomyces cerevisiae STE6 gene product: a novel pathway for protein export in eukaryotic cells. *EMBO Journal.* 8:3973–3984.

Kuchler, K., and J. Thorner. 1990. Membrane translocation of proteins without hydrophobic signal peptides. *Current Opinion in Cell Biology.* 2:617–624.

Kunkel, T. A., J. D. Roberts, and R. A. Zakour. 1987. Rapid and efficient site-specific mutagenesis without phenotypic selection. *Methods in Enzymology.* 154:367–382.

Kurjan, J. 1992. Pheromone response in yeast. *Annual Review of Biochemistry.* 61:1097–1129.

Mackman, N., J.-M. Nicaud, L. Gray, and I. B. Holland. 1986. Secretion of haemolysin by Escherichia coli. *Current Topics in Microbiology and Immunology.* 125:159–181.

McGrath, J. P., and A. Varshavsky. 1989. The yeast STE6 gene encodes a homologue of the mammalian multidrug resistance P-glycoprotein. *Nature.* 340:400–404.

Michaelis, S. 1992. STE6, the yeast **a**-factor transporter. 1993. *Seminars in Cell Biology.* In press.

Michaelis, S., and I. Herskowitz. 1988. The **a**-factor pheromone of Saccharomyces cerevisiae is essential for mating. *Molecular and Cellular Biology.* 8:1309–1318.

Mimmack, M. L., M. P. Gallagher, S. R. Pearce, S. C. Hyde, I. R. Booth, and C. F. Higgins. 1989. Energy coupling to periplasmic binding protein-dependent transport systems: stoichiometry of ATP hydrolysis during transport in vivo. *Proceedings of the National Academy of Sciences, USA.* 86:8257–8261.

Mimura, C. S., S. R. Holbrook, and G. F.-L. Ames. 1991. Structural model of the nucleotide-binding conserved component of periplasmic permeases. *Proceedings of the National Academy of Sciences, USA.* 88:84–88.

Parham, P. 1992. Flying the first class flag. *Nature.* 357:193–194.

Raymond, M., P. Gros, M. Whiteway, and D. Y. Thomas. 1992. Functional complementation of yeast ste6 by a mammalian multidrug resistance mdr gene. *Science.* 256:232–234.

Rine, J. 1979. Regulation of transposition of cryptic mating type genes in Saccharomyces cerevisiae. PhD dissertation. University of Oregon, Eugene, OR. 265 pp.

Riordan, J. R., J. M. Rommens, B.-S. Kerem, N. Alon, R. Rozmahel, Z. Grzelczak, J. Zielenski, S. Lok, N. Plavsic, J.-L. Chou et al. 1989. Identification of the cystic fibrosis gene: cloning and characterization of complementary DNA. *Science.* 245:1066–1073.

Saraste, M., P. R. Sibbald, and A. Wittinghofer. 1990. The P-loop: a common motif in ATP- and GTP-binding proteins. *Trends in Biochemical Science.* 15:430–434.

Sikorski, R. S., and P. Hieter. 1989. A system of shuttle vectors and yeast host strains designed for efficient manipulation of DNA in Saccharomyces cerevisiae. *Genetics.* 122:19–27.

Spies, T., V. Cerundolo, M. Colonna, P. Creswell, A. Townsend, and R. DeMars. 1992. Presentation of viral antigen by MHC class I molecules is dependent on a putative peptide transporter heterodimer. *Nature.* 355:644–646.

Thomas, P. J., P. Shenbagamurthi, J. Sondek, J. M. Hullihen, and P. L. Pedersen. 1992. The cystic fibrosis transmembrane conductance regulator: effects of the most common cystic fibrosis-causing mutation on the secondary structure and stability of a synthetic peptide. *Journal of Biological Chemistry.* 267:5727–5730.

Walker, J. E., M. Saraste, M. J. Runswick, and N. J. Gay. 1982. Distantly related sequences in the alpha- and beta-subunits of ATP synthase, myosin, kinases and other ATP-requiring enzymes and a common nucleotide binding fold. *EMBO Journal.* 1:945–951.

Wang, M., E. Balzi, L. Van Dyck, J. Golin, and A. Goffeau. 1991. Sequencing of the yeast multidrug resistance PDR5 gene encoding a putative pump for drug efflux. *Yeast.* 8:5528.

Wilson, K. L., and I. Herskowitz. 1984. Negative regulation of STE6 gene expression by the alpha2 product of Saccharomyces cerevisiae. *Molecular and Cellular Biology.* 4:2420–2427.

Wilson, K. L., and I. Herskowitz. 1987. STE16, a new gene required for pheromone production by **a** cells of Saccharomyces cerevisiae. *Genetics.* 155:441–449.

Coupled Ion Exchangers

Chapter 11

The Evolution of Membrane Carriers

Peter C. Maloney and T. Hastings Wilson

Department of Physiology, Johns Hopkins Medical School, Baltimore, Maryland 21205; and Department of Physiology, Harvard Medical School, Boston, Massachusetts 02115

Molecular Biology and Function of Carrier Proteins © 1993 by The Rockefeller University Press

Introduction

As a preface to these chapters on antiport systems, we would like to discuss the questions of how and why these (and other) membrane carriers developed over evolutionary time. We have speculated on this topic before (Wilson and Maloney, 1976; Maloney and Wilson, 1985; Maloney, 1991), and the exercise is worth repeating now that advances in molecular and cell biology have made so much new information available. As it happens, the impact of these new findings can be well illustrated by the literature on exchange carriers, and we have placed a special emphasis on such examples. Nevertheless, the discussion clearly bears on general issues in the evolution of membrane transport.

Early Cells Faced an Osmotic Crisis

As before (Wilson and Maloney, 1976; Maloney and Wilson, 1985), we would like to suggest that the origins of membrane transport coincided with the shift of replicating systems from an acellular to a cellular form of organization. At this time, we argue, use of a semipermeable lipid bilayer to separate outside from inside brought living systems into an unavoidable conflict with physical chemistry, a conflict that expressed itself as an "osmotic crisis" and that required as a solution the regulation of cell volume. We view this as the selective pressure driving subsequent evolution of membrane transport.

Active Transport Could Have Regulated Cell Volume in Primitive Cells

The diagram of Fig. 1 illustrates the basic argument. A compartmentalized form of life made it possible for early cells to sequester scarce reagents and products (proteins, nucleic acids, etc.) so these materials could interact productively within reasonable time. But the lipid bilayer that limited the escape of large molecules did not restrict the eventual movement of small ones, the internal and external ions (Na, Cl, etc.). As a result, there arose an inevitable tendency for the *net* entry of environmental salt and water, as these permeant species moved in the direction of both electrochemical and osmotic equilibria. Since these equilibria cannot coexist under such conditions, the inevitable result would have been cell lysis.

There are only a few solutions to this kind of problem. (*a*) The presence of an internal gel (Lechene, 1985), invention of an internal cytoskeleton, or use of an external cell wall could have sustained osmotic gradients and allowed ions to attain their equilibrium positions without cell lysis. (*b*) Alternatively, membrane transport could have drawn on metabolic energy to extrude the ions moving inward, sacrificing ionic equilibrium for osmotic stasis. While both solutions are evident in contemporary systems, we believe the latter was the earlier one, and that primitive cells, without access to the complex machinery required for construction of cytoskeletons and cell walls, used the energy from metabolism to actively extrude incoming salts.

Fig. 1 illustrates membrane events that could have accomplished these goals. Specifically, we suggest that early cells regulated cell volume by relying on a proton pump working in parallel with an ion exchange system. The membrane potential (negative inside) established by the pump would have excluded extracellular anions (mainly Cl), while the transmembrane pH gradient (alkaline inside) would have allowed an exchange of H with the cations that had leaked inward (mainly Na). Together, then, the membrane potential and pH gradient (the proton-motive force) created by an early proton pump could have provided the means to prevent net entry of extracellular salt and water, avoiding cell lysis.

Why Did Early Cells Rely on a Proton Circulation?

Several things suggest to us that the osmotic crisis was solved by invention of a proton pump and proton-linked carriers. (*a*) We know, for example, that proton pumps dominate the membrane energetics of contemporary cells, whether prokaryote or eukaryote, plant or animal (Maloney and Wilson, 1985). Indeed, only higher animal cells lack proton pumps on their plasma membranes. Such cells retain the use of proton pumps on their internal endomembrane system, while the plasma membrane has a Na pump in place of the proton pump found in other complex cells (lower eukaryotes and plants). (*b*) Biological materials have a natural affinity and selectivity

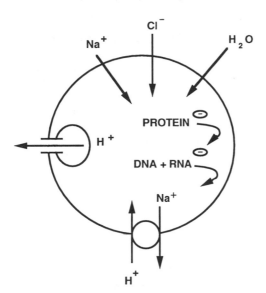

Figure 1. The osmotic crisis faced by the "first" cell. Negatively charged macromolecules within the cell exert colloid osmotic and Donnan forces which result in the entry of extracellular NaCl and water. To prevent swelling and lysis, this cell expends energy to pump protons and establish a proton-motive force. This prevents entry of the negatively charged external anion and allows outward movement of Na^+ by exchange with H^+ (Wilson and Maloney, 1976).

for protons because of the abundance of carboxyl and amino groups that engage in both ionic and covalent interactions with H; these ever-present buffers also ensure a steady supply of protons, even for small compartments at neutral pH. If only for these reasons, one must seriously consider a role for protons in the first transport mechanisms. (*c*) Finally, recent experimental work, to be summarized below, suggests that construction of a proton pump was quite easy. Most contemporary cells have rather complex proton pumps, but there are now several examples in which the equivalent reaction arises in a natural way, from the combined activity of physically separate transport and metabolic events. One can now avoid invoking complex proton pumps in simple cells.

A Proton Pump Can Arise by Linking Transport to Metabolism

One possible way to form a proton-motive force is illustrated by Fig. 2. In this case, we have taken as a starting point the mechanism proposed by Stillwell (1980) to link internal protein synthesis and external amino acids. Stillwell imagined that amino acids were made more permeant by their reversible combination, through a Schiff's base, with an unspecified hydrophobic partner. Note that formation of the Schiff's base derives from a zwitterionic amino acid and therefore releases a proton. Note also that the inward diffusion of the lipid-soluble carboxylate generates a membrane potential (negative inside). And finally, note that reversal of the original reaction generates internal alkalinity as the zwitterion is regenerated. It should be clear, then, that in a formal sense this overall cycle represents the movement of protons from inside to outside. More important, when this cycle is made unidirectional by consumption of the amino acid during protein synthesis, one has the equivalent of a proton pump driven by ongoing metabolic activity.

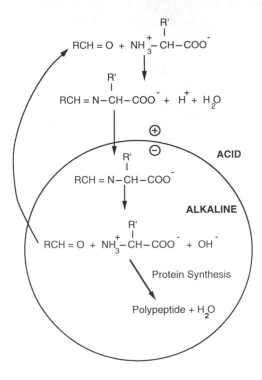

Figure 2. Generation of a proton-motive force by reactions involving the facilitated entry of amino acids. An amino acid outside the cell combines with a long chain aliphatic aldehyde to form a lipid-soluble Schiff's base complex which can diffuse into the cell. The net entry of negative charge on the complex establishes a membrane potential. Regeneration of the free aldehyde and amino acid leads to generation of a pH gradient. A proton pump arises when the reaction cycle is made unidirectional by outward diffusion of the aldehyde and consumption of the amino acid during protein synthesis. (Modified from Stillwell, 1980.)

This kind of model (Fig. 2) was first suggested as a way to understand how proto-cells gained access to extracellular amino acids (Stillwell, 1980). Viewed in a different context, however, it offers an answer to the question of how early cells might have pumped protons. And in addressing this second issue, the model points to the feasibility of an early structural link between metabolism and ion transport, one in which the active transport of protons (but not of other ions) arises as a natural consequence of the chemical and biochemical reactions around early cells. This seems a sensible way to think of the origin of the earliest ion pumps: that they were organized as "virtual" or "metabolic" or "indirect" proton pumps.

Indirect Proton Pumps Are Found in Bacteria

We have confidence in the suitability of such structural links between transport and metabolism because indirect proton pumps are now known in a number of bacterial systems. The best characterized of these, and the first to have been recognized as such, is found in the Gram-negative anaerobe, *Oxalobacter formigenes* (Anantharam et al., 1989). As outlined in Fig. 3, in this cell a proton-motive force arises in an unexpected way from the sequential operation of a simple anion exchange carrier and a cytosolic enzyme system. Thus, entry of divalent oxalate is followed by its decarboxylation in a proton-consuming reaction that generates monovalent formate and carbon dioxide. To complete the cycle, efflux of monovalent formate is then used as the countersubstrate to fuel the next turnover. (Influx of monovalent oxalate and outward diffusion of formic acid accomplishes the same thing, but the experimental

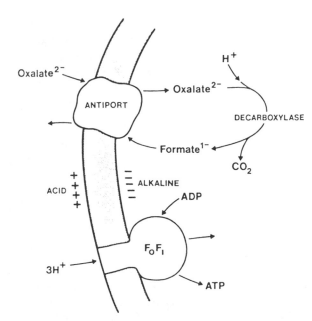

Figure 3. Origin of a proton-motive force relies on the entry and metabolism of oxalate in *O. formigenes*. The carrier-mediated entry of divalent oxalate occurs by exchange with monovalent formate, giving rise to a membrane potential (inside negative). The decarboxylation of oxalate consumes an internal proton during production of formate, and this both establishes a pH gradient (inside alkaline) and provides the countersubstrate for the next cycle (Anantharam et al., 1989).

evidence points to an anion exchange [Maloney et al., 1992].) Since entry of negative charge is stoichiometric with disappearance of an internal proton, each cycle of influx/decarboxylation/efflux is associated with the equivalent of one H^+ pumped outward. For this reason, *O. formigenes* may be said to carry out "decarboxylative" phosphorylation rather than oxidative phosphorylation (Fig. 3).

Biochemical evidence supporting this scenario (Fig. 3) is quite good and can be illustrated by work using the purified anion exchange carrier, OxlT (Fig. 4). In this case, proteoliposomes were loaded with formate under conditions where a potassium gradient was oriented either inward (squares) or outward (circles). Of itself, this had no effect on the incorporation of external oxalate in exchange for internal formate. However, when a potassium diffusion potential was generated by adding the ionophore, valinomycin, oxalate movement was accelerated or inhibited according to whether the internal potential was positive or negative, just as expected of the electrogenic exchange of divalent oxalate and monovalent formate.

This kind of work (Fig. 4) demonstrates directly that OxlT has the properties required to sustain a structural link between membrane transport and intracellular metabolism (Fig. 3). In the same fashion, a recent study of malate:lactate exchange in Gram-positive cells (Poolman et al., 1991) shows that malolactic fermentation has a similar overall organization: (*a*) entry of divalent substrate (malate); (*b*) decarboxylation, consuming a proton and generating a monovalent product (and carbon dioxide); and (*c*) efflux of monovalent product (lactate). In fact, it seems likely that this sort of indirect proton pump will be widely spread among bacteria. Of course, such pumps need not be based on anion exchange. Instead, the paradigm represented by *O. formigenes* (Fig. 3) requires only that there be entry of negative charge

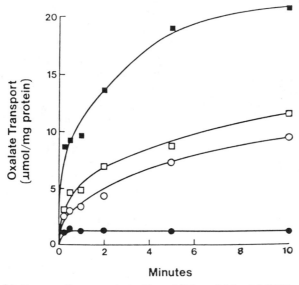

Figure 4. The electrogenic nature of oxalate:formate exchange. The purified oxalate: formate exchange carrier, OxlT, was reconstituted into proteoliposomes containing 200 mM potassium formate ($\bigcirc$, $\bullet$) or 200 mM *N*-methylglucamine formate ($\square$, $\blacksquare$). These proteoliposomes were placed in assay buffers containing, respectively, 100 mM *N*-methylglucamine sulfate or 100 mM potassium sulfate to establish an outwardly or inwardly directed potassium gradient. After addition of 1 μM valinoymcin (*filled symbols*) or the equivalent amount of ethanol (*open symbols*), the reaction was started by addition of 0.1 mM [^{14}C]oxalate. (Reproduced from *Journal of Biological Chemistry,* 1992, 267:10537–10542, by copyright permission of The American Society for Biochemistry and Molecular Biology.)

(or exit of positive charge) stoichiometric with disappearance of an internal proton (generation of internal hydroxyl). Such indirect pumps can be constructed by any appropriate set of vectorial (transport) and scalar (metabolic) reactions, regardless of their complexity.

While these indirect proton pumps are clearly of interest for their roles in bacterial cell biology (eukaryotes, too, may have such virtual pumps), they are of particular interest in an evolutionary context, inasmuch they directly support the idea that active transport might arise by a cooperativity of membrane and cytosolic events. Perhaps just as relevant, such ion transport has an automatic specificity for protons, making it realistic to propose these as substrates of the first ion pump (Fig. 1).

The Origins of Membrane Carriers

Having speculated on the origins of the first ion pumps, it is appropriate to consider the parallel development of membrane carriers, once again using contemporary examples to support speculation. There are, in fact, a large number of known carriers: physiological work has identified 500 or so different carrier-mediated reactions (Stein, 1986), and sequence analysis has yielded ~ 100 case studies, some of which have not yet been studied at the physiological level (Maloney, 1991; Maloney, P. C., unpublished observations). Unfortunately, biochemical analysis is still restricted to a few carriers, and no single example is known in any real structural detail. Nevertheless, the ensemble data base is large enough to prompt educated guesses about the evolution of this superfamily.

Carriers Have Large Numbers of Transmembrane Alpha Helices

Even a cursory scan of carrier protein sequences suggests that the transmembrane alpha helix is the most commonly found structural element, and this simple conclusion is consistent with all of the theoretical and experimental studies at hand: the various predictive algorithms that assess secondary structure hydropathy (see Kyte and Doolittle, 1982; Engelman et al., 1986), CD spectroscopy (Foster et al., 1983; Chin et al., 1986), and inter-residue distance as assessed by the propensity to form disulfide bridges when pairs of cysteines are introduced by mutagenesis (Pakula and Simon, 1992). Overall, we think it fair to state that the "typical" membrane carrier is mostly (60–70% or more) alpha helical, and that these helices are sufficiently long (20–25 residues) and of sufficient hydrophobicity to span the 40-A lipid bilayer. One should add, of course, that these things are known to be true for the transmembrane segments of crystallized membrane proteins such as the photosynthetic reaction center (Deisenhofer et al., 1985), bacteriorhodopsin (Henderson et al., 1990), and diphtheria toxin (Choe ct al., 1992). Although these specific proteins are not carriers, they clearly contain the structural features we associate with authentic carriers.

Given these findings, it is commonly assumed that the transmembrane alpha-helix is the main *structural* element of membrane carriers. This has, understandably, led to the unstated proposition that the alpha helix is also the main *functional* element in these proteins, but this latter speculation is much harder to defend. It might be worth keeping an open mind on this second issue, if only to honor the current models of ion channel structure, in which the role of an eight-stranded beta barrel is emphasized (see Miller, 1991), despite the fact that even for ion channels the dominant structural element is (in almost all cases) the transmembrane alpha helix.

The Sequences of a Few Carriers Suggest a Pattern of Evolution Involving Gene Duplication and Fusion

A few contemporary carriers offer especially interesting sequences in the setting of an evolutionary discussion. For example, we suppose primitive carriers were of rather simple structure and limited sequence diversity. Are simple carriers presently known? Perhaps. Among today's examples, the simplest is clearly represented by the proteolipid subunit (subunit c) of the F-type proton-translocating ATPase, FoF1.

Subunit c contains only a single pair of transmembrane segments (alpha helices) connected by a short intracellular loop. This simple hairpin is then replicated many times (9–12) to form an aggregate which of itself or in combination with the remainder of the Fo sector forms the proton-translocating carrier/channel within the enzyme (Fillingame, 1990). Accordingly, one might suggest that something as simple as subunit c could have been the first membrane carrier; that is, identical pairs of transmembrane helices associated so as to yield multiple copies per functional unit.

Still another carrier gives reason to believe such a hairpin structure actually served as precursor to a more elaborate contemporary version. The idea is nicely illustrated in the work of Klingenberg and his collaborators (1980), who have examined the mitochondrial ADP:ATP carrier. This polypeptide, which is some 300 residues in length and which contains six transmembrane alpha helices, can be subdivided into three domains, each comprising one-third of the molecule. Each domain contains a pair of transmembrane alpha helices connected by a short hydrophilic loop, and each of these hairpin domains is related to the others by significant sequence homology: nearly 40% of the time, identical (or similar) residues are found at the same position in at least two of the three domains (Fig. 5). Similar

| = same residue in 3/3 regions   16%

● = same residue in 2/3 regions   26%

Figure 5. Internal sequence homologies in the mitochondrial ATP:ADP carrier. Sequence analysis indicates this carrier has three internal domains, each ∼ 100 residues in length. These domains are shown here, proceeding from the NH_2 terminus (*upper left*) to the COOH terminus (lower right) of the protein. Vertical lines indicate positions in which the residues are similar or identical in all three domains; the solid circles show positions containing identical (or similar) residues in two of three domains (Aquila et al., 1987).

observations have been made for several other mitochondrial carriers (Runswick et al., 1987).

The simplest way to understand these internal homologies is to suppose that in historic time such carriers were constructed as multimers of identical domains (pairs of helices), and that evolutionary progression involved gene duplication, gene triplication, and then gene fusion. As a result, there was both a reduction in subunit number and a corresponding increase in sequence complexity, presumably allowing for development of specificity. (The proteolipid also illustrates the argument, since subunit c of V-type FoF1 arose from gene duplication and fusion [Nelson, 1992].) The overall concept is further outlined by Fig. 6, which now incorporates the experimental finding that the ADP:ATP carrier exists as a homodimer (Klingenberg et al., 1980).

The fact that the adenine nucleotide carrier exists as a dimer suggests that we might extend the argument concerning gene duplication and fusion to anticipate examples in which the polypeptide represents, in effect, a covalent heterodimer containing a total of 6 hairpin domains and 12 transmembrane helices. Are there such examples now known? There is, of course, no example in which one finds six identical helical pairs, but perhaps that's too much to expect. On the other hand, there are a number of carriers whose sequences are consistent with the general idea. Take, for example, the internal homology (and complementation) exhibited by the MgTetracycline:H antiporter (Rubin et al., 1990). That protein, which is approximately twice the size of the nucleotide exchanger, shows an evident homology between its NH_2- and COOH-terminal halves, and while such internal similarities are not found in all carriers, they are reasonably frequent, as Henderson and others have noted for the monosaccharide carriers (Baldwin and Henderson, 1989). Others have noted similarly repeated domains in eukaryote proteins, the most prominent example being the multiple drug resistance protein, MDR (Pastan ct al., 1991). We therefore assume that these findings reflect a general pattern, one that suggests an evolutionary history involving gene duplication and fusion, as discussed before. In cases where internal homology is not evident, we assume heterologous rather than homologous gene fusions.

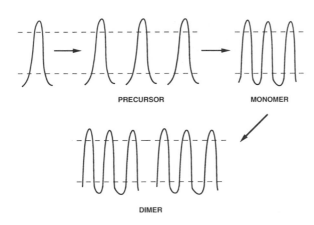

PRECURSOR   MONOMER

DIMER

Figure 6. Hypothetical evolutionary development of a membrane carrier protein. We suggest early transport proteins (carriers, possibly channels) were multimeric (hexamer?) structures with a subunit having a pair of trans-membrane alpha helices. Subsequent gene duplication, triplication, and fusion gave rise to a new carrier protein capable of functioning as a dimer. As noted in the text, a second gene duplication/ fusion event would generate a carrier able to work as a monomer.

This argument naturally leads one to question whether such events have repeated themselves on a still larger scale: arc there carriers in which gene duplication and fusion events have led to even larger constructs, having, say, 9 or 12 hairpin domains (18 or 24 helices)? Presently, the answer to this question is negative; all known carriers show sequences placing them within the boundaries already discussed (sec below). Instead, when one encounters higher domain multiplicities (helix numbers of 18, 24, or higher), the protein appears to function as a channel or junction rather than as a carrier (see Maloney, 1991).

Carriers Belong to a Superfamily Having a Common Overall Structure

We believe the evolutionary progression outlined here adequately describes the origin and development of membrane carriers. Moreover, this simple progression is consistent with the idea (Maloney, 1991) that carriers form a cohesive group (a superfamily) whose members differ in the details required to confer selectivity and specificity, but not in aspects of overall structure.

The likely two-dimensional structure of such carriers is illustrated by UhpT, the bacterial Pi-linked sugar 6-phosphate antiporter (Fig. 7), an example that is suited to the present discussion for several reasons: (*a*) Certainly, UhpT shows the general features we now expect to see in membrane carriers. That is, the bulk of the molecule is within the membrane (or within the cytoplasm); there are multiple transmembrane segments (usually 10 or 12) segregated into two groups by a large central cytoplasmic loop. (*b*) In addition, work with UhpT illustrates the application of genetic tools. As

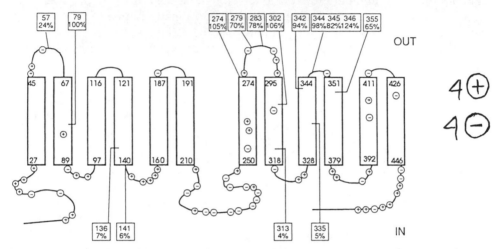

Figure 7. A two-dimensional topological model of the Pi-linked sugar phosphate antiporter (UhpT) from *E. coli.* The likely secondary structure of UhpT is derived by considering three factors: (*a*) the analysis of hydropathy; (*b*) the "positive-inside" rule (von Heijne, 1986), which assigns clusters of positively charged residues to the inner cytoplasmic border; and (*c*) the results of fusions to alkaline phosphatase, a reporter whose activity reflects its internal or external location. In the diagram shown, the boxes indicate the phosphatase activity of UhpT-AP fusion proteins relative to wild-type AP. Low AP activity indicates that the fusion occurs at a residue originally found within the cytoplasm; high AP activity reflects an external position. (Reproduced from *Journal of Bacteriology,* 1990, 172:1688–1693, by copyright permission of The American Society for Microbiology.)

discussed by Boyd et al. (this volume), such an approach can document the presence of multiple transmembrane segments by using gene fusion technology to attach various reporter groups at different positions along or within the carrier sequence. Studies with UhpT have generated 20 such fusions (Lloyd and Kadner, 1990; Island et al., 1992), which together define most of the 12 transmembrane segments shown here (Fig. 7). (*c*) UhpT is of interest for its physiological behavior as well. Thus, in the neutral exchange of phosphate and sugar phosphate, UhpT displays an unusual 2:1 stoichiometry, as two monovalent phosphate anions move against a single divalent sugar phosphate (Maloney et al., 1990). This stoichiometry has been taken to reflect the presence of paired binding sites cooperatively linked to accept either two monovalent substrates or a single divalent passenger. Clearly, one may reasonably speculate that such paired sites echo an evolutionary history in which two independent subunits operated together, each contributing a single binding site. (*d*) Perhaps most important, the biochemical study of UhpT has significantly contributed to the idea that all carriers are structured in essentially the same way. This suggestion was prompted by the combined results of biochemistry and molecular biology (Maloney, 1989, 1991; Ambudkar et al., 1990), but since only a few carriers have been examined in the necessary biochemical detail, the idea must still be considered speculative. Even so, because the idea offers a simple summary of all current information, it is worth discussion.

It is perhaps easiest to begin by noting the result of biochemical studies with UhpT, a bacterial anion exchange protein, and with band 3, the anion exchanger of the red cell membrane. The striking finding is that each of these functions as a monomer, in sharp contrast to the dimeric nature of the adenine nucleotide carrier (see above). The significance of such contrasts is highlighted by the data in Table I, which shows the minimal functioning unit of the seven membrane carriers now analyzed in enough detail to address the issue. Note that the biochemical analysis

TABLE I
The Minimal Functioning Units of Seven Membrane Carriers

Example	Functional unit	Helix number
LacY	Monomer	12
UhpT	Monomer	12
GLUT1	Monomer	12
Band 3	Monomer	12
ADP:ATP	Dimer	6
UCP	Dimer	6
Pi:3PGA	Dimer	7

Updated from Ambudkar et al. (1990).

distinguishes two types of carriers. On the one hand, there are those that work as monomers, a group now including the lactose (LacY) and sugar phosphate (UhpT) carriers of *Escherichia coli,* and the glucose facilitator (GLUT1) and anion carriers (band 3) of the red cell. Among these, the evidence is unambiguous for band 3, where proteoliposomes containing single monomers can be physically isolated and tested for function (Lindenthal and Schubert, 1991). Nearly as good is the evidence for UhpT and LacY, where somewhat fewer direct criteria are used to infer that function is a result of single monomers in single proteoliposomes (Costello et al., 1987; Ambudkar et al., 1990). The evidence for GLUT1 is perhaps the weakest, but consistent with the others nevertheless. The second class distinguished by biochemical study contains carriers that appear as dimers—the nucleotide exchange carriers (ATP:ADP), the uncoupling protein (UCP) of brown fat mitochondria, and the phosphate:phosphoglycerate antiporter of chloroplasts. The evidence in these cases is mostly from analytical ultracentrifugational analysis of purified proteins in detergent micelles (Klingenberg et al., 1980; Wagner et al., 1989).

Table I also summarizes the molecular biology of these carriers by noting their helix numbers, that is, the number of transmembrane alpha helices inferred from hydropathy analysis (or in some cases by more complete study; cf. Fig. 7). The central observation is that the two biochemically distinct groups (monomers and dimers) can also be identified by a corresponding twofold difference in helix numbers. As a result, all seven cases show a functional unit with a helix number of ~12.

We believe this correlation between biochemistry and molecular biology (Table I) is significant and that it forms the structural basis for delineation of the superfamily of membrane carriers. At the least, a survey of the helix numbers of known carriers is consistent with the argument (Maloney, 1991; Maloney, P. C., unpublished observations). Thus, hydropathy plots identify two groups of carriers. The smaller group (~20 cases) contains carriers with helix numbers centering on 6; these are usually found in mitochondria and chloroplasts. (Since this group contains many isoforms of the adenine nucleotide exchanger, this correlation may be fortuitous.) In the larger group (~80 cases) helix numbers almost always center on 10–12. Based on the correlation shown earlier (Table I), one might therefore suppose that those in the former group operate as dimers, while those in the latter group are functional monomers.

If the correlation shown by Table I continues to hold as more carriers are subjected to biochemical analysis, one important outcome will be that we can recall earlier thoughts on how substrates might move through membrane proteins. Both Klingenberg (1981) and Kyte (1981) argued that transport systems should be oligomeric so that the volume entrapped between subunits would become the transport pathway. The reasoning fits with the finding of dimeric carriers but seems contradicted by those that work as monomers. If, however, these monomers act as a virtual dimer (Maloney, 1989) by virtue of their evolutionary history of gene duplication and fusion, then theory would again match experiment.

Acknowledgments

Work in the authors' laboratories is supported by grants from the United States Public Health Service (GM-24195 and DK-44015 to P. C. Maloney, and DK-05735 to T. H. Wilson), and the National Science Foundation (DCB-8905130 to P. C. Maloney and DCB-9017255 to T. H. Wilson).

References

Ambudkar, S. V., V. Anantharam, and P. C. Maloney. 1990. UhpT, the sugar phosphate antiport of Escherichia coli, functions as a monomer. *Journal of Biological Chemistry.* 265:12287–12292.

Anantharam, V., M. J. Allison, and P. C. Maloney. 1989. Oxalate:formate exchange. The basis for energy coupling in Oxalobacter. *Journal of Biological Chemistry.* 264:7244–7250.

Aquila, H., T. A. Link, and M. Klingenberg. 1987. Solute carriers involved in energy transfer of mitochondria form a homologous protein family. *FEBS Letters.* 212:1–9.

Baldwin, S. A., and P. J. F. Henderson. 1989. Homologies between sugar transporters from eukaryotes and prokaryotes. *Annual Review of Physiology.* 51:459–471.

Chin, J. J., E. K. Y. Jung, and C. Y. Jung. 1986. Structural basis of human erythrocyte glucose transporter function in reconstituted vesicles: α-helix orientation. *Journal of Biological Chemistry.* 261:7101–7104.

Choe, S., M. J. Bennett, G. Fujii, M. G. Curmi, K. A. Kantardjieff, R. J. Collier, and D. Eisenberg. 1992. The crystal structure of diphtheria toxin. *Nature.* 357:216–221.

Costello, M. J., J. Escaig, K. Matsushita, P. V. Viitanen, D. R. Menick, and H. R. Kaback. 1987. Purified *lac* permease and cytochrome *o* oxidase are functional as monomers. *Journal of Biological Chemistry.* 262:17072–17082.

Deisenhofer, J., O. Epp, K. Miki, R. Huber, and H. Michel. 1985. Structure of the protein subunits in the photosynthetic reaction centre of Rhodopseudomonas viridis at 3A resolution. *Nature.* 318:618–624.

Engelman, D. M., T. A. Steitz, and A. Goldman. 1986. Identifying nonpolar transbilayer helices in amino acid sequences of membrane proteins. *Annual Review of Biophysics and Biophysical Chemistry.* 15:321–353.

Fillingame, R. H. 1990. Molecular mechanics of ATP synthesis by F1F0-type H+ transporting ATP synthases. *In* The Bacteria. T. A. Krulwich, editor. Academic Press, Inc., San Diego. 345–392.

Foster, D. L., M. Boublik, and H. R. Kaback. 1983. Structure of the lac carrier protein of *Escherichia coli. Journal of Biological Chemistry.* 258:31–34.

Henderson, R., J. M. Baldwin, T. A. Ceska, F. Zemlin, E. Beckmann, and K. H. Downing. 1990. Model for the structure of bacteriorhodopsin based on high-resolution electron cryo-microscopy. *Journal of Molecular Biology.* 213:899–929.

Island, M. D., B.-Y. Wei, and R. J. Kadner. 1992. Structure and function of the uhp genes for the sugar phosphate transport system in Escherichia coli and Salmonella typhimurium. *Journal of Bacteriology.* 174:2754–2762.

Klingenberg, M. 1981. Membrane protein oligomeric structure and transport function. *Nature.* 290:449–454.

Klingenberg, M., H. Hackenberg, R. Kramer, C. S. Lin, and H. Aquila. 1980. Two transport proteins from mitochondria. I. Mechanistic aspects of asymmetry of the ADP/ATP translocator. I. The uncoupling protein of brown adipose tissue mitochondria. *Annals of the New York Academy of Sciences.* 358:83–95.

Kyte, J. 1981. Molecular considerations relevant to the mechanism of active transport. *Nature.* 292:201–204.

Kyte, J., and R. F. Doolittle. 1982. A simple method for displaying the hydropathic character of a protein. *Journal of Molecular Biology.* 157:105–132.

Lechene, C. 1985. Cellular volume and cytoplasmic gel. *Biology of the Cell.* 55:177–180.

Lindenthal, L., and D. Schubert. 1991. Monomeric erythrocyte band 3 protein transports anion. *Proceedings of the National Academy of Sciences, USA.* 88:6540–6544.

Lloyd, A. D., and R. J. Kadner. 1990. Topology of the *Escherichia coli uhpT* sugar-phosphate transporter analyzed by using Tn*phoA* fusions. *Journal of Bacteriology.* 172:1688–1693.

Maloney, P. C. 1989. Resolution and reconstitution of anion exchange reactions. *Philosophical Transactions of the Royal Society of London, Series B.* 326:437–454.

Maloney, P. C. 1991. A consensus structure for membrane transport. *Research in Microbiology.* 141:374–383.

Maloney, P. C., A. V. Ambudkar, V. Anantharam, L. A. Sonna, and A. Varadhachary. 1990. Anion-exchange mechanisms in bacteria. *Microbiological Reviews.* 54:1–17.

Maloney, P. C., V. Anantharam, and M. J. Allison. 1992. Measurement of the substrate dissociation constant of a solubilized membrane carrier. *Journal of Biological Chemistry.* 267:10531–10536.

Maloney, P. C., and T. H. Wilson. 1985. The evolution of ion pumps. *BioScience.* 35:43–48.

Miller, C. 1991. 1990: annus mirabilis of potassium channels. *Science.* 252:1092–1096.

Nelson, N. 1992. Evolution of organellar proton-ATPases. *Biochimica et Biophysica Acta.* 1100:109–124.

Pakula, A. A., and M. I. Simon. 1992. Determination of transmembrane protein structure by disulfide cross-linking: the *Escherichia coli* Tar receptor. *Proceedings of the National Academy of Sciences, USA.* 89:4144–4148.

Pastan, I., M. C. Willingham, and M. Gottesman. 1991. Molecular manipulations of the multi-drug transporter: a new role of transgenic mice. *FASEB Journal.* 5:2523–2528.

Poolman, B., D. Molenaar, E. J. Smid, T. Ubbink, T. Abee, P. P. Renault, and W. N. Konings. 1991. *Journal of Bacteriology.* 173:6030–6037.

Ruan, Z.-S., V. Anantharam, I. C. Crawford, S. V. Ambudkar, S. Y. Rhee, M. J. Allison, and P. C. Maloney. 1992. Identification, purification and reconstitution of OxlT, the oxalate: formate antiport protein of Oxalobacter formigenes. *Journal of Biological Chemistry.* 267:10537–10542.

Rubin, R., A., S. B. Levy, R. L. Heinrikson, and F. J. Kezdy. 1990. Gene duplication in the evolution of the two complementing domains of Gram-negative bacterial tetracycline efflux proteins. *Gene.* 87:7–13.

Runswick, M. J., S. J. Powell, P. Nyren, and J. E. Walker. 1987. Sequence of the bovine mitochondrial phosphate carrier protein: structural relationship to ADP/ATP translocase and the brown fat mitochondria uncoupling protein. *EMBO Journal.* 6:1367–1373.

Stein, W. D. 1986. Transport and Diffusion across Cell Membranes. Academic Press, Inc., San Diego. 685 pp.

Stillwell, W. 1980. Facilitated diffusion as a method for selective accumulation of materials from the primordial oceans by a lipid-vesicle protocell. *Origins of Life.* 10:277–292.

von Heijne, G. 1986. The distribution of positively charged residues in bacterial inner membrane proteins correlates with the trans-membrane topology. *EMBO Journal.* 5:3021–3027.

Wagner, R., E. C. Apley, G. Gross, and U. I. Flugge. 1989. The rotational diffusion of chloroplast phosphate translocator and of lipid molecules in bilayer membranes. *European Journal of Biochemistry.* 182:165–173.

Wilson, T. H., and P. C. Maloney. 1976. Speculations on the evolution of ion transport mechanisms. *Federation Proceedings.* 35:2174–2179.

Chapter 12

Molecular Characterization of the Erythrocyte Chloride-Bicarbonate Exchanger

Reinhart A. F. Reithmeier, Carolina Landolt-Marticorena, Joseph R. Casey, Vivian E. Sarabia, and Jing Wang

MRC Group in Membrane Biology, Department of Medicine and Department of Biochemistry, University of Toronto, Toronto, Ontario, Canada, M5S 1A8

Band 3 Is a Multifunctional Protein

The basic structural unit of band 3 (M_r 95,000) in the erythrocyte membrane is a dimer; however, the individual subunits within the oligomer are capable of carrying out anion transport independently. The oligomeric structure allows for allosteric regulation or modulation of transport, perhaps even to the extent of changing transport kinetics from ping-pong (uncoupled) to simultaneous (coupled). The transport function (Jennings, 1989) is carried out by the membrane domain (M_r 52,000), which is also dimeric. The cytosolic domain of band 3 (M_r 43,000) carries out a plethora of functions (Low, 1986) largely unrelated to anion transport. This suggests that band 3 may have arisen as the result of gene fusion, with the two protein domains corresponding to ancestral genes. Alternatively, the cytosolic domain may have originally regulated anion transport, with this function having been lost through evolution of the erythrocyte form of band 3. The cytosolic domain anchors the cytoskeleton to the membrane, regulates glycolysis through phosphorylation/dephosphorylation of tyrosine 8, and binds hemoglobin (Low, 1986). Band 3 contains a third domain, the N-linked oligosaccharide chain exposed on the cell surface. This elaborate structure is a red cell antigen and probably plays a role in senescence (Beppu et al., 1992). The multifunctional nature of band 3 suggests that a single gene may encode a membrane protein that has domains (membrane and cytosolic) with unrelated functions.

Sequence Analyses

The amino acid sequence of a membrane protein provides a wealth of information which, if properly decoded, will eventually lead to a prediction of the three-dimensional structure of the protein. One example of such an analysis that is routinely applied to membrane proteins is the search for hydrophobic stretches of amino acids that are of sufficient length to span a lipid bilayer. Such hydropathy profiles lead to models of membrane protein folding that can then be tested experimentally. For band 3, the hydropathy analysis reveals the presence of 10 hydrophobic peaks, some of which are long enough to span the membrane twice. Experimental evidence (Tanner, 1989) has shown that band 3 may contain up to 14 transmembrane segments. We have used an immunological approach to study the folding pattern of band 3 in the membrane. Polyclonal antibodies were raised against synthetic peptides corresponding to hydrophilic sequences predicted to face the cytoplasmic side of the membrane. An antibody against the extreme carboxyl terminus was used to localize this portion of band 3 to the cytosol (Lieberman and Reithmeier, 1988). Since the amino terminus of band 3 also faces the cytosol, band 3 spans the erythrocyte membrane an even number of times. Antibodies directed toward predicted internal loops of protein failed to react with the native protein in the membrane and therefore could not be used for localization studies. Either the conformation of the loops is different in the protein and the peptide or the sequences are simply not accessible to the very large antibody probe.

We have completed a survey of the sequences of single span membrane proteins and have found that the transmembrane segments of human type I plasma membrane proteins (Nout, Cin) are not simply random sequences of exclusively hydrophobic amino acids (Landolt-Marticorena et al., 1993). The sequence can be organized into a motif that has implications for membrane protein biosynthesis, folding, and stability. The single transmembrane segment of glycophorin A (Nt-EPE**ITLIIF-GVMAGVIGTILLISYGI**RRLIKK-Ct; hydrophobic domain in bold) provides an excellent example of such a motif. Single span transmembrane segments are enriched in Ile, Val, Leu, Phe, Ala, and Gly. The first four amino acids ensure the hydrophobic nature of the segment, Ile and Val displaying β-branched side-chains to the bilayer. Alanine is an excellent helix former, while glycine is the poorest. This suggests that glycine residues within transmembrane segments must play a special role. The motif begins with a residue designed to initiate helix formation (Pro) at the amino-terminal flanking domain. The hydrophobic segment contains β-branched amino acids at the amino terminus, while the carboxyl terminus is enriched in leucine residues. Aromatic residues (Tyr) are found at the boundary region between the hydrophobic domain and the aqueous domain. The transmembrane segments contain a number of glycine residues that are disposed to one side of the helix. These small residues provide the site of interaction to stabilize the glycophorin dimer. Glycine residues may also impart a degree of flexibility to the transmembrane segment, allowing the formation of higher order structures. The carboxyl-terminal flank is enriched in basic residues, which act to retain the carboxyl terminus to the cytoplasmic side of the membrane during biosynthesis.

Type II single span membrane proteins contain some of these elements; however, a rigorous survey was not possible due to the limited number of sequences available. Aromatic residues again are localized to the interfacial region. Basic residues are found at the amino terminus, again facing the cytoplasm. For example, proline residues near the amino terminus of transmembrane segments may initiate helix formation.

Oligosaccharide Chains

Oligosaccharide chains may be linked to asparagine residues contained within the sequon Asn-X-Ser/Thr, where X is any amino acid other than proline (Reithmeier and Deber, 1992). These sites provide markers for the outside of the cell, since N-linked carbohydrate is found on the cell surface. Unfortunately, many of the sequons are not glycosylated since they are located on the cytoplasmic side of the membrane. Band 3 provides such an example. Asparagines 593 and 642 are contained within consensus sequons, but they are separated by a single hydrophobic segment and must therefore reside on opposite sides of the membrane. Protein chemical studies have shown that asparagine 642 is a site of N-linked glycosylation and is located on the cell exterior (Jay and Cantley, 1986; Tanner et al., 1988). Interestingly, the other members of the band 3 gene family (Alper, 1991), AE2 and AE3, do not contain a consensus glycosylation site at the same position, but rather contain consensus glycosylation sites on the preceding extracellular loop of protein.

In a survey of multispanning membrane proteins we have found that glycosylation sites are restricted to a single extracellular loop of protein (Landolt-Marticorena, C., and R. A. F. Reithmeier, manuscript in preparation). This loop may contain multiple glycosylation sites, but other glycosylation sites are rarely permitted in other extracellular domains. Why glycosylation does not occur on multiple loops of protein is unknown. It may be that during biosynthesis, once one loop of protein occupies the glycosyltransferase enzyme in the endoplasmic reticulum, no other loop has access to the enzyme.

Most multispan membrane proteins localized to the plasma membrane contain at least one N-linked oligosaccharide chain or are associated with another polypeptide that is glycosylated. This suggests that oligosaccharide chains must fulfill some essential role in the biosynthesis, structure, or function of these membrane proteins. On the other hand, nature may simply be using plasma membrane protein as a vehicle to transport exotic cell surface antigens to the cell surface. In some cases, oligosaccharide chains may aid in the folding process since the oligosaccharide chain is attached to the protein in the endoplasmic reticulum before synthesis is completed. The targeting role for oligosaccharide seems to be limited to phosphomannose residues, which direct proteins to the lysosome.

The role of the single oligosaccharide chain attached to band 3 was studied after enzymatic deglycosylation (Casey et al., 1992). It was found that deglycosylation decreased the solubility of the protein and promoted aggregation. No change in oligomeric structure, detergent binding, or secondary structure was observed. Deglycosylation of band 3 did not alter its ability to transport anions or to bind inhibitors. The oligosaccharide chain of band 3 may be thought of as a separate domain of the protein that functions as a blood group antigen (I), being relatively independent from the other functions of band 3.

Helix–Helix Interactions

Circular dichroism studies of purified band 3 in reconstituted lipid vesicles and in detergent ($C_{12}E_8$) solution have shown that the secondary structure of the protein is maintained in this detergent (Reithmeier et al., 1989). Detergents with short alkyl chains (octyl glucoside), in contrast, denature band 3 with significant changes in secondary structure (Werner and Reithmeier, 1985). Exhaustive proteolysis of ghost membranes has shown that the transmembrane segments of band 3 that are protected from proteolysis are entirely α-helical in conformation (Oikawa et al., 1985). Band 3 transmembrane sequences with known topography may be modeled as α-helices.

The membrane-associated sequences of membrane proteins are often highly conserved. This is unlikely to be due to specific protein–lipid interactions. Indeed, analysis of the structure of the photoreaction center (Rees et al., 1989) has shown that the lipid-facing side chains are poorly conserved. The high degree of conservation of transmembrane segments is therefore due to the constraints imposed by protein–protein interactions. Since membrane-spanning segments are α-helical in conformation, helix–helix interactions play a central role in the structure of intrinsic membrane proteins (Lemmon and Engelman, 1992). Transmembrane segments that are linked by short loops or turns are probably adjacent to one another in the final structure. This is certainly the case for bacteriorhodopsin. The packing of antiparallel helices is well characterized in soluble proteins, and similar modes of packing may exist for membrane proteins. In soluble proteins helix–helix interactions are mediated by hydrophobic surfaces. The packing of transmembrane helices involves similar interactions, but with more hydrophobic surface of the helices being displayed to the bilayer.

An examination of the transmembrane segments of single span membrane proteins like glycophorin reveals an asymmetric helix. One face is very hydrophobic and probably faces the bilayer, while the opposite face contains a cluster of glycine residues and residues with small side chains like alanine. We propose that glycine residues in transmembrane helices provide sites of close contact between transmembrane segments. Glycine residues also impart to helical segments a degree of flexibility. This hypothesis was tested using the glycophorin dimer as a model. Energy minimization and dynamics showed that the transmembrane segment of the glycophorin dimer involved extensive interactions between the glycine faces of the helix. Energy minimization studies have shown that a number of low energy structures are possible. They all, however, involve interactions between the glycine face of glycophorin. The final structure was a left-handed coiled coil. The lack of side chains at the glycine position allowed an intimate backbone–backbone interaction. Glycine residues are enriched in the transmembrane segments of type I single span membrane proteins, consistent with its role as mediated helix–helix contacts. A similar coiled coil structure was observed between the transmembrane segments of HLA class II molecules, which consist of an α and a β subunit.

The modeling of multi-spanning membrane proteins presents additional challenges: Which helices interact? What is the conformation of the segments? Are all transmembrane segments helical? Transmembrane segments that are not in a hydrophobic environment (e.g., not facing the lipid bilayer) need not be helical in conformation. The transmembrane segments of transport proteins are enriched in proline residues (Brandl and Deber, 1986), which may disrupt the helical segment. Proline residues can bend helices, providing space for transporter substrates. This amino acid is commonly found at the amino terminus of helices in soluble proteins, where it plays a role in helix initiation (Richardson and Richardson, 1988). Since helices are dipoles, a slight positive charge exists at the amino termini of helices (Hol et al., 1978). A number of helices can provide a binding site for anions at the amino-terminal ends. The limits of the helix may be difficult to define; however, the presence of proline residues as helix initiators may help define helix ends.

Oligomeric Structure

Band 3 exists as a mixture of dimers (70%) and tetramers in the erythrocyte membrane and in some detergent solutions (Casey and Reithmeier, 1991). The various oligomeric forms of band 3 express different functions. The individual subunits within band 3 oligomers are fully capable of anion transport; however, an oligomeric structure allows cooperative interactions to occur between subunits. Tetramers of band 3 bind to the cytoskeleton. The membrane domain is exclusively dimeric, showing that the tetramer is held together by interactions involving the cytosolic domain. Higher oligomeric forms of band 3 form the senescence antigen and are involved in red cell aging (Low et al., 1985). In this role band 3 acts as a transmembrane signaling molecule, relaying the denaturation of hemoglobin to the cell surface via band 3 aggregates that provide external binding sites for antibodies involved in red cell removal.

Band 3 in Ovalocytes

Band 3 from red cells of individuals with Southeast Asian ovalocytosis (SAO) contain a very interesting deletion, with nine amino acids located at the boundary position of the cytosol and the first transmembrane segment missing in SAO band 3 (Jarolim et al., 1991). No homozygous carriers of the SAO gene have been found, suggesting that this mutation is lethal and that the gene for band 3 (AE1) is essential either in the context of the red cell or in the kidney, where an alternatively spliced form of the gene is expressed (Brosius et al., 1989). This deletion alters the interaction of band 3 with the cytoskeleton, resulting in the unusual red cell shape. SAO is also associated with resistance to malarial infection (Kidson et al., 1981). We have noted changes in the membrane domain as well. First, SAO band 3 is incapable of binding stilbene disulfonates (Kidson et al., 1981; Schofield et al., 1992; Sarabia, V., J. Casey, and R. A. F. Reithmeier, manuscript submitted for publication). This suggests that the first transmembrane segment is involved directly in stilbene binding, or that the structure of the protein has been dramatically altered by this deletion. It is important to note that the stilbene site is accessible from the outside, while the mutation is on the opposite side of the membrane. Recent results have shown that SAO band 3 is inactive in transport (Schofield et al., 1992).

A second alteration occurs in oligosaccharide processing. As mentioned above, band 3 contains a polylactosaminyl oligosaccharide structure that is attached in the *trans*-Golgi. This sugar structure is commonly found on lysosomal enzymes and its presence on band 3 remains a mystery. SAO band 3 entirely escapes this modification (Sarabia, V., J. Casey, and R. A. F. Reithmeier, manuscript submitted for publication). This suggests that SAO band 3 proceeds quickly through the *trans*-Golgi, or that the structure of the protein has been changed such that the recognition elements for the lactosaminyl transfer have been destroyed. The fact that SAO band 3 is made in equal amounts to normal band 3 suggests that the synthesis and movement of the protein out of the endoplasmic reticulum has not been impaired by the mutation. Since band 3 is oligomeric, it is also possible that SAO band 3 forms heterodimers with normal band 3 and piggybacks its way to the cell surface.

Acknowledgments

Research was supported by the Medical Research Council of Canada and the Department of Medicine, University of Toronto.

References

Alper, S. L. 1991. The Band 3-related anion exchanger family. *Annual Review of Physiology.* 53:549–564.

Beppu, M., A. Mizukami, K. Ando, and K. Kikugawa. 1992. Antigenic determinants of senescent antigen of human erythrocytes are located in sialylated carbohydrate chains of Band 3 glycoprotein. *Journal of Biological Chemistry.* 267:14691–14696.

Brandl, C. J., and C. M. Deber. 1986. Hypothesis about the function of membrane-buried proline residues in transport proteins. *Proceedings of the National Academy of Sciences, USA.* 83:917–921.

Brosius, F. C., III, S. L. Alper, A. M. Garcia, and H. F. Lodish. 1989. The major kidney Band 3 gene transcript predicts an amino-terminal truncated Band 3 polypeptide. *Journal of Biological Chemistry.* 264:7784–7787.

Casey, J. R., C. A. Pirraglia, and R. A. F. Reithmeier. 1992. Enzymatic deglycosylation of human Band 3, the anion transport protein of the erythrocyte membrane: effect on protein structure and transport properties. *Journal of Biological Chemistry.* 267:11940–11948.

Casey, J. R., and R. A. F. Reithmeier. 1991. Analysis of the oligomeric state of Band 3, the anion transport protein of the human erythrocyte membrane, by high performance size exclusion liquid chromatography: oligomeric stability and origin of heterogeneity. *Journal of Biological Chemistry.* 266:15726–15737.

Hol, W. G. J., P. T. van Duijnen, and H. J. C. Berendsen. 1978. The α-helix dipole and the properties of proteins. *Nature.* 273:443–447.

Jarolim, P., J. Palek, D. Amato, K. Hassan, P. Sapak, G. T. Nurse, H. L. Rubin, S. Zhai, K. E. Sahr, and S.-C. Liu. 1991. Deletion in erythrocyte Band 3 gene in malaria resistant southeast asian ovalocytes. *Proceedings of the National Academy of Sciences, USA.* 88:11022–11026.

Jay, D., and L. Cantley. 1986. Structural aspects of the red cell anion exchange protein. *Annual Review of Biochemistry.* 55:511–538.

Jennings, M. L. 1989. Structure and function of the red blood cell anion transport protein. *Annual Review of Biophysics and Biophysical Chemistry.* 18:397–430.

Kidson, C., G. Lamont, A. Saul, and G. T. Nurse. 1981. Ovalocytic erythrocytes from Melanesians are resistant to invasion by malaria parasites in culture. *Proceedings of the National Academy of Sciences, USA.* 78:5829–5832.

Kyte, J., and R. F. Doolittle. 1982. A simple method for displaying the hydropathic character of a protein. *Journal of Molecular Biology.* 157:105–132.

Landolt-Marticorena, C., K. A. Williams, C. M. Deber, and R. A. F. Reithmeier. 1993. Non-random distribution of amino acids in the transmembrane segments of human type I single span membrane proteins. *Journal of Molecular Biology.* In press.

Lemmon, M. A., and D. M. Engelman. 1992. Helix-helix interactions inside lipid bilayers. *Current Opinion in Structural Biology.* 2:511–518.

Lieberman, D. D., and R. A. F. Reithmeier. 1988. Localization of the carboxyl terminus of Band 3 to the cytoplasmic side of the erythrocyte membrane using antibodies raised against a synthetic peptide. *Journal of Biological Chemistry.* 263:10022–10028.

Low, P. S. 1986. Structure and function of the cytoplasmic domain of Band 3: center of erythrocyte membrane-peripheral protein interactions. *Biochimica et Biophysica Acta.* 864:145–167.

Low, P. S., S. M. Waugh, and D. Drenckhahn. 1985. The role of hemoglobin denaturation and Band 3 clustering in red cell aging. *Science.* 227:531–533.

Oikawa, K., D. M. Lieberman, and R. A. F. Reithmeier. 1985. Conformation and stability of the anion transport protein of human erythrocyte membranes. *Biochemistry.* 24:2843–2848.

Rees, D. C., H. Komiya, T. O. Yeates, J. P. Allen, and G. Feher. 1989. The bacterial photoreaction center as a model for membrane proteins. *Annual Review of Biochemistry.* 58:607–633.

Reithmeier, R. A. F., and C. M. Deber. 1992. Intrinsic membrane protein structure: principles and prediction. *In* The Structure of Biological Membranes. P. Yeagle, editor. CRC Press, Inc., Boca Raton, FL. 337–393.

Reithmeier, R. A. F., D. M. Lieberman, J. R. Casey, S. W. Pimplikar, S. W. Werner, H. See, and C. A. Pirraglia. 1989. Structure and function of the Band 3 Cl^-/HCO_3^- transporter. *Annals of the New York Academy of Sciences.* 572:75–83.

Richardson, J. S., and D. C. Richardson. 1988. Amino acid preferences for specific locations at the ends of α-helices. *Science.* 240:1648–1652.

Schofield, A. E., D. M. Reardon, and M. J. A. Tanner. 1992. Defective anion transport activity of the abnormal band 3 in hereditary ovalocytic blood cells. *Nature.* 355:836–838.

Tanner, M. J. A. 1989. Proteolytic cleavage of the anion transporter and its orientation in the membrane. *Methods in Enzymology.* 173:423–432.

Tanner, M. J. A., P. G. Martin, and S. High. 1988. The complete amino acid sequence of the human erythrocyte membrane anion-transport protein deduced from the cDNA sequence. *Biochemical Journal.* 256:703–712.

Werner, P. K., and R. A. F. Reithmeier. 1985. Molecular characterization of human erythrocyte anion transport protein in octyl glucoside. *Biochemistry.* 24:6375–6381.

Chapter 13

Molecular Biology and Hormonal Regulation of Vertebrate Na$^+$/H$^+$ Exchanger Isoforms

Laurent Counillon and Jacques Pouyssegur

Centre de Biochimie-Centre National de la Recherche Scientifique, Université de Nice, 06018 Nice, France

Molecular Biology and Function of Carrier Proteins © 1993 by The Rockefeller University Press

Introduction

Eukaryotic Na^+/H^+ antiporters are integral plasma membrane proteins which exert important functions in the maintenance of acid–base balance and ionic homeostasis. These ion transporters use the energy of the transmembrane gradient of sodium to catalyze the exchange of one extracellular sodium ion for one intracellular proton.

The most widely studied Na^+/H^+ antiporter is the Na^+/H^+ exchanger 1 (NHE1) isoform, which appears to be expressed in every cell system where it has been investigated (for review, see Grinstein, 1988). This exchanger is involved in the regulation of intracellular pH and cell volume (Hoffmann and Simonsen, 1989) and is blocked by the compound amiloride. The discovery that NHE1 is rapidly turned on by all known mitogens has initiated many studies trying to elucidate both the contribution of this antiporter in the control of cell proliferation and the cellular signaling pathways leading to NHE1 activation (for review, see Grinstein et al., 1989). Besides the ubiquituously expressed Na^+/H^+ exchanger, several isoforms possessing a lower sensitivity to amiloride have been identified (Orlowski et al., 1992; Tse et al., 1992). NHE2 and NHE3 are found to be apically expressed in small intestine and kidney proximal tubule epithelia, where they perform more specialized ion transport functions such as sodium reabsorption linked to acid secretion. NHE4, which is highly expressed in the gastrointestinal tract, may be involved in pH regulation upon stomach acid secretion.

Because of their importance in numerous physiological functions, the Na^+/H^+ exchangers have been suspected to exert a contribution in several pathological processes such as essential hypertension (Canessa et al., 1988) or postischemic arrythmia and cell death occurring after a heart attack (Lazdunski et al., 1985).

During the past decade our attention has been focused on the biochemical properties of NHE1 (Paris and Pouysségur, 1983), its molecular mechanism of activation by growth factors (Paris and Pouysségur, 1984), the identification of its primary structure, and the cloning of its gene (Sardet et al., 1989; Miller et al., 1991). We are presently trying to elucidate the relations between the primary structure and the functions of the Na^+/H^+ antiporters. After an overview describing the basic features of Na^+/H^+ exchange, this review article will expose recent data concerning the structure of Na^+/H^+ exchangers, the mechanism of growth factor activation of the NHE1 antiporter, the identification of its amiloride binding site, and the recent progress on the other isoforms of Na^+/H^+ exchanger.

Biochemical Properties

Na^+/H^+ antiporters use the energy provided by the inwardly directed electrochemical gradient of sodium to perform ion translocation across the plasma membrane (Fig. 1). This exchange therefore does not need direct ATP hydrolysis to occur, but depends instead on the cellular energy level, because the Na^+ transmembrane gradient has to be maintained by the Na^+/K^+ATPase, and since a basal state of phosphorylation of the antiporter is required for its function (Cassel et al., 1986; Wakabayashi et al., 1992). This electroneutral ion exchange (1:1 stoichiometry) can function in both directions, depending on the orientation of the Na^+ transmembrane gradient. Interestingly, other cations can compete with sodium on the external transport site of the antiporter: Na^+/H^+ exchangers can mediate Li^+/H^+ as well as H^+/Li^+ exchange (Kinsella and Aronson, 1981; Paris and Pouysségur, 1983), and external protons have also been reported to inhibit the antiporter by competing with external sodium (Fig. 1). Thus it has been hypothesized that a protonatable amino acid, probably a histidine residue, could play an important role in external sodium binding or translocation (Grillo and Aronson, 1986). Although the complete revers-ibility of the ion exchange had suggested a certain degree of symmetry of the antiporter, the behavior of this transporter in respect to internal protons and to amiloride action clearly points to an asymmetric functioning. The Na$^+$/H$^+$ exchanger, which is barely measurable at physiological intracellular pH, becomes allosterically activated as the internal pH decreases. Physiologically, this results in an accelerated extrusion of the intracellular protons, and kinetically, this indicates that two distinct internal proton binding sites are interacting cooperatively. This hypothesis, first proposed by Aronson and Nee (1982) suggests that the binding of a first proton to an internal protonatable regulatory site, or proton sensor site, would increase the affinity of the proton internal transport site, thereby enhancing the activation of the antiporter when the cytoplasm becomes more acidic (Fig. 1).

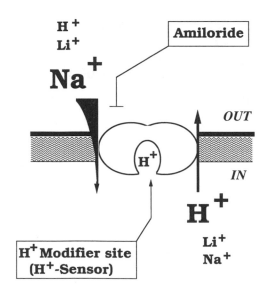

Figure 1. The basic features of the Na$^+$/H$^+$ exchangers.

Like most of the members of the sodium transmembrane transporter family, the Na$^+$/H$^+$ exchange is blocked by guanidinium ions, with a K_i that is close to the K_m of external sodium (Frelin et al., 1986). Among all the guanidinium derivatives that have been tested for their ability to inhibit the antiporter, the diuretic molecule amiloride appears to be the prototype of the most specific inhibitor (Benos, 1982) (Fig. 1). Since amiloride is a purely competitive inhibitor of external sodium with a K_i for the NHE1 antiporter of 3 μM in sodium-free medium (L'Allemain et al., 1984), millimolar concentrations of this compound are necessary to perform studies where physiological external Na$^+$ concentrations have to be maintained. Hence, to avoid the side effects of these high amiloride concentrations, substitutions on this molecule have been performed in order to obtain Na$^+$/H$^+$ exchanger inhibitors having higher affinity (for review, see Kleyman and Cragoe, 1988). The 5-(N-alkyl) substituted derivatives of amiloride appear to be ~ 100-fold more potent than amiloride itself in their ability to inhibit the NHE1 isoform (K_i = 50 nM for the 5-[N-methyl, N-propyl] amiloride), suggesting that the 5-amino moiety of the inhibitor is contacting a hydrophobic pocket within the NHE1 antiporter. These pharmacological agents have allowed the identification of Na^+/H^+ antiporter isoforms specifically expressed in tissues involved in sodium transport such as small intestine and kidney. These isoforms, which are located on the apical face of epithelial cells, display a lower sensitivity to amiloride and 5-amino substituted derivatives when compared with the NHE1 isoform (Haggerty et al., 1988). Moreover, a totally amiloride-insensitive Na^+/H^+ exchanger has been reported to operate in cultured hippocampal neurons (Raley-Susman et al., 1991).

Structural Features of the Na^+/H^+ Antiporters
Primary, Secondary, and Quaternary Structure of NHE1

The eukaryotic Na^+/H^+ exchangers are expressed at the plasma membrane in relatively low abundance (Dixon et al., 1987), which makes their study more difficult. In addition, inhibitors such as amiloride, which bind with relatively low affinity, are not particularly specific, being able to interact with many other extracellular and intracellular proteins (Kleyman and Cragoe, 1988). Hence, instead of using classical protein purification techniques to characterize the structure of the antiporter, we preferred to gain access to the primary sequence of the human NHE1 exchanger by the use of a genetic approach (Pouysségur, 1985). Briefly, this genetic approach was based on: (*a*) the selection of mouse fibroblast mutants deficient in Na^+/H^+ antiporter (Pouysségur et al., 1984), (*b*) functional complementation of this antiporter-deficient cell line using human genomic DNA (Franchi et al., 1986*b*), and (*c*) cloning of the human cDNA restoring the function (Sardet et al., 1989).

The expression of this cDNA in the null cells restored all the features of NHE1 Na^+/H^+ exchange, and its complete sequencing revealed an open reading frame coding for a putative protein of 815 amino acids (predicted molecular mass, 90,704 kD). The hydropathy plot of this primary sequence, as determined both by the method of Engelman et al. (1986) and by the method of Kyte and Doolittle (1982), showed that the encoded protein possessed a dichotomic structure. The secondary structure of the antiporter appeared to consist of an amphipathic NH_2-terminal domain of 500 amino acids composed of 10–12 α-helical putative transmembrane segments, and a 315–amino acid-long COOH-terminal hydrophilic cytoplasmic domain (Fig. 2). As discussed further, the NH_2-terminal transmembrane domain possesses all the information to catalyze the ion exchange, while the cytoplasmic domain, although not required for ion translocation, is necessary for mediating hormonal regulation (Wakabayashi et al., 1992).

To isolate the protein encoded by the cloned cDNA, a fusion protein was engineered using the last 157 amino acids of the COOH-terminal domain and the *Escherichia coli* β-galactosidase. Antibodies raised against this fusion protein were able to specifically immunoprecipitate a protein with an apparent molecular mass of 110 kD, as well as being able to recognize the same band in immunoblots. This antiserum was not able to detect any protein in cells deficient in Na^+/H^+ exchange activity. The molecular weight of the immunologically characterized antiporter was shown to be greater than the molecular weight deduced from the cDNA sequence because of the glycosylation of the protein. Treatment with neuraminidase and endoglycosidase F reduced the apparent size of the protein to the expected molecu-

lar weight under denaturing SDS-PAGE conditions. In ^{32}P-labeled cells, immunoprecipitation of the antiporter showed that this 110-kD protein is constitutively phosphorylated, but that its degree of phosphorylation can vary with the physiological state of
the cells. Additionally, when used in immunofluorescence, this antiserum could label
cells expressing the antiporter only when they were permeabilized, showing that, as
previously predicted, the hydrophilic COOH-terminal domain is indeed intracellular
(Sardet et al., 1990).

Several ion exchangers, such as the Cl$^-$/HCO$_3^-$ exchanger, have been hypothesized to function as dimers (Jennings, 1989). Some experimental results also strongly
suggest that the functional unit of the Na$^+$/H$^+$ exchanger could be oligomeric. First,
pre–steady state kinetics on brush border vesicles showed that the sodium transport
catalyzed by the Na$^+$/H$^+$ antiporter (probably a distinct isoform from NHE1)

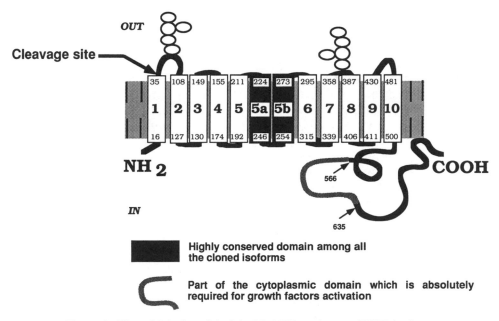

Figure 2. Hypothetical model of the Na$^+$/H$^+$ exchanger NHE1 isoform.

exhibits allosteric features during the first turnover of the antiporter (Otsu et al.,
1989). The authors suggested that this cooperative behavior (which is followed by a
purely Michaelian steady-state phase) indicates that two external sodium sites are
interacting cooperatively during the first cycle of the antiporter. They concluded that
the functional unit of the Na$^+$/H$^+$ exchanger could be a homodimer, transporting
ions across the plasma membrane by a "flip-flop" mechanism (Lazdunski, 1972).
Second, after an immunoprecipitation of the NHE1 isoform in ^{32}P-labeled cells, in
addition to the major 110-kD band, a fainter band of ~200 kD appears even under
SDS-PAGE denaturing conditions. When intact cells are pretreated with the
crosslinking agent disuccimidyl suberate, almost all the material present in the
110-kD band is shifted to the 200-kD band (Fafournoux, P., manuscript in preparation). Although these results strongly suggest that the Na$^+$/H$^+$ antiporter could exist
as an homodimer in the plasma membrane, they cannot determine whether an

oligomeric state is absolutely required for the function of the antiporter. However, the hypothesis that the functional unit is a dimer leads to the prediction that the expression of an inactive Na^+/H^+ antiporter into fibroblasts would titrate the endogenous exchange activity. We are currently using this strategy to answer the above question.

The Na^+/H^+ Antiporter Isoforms

The ubiquituously expressed exchanger NHE1 has been the most widely studied antiporter since it is present in every cell system, where it achieves basic functions that are absolutely required for cell life. A rapidly expanding field in the studies of Na^+/H^+ exchange is the search for and molecular characterization of new members of the Na^+/H^+ antiporter family. The antiporter isoforms have been characterized first by their differing pharmacological profile for amiloride and its 5-amino substituted derivatives (Haggerty et al., 1988; Raley-Susman et al., 1991; Ramamoorthy et al., 1991). They have been discovered in various tissues, such as small intestine or kidney proximal tubule epithelia, where they are involved in specialized ion transport. For example, isoforms displaying a decreased affinity for amiloride 5-*N* substituted derivatives are specifically expressed at the apical but not at the basolateral face of the kidney proximal tubule. They are responsible for $\sim 70\%$ of the sodium and bicarbonate reabsorption coupled to acid secretion by the kidney.

To characterize these isoforms at a molecular level, the NHE1 cDNA has been used as a probe to screen cDNA libraries made from intestine and kidney by low stringency hybridization. Using this method, three cDNAs coding for distinct isoforms referred as NHE2 (Tse, C. M., manuscript in preparation), NHE3 (Orlowski et al., 1992; Tse et al., 1992), and NHE4 (Orlowski et al., 1992) have been cloned. Since all the NHE1 cDNAs cloned from different species (human: Sardet et al., 1989; hamster: Counillon, L., manuscript in preparation; pig: Reilly et al., 1991; rat: Orlowski et al., 1992; and rabbit: Tse et al., 1991) have been shown to share a very high sequence homology ($> 90\%$), the sequences of these apparently more specialized isoforms have provided an "evolutionary shortcut" revealing the conserved essential functional domains of the antiporter. All these isoforms share the same hydropathy plot as the NHE1 isoform, with an amino acid homology of $\sim 50\%$ in the transmembrane domain and $\sim 30\%$ in the cytoplasmic domain for the NHE3 and NHE4 isoforms, and 60 and 35% for the NHE2 isoform (Fig. 3), respectively.

The analysis of sequence homology between the cloned isoforms has been very helpful in identifying a highly conserved domain of the antiporter, which is probably absolutely required for Na^+/H^+ exchange. This part of the molecule corresponds to the 5a and 5b putative transmembrane segments in the 12 transmembrane helices topological model of the antiporter. Interestingly, these two hydrophobic putative transmembrane domains contain several negatively charged residues which could be engaged in ion transport or proton sensitivity. Several other charged amino acids present in the other transmembrane domains are also conserved among all the isoforms and should be important for the ionic transport, and/or in the tertiary or quaternary structure of the exchanger. Site-directed mutagenesis studies are now in progress to elucidate their function.

By contrast, the sequence of the first putative transmembrane domain of NHE1, which was originally hypothesized to be a signal peptide, is not conserved among the isoforms, suggesting that this part of the molecule is not essential for the function of the antiporters. The presence of a signal peptide consensus cleavage site conserved among all the NHE1 isoforms and several experimental results suggest that this hydrophobic segment is cleaved during the protein processing. We have introduced a sequence coding for an artificial epitope at the NH$_2$ terminus of the human NHE1,

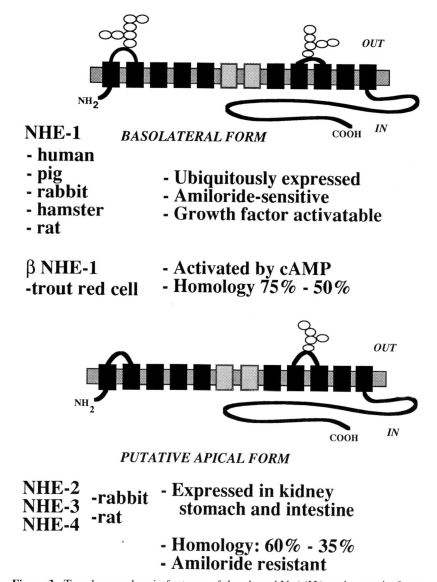

NHE-1
- **human**
- **pig**
- **rabbit**
- **hamster**
- **rat**

BASOLATERAL FORM

- **Ubiquitously expressed**
- **Amiloride-sensitive**
- **Growth factor activatable**

β NHE-1
-trout red cell

- **Activated by cAMP**
- **Homology 75% - 50%**

PUTATIVE APICAL FORM

NHE-2
NHE-3   **-rabbit**
NHE-4   **-rat**

- **Expressed in kidney stomach and intestine**

- **Homology: 60% - 35%**
- **Amiloride resistant**

Figure 3. Topology and main features of the cloned Na$^+$/H$^+$ exchanger isoforms.

and alternatively, downstream of the first transmembrane segment. In Western blots, specific antibodies against this epitope recognize the second but not the first engineered protein after expression in antiporter-deficient cells (Fafournoux, P., unpublished results). These results strongly suggest that the first putative transmembrane segment is removed from the protein during its processing.

The first glycosylation site predicted in the topological model of NHE1 is not conserved among all the isoforms, suggesting that it is not necessary for the function of the protein. This has been confirmed in our laboratory: the disruption of the glycosylation site in the NHE1 sequence gives a totally functional protein with a lower apparent molecular weight (Fafournoux, P., unpublished results). These data show that this site is effectively glycosylated but that it is not necessary for the function of the antiporter.

The Pharmacological Profile of the Cloned Na$^+$/H$^+$ Antiporter Isoforms

It was also necessary to determine if these recently cloned cDNAs were coding for functional isoforms of Na$^+$/H$^+$ antiporters, and if their affinity for amiloride and its derivatives was similar to the pharmacological profile of the isoforms that had been initially characterized in epithelial tissues. Thus, we used our antiporter-deficient PS120 cell line to express the cDNAs coding for the NHE2 and NHE3 isoforms. The cells where these two cDNAs had been transfected exhibited a fully functional antiporter activity, and we used inhibition of ^{22}Na$^+$ uptake initial rates to determine the pharmacological profile of these isoforms for amiloride and its 5-amino substituted derivatives (Counillon, L., manuscript in preparation). When compared with NHE1, the NHE2 isoform exhibits a 10-fold decrease in affinity for the 5-*N* substituted derivatives of amiloride (5-*N*-dimethyl amiloride [DMA], and 5-[*N*-methyl, *N*-propyl] amiloride [MPA]), but not for amiloride itself. By contrast, the NHE3 isoform exhibits a much stronger resistant phenotype toward both amiloride and DMA or MPA. We therefore conclude that NHE2 and NHE3 cDNAs are encoding functional Na$^+$/H$^+$ antiporters which are likely to be the same as the ones that have been pharmacologically characterized in epithelia. Unfortunately, it was impossible to express the NHE4 cDNA in fibroblasts, and therefore no pharmacological information is yet available for this isoform.

Identification of Amino Acids Involved in the Amiloride Binding of the Na$^+$/H$^+$ Exchangers

Another piece of work performed to elucidate structure–function relationship in the Na$^+$/H$^+$ exchanger was a genetic approach that has lead to the identification of amino acids involved in amiloride binding (Counillon, L., manuscript in preparation). As amiloride is a pure competitor of external sodium, it was initially hoped that the mapping of residues interacting with amiloride in the primary structure of the antiporter could be used as a starting point to identify sodium interaction sites. Briefly, we have isolated, by specific proton-killing techniques, a fibroblast cell line overexpressing a mutated Na$^+$/H$^+$ exchanger possessing a decreased affinity for amiloride and 5-amino substituted derivatives (Franchi et al., 1986a). The cloning of the cDNA coding for this antiporter and the comparison of its sequence to the wild-type one allowed us to identify one point mutation in the exchanger's gene (cytosine to thymidine transition) changing a leucine to a phenylalanine at position 167 in a highly conserved part of the fourth putative transmembrane domain of the antiporter (Fig. 4). The fact that a point mutation of this residue has been shown to be necessary and sufficient to account for the phenotype of the amiloride-resistant cells strongly suggests that this amino acid is interacting with amiloride. Interestingly, the cloned isoforms (NHE2 to NHE4) exhibit sequence differences for residues surrounding the previously identified one (Fig. 4). To determine if these changes

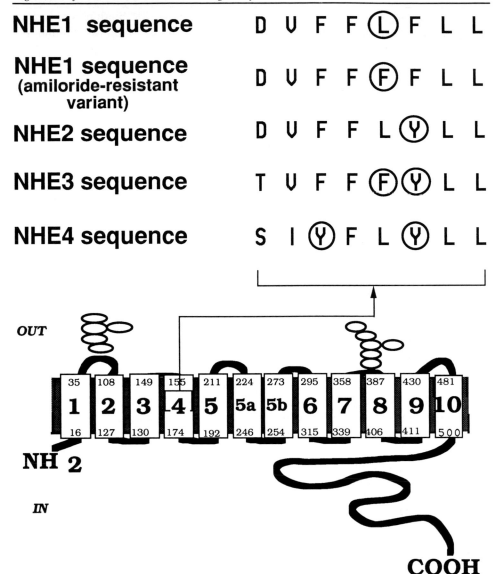

Figure 4. Comparison of the sequences of the amiloride binding region of the cloned Na$^+$/H$^+$ exchanger isoforms.

were responsible for the different pharmacological profiles of these isoforms, we introduced them by site-directed mutagenesis in the NHE1 cDNA. Expression of these mutated cDNAs in antiporter-deficient cells revealed that a phenylalanine to tyrosine change at position 168 (NHE2) does not produce any pharmacological modification, while the same change, when introduced at position 165 (NHE4), reduces both NHE1 affinity for amiloride and V_{max} for sodium transport. Interestingly, these substitutions were unable to reproduce the same pharmacological profiles as in the original isoforms, indicating that other amino acids of the Na$^+$/H$^+$ exchangers are contacting amiloride.

The finding of these two residues, involved in amiloride binding and sodium transport, is now used as a starting point to fully identify the amiloride binding site and to study sodium translocation. To achieve this, we are using a variety of techniques including protein computer modeling, chimeric constructions between amiloride-sensitive and -resistant isoforms, and site-directed mutagenesis.

Growth Factor Activation of NHE1 Molecular Mechanism
The Different Events Activating the Na$^+$/H$^+$ Exchanger

Addition of growth factors to quiescent fibroblasts results in a rapid cytoplasmic alkalinization of 0.2–0.3 pH units in the absence of bicarbonate. This pH increase has been shown to be due to the activation of the Na$^+$/H$^+$ antiporter since it is (*a*) coupled and dependent on an entry of sodium, (*b*) totally amiloride sensitive (for review see Grinstein et al., 1989), and (*c*) absent in PS120 cells lacking NHE1. In the absence of bicarbonate, mutant fibroblasts devoid of Na$^+$/H$^+$ exchanger (PS120) cannot reinitiate DNA synthesis if the external pH is lower than 7.2 pH units. A simple increase of 0.2 units of the internal pH, mimicking the action of the Na$^+$/H$^+$ antiporter, is sufficient to restore the ability of the cells to divide in the presence of mitogens (Pouysségur et al., 1984).

Many other external stimuli, such as sperm upon fertilization, hormones, neurotransmitters, chemotactic peptides, and mitogenic lectins (Pouysségur, 1985), as well as osmotic shocks and cell spreading (Margolis et al., 1988; Schwartz et al., 1989) can activate the antiporter system apparently in the same manner. The amiloride-sensitive Na$^+$/H$^+$ exchanger can thus be considered as the target of many intracellular signaling pathways of different natures. Therefore, one of the major purposes of the studies performed by our group was to understand the mechanism of activation of the antiporter by such diverse extracellular signals, and to see whether a converging route for activation prevails.

Briefly, the different events that can activate the exchanger can be grouped into three categories: (*a*) activation of receptors possessing a tyrosine kinase activity (for review see Ullrich and Schlessinger, 1990), (*b*) activation of G protein–coupled receptors (for review see Seuwen and Pouysségur, 1992), and (*c*) modifications of the cell shape. The mechanisms of activation due to the third category of stimuli have been at present only poorly investigated. Several hypotheses concerning the direct activation of the antiporter by cytoskeletal elements physically linked to its cytoplasmic loop have been proposed, but remain to be tested. By contrast, significant progress has been made in the two other signaling pathways, since both are activated by the above-mentioned two classes of growth factors.

In our fibroblast cell system (CCL39 cell line), the most potent mitogen is α-thrombin. This serine protease, which plays a major role in the blood coagulation process, triggers fibroblast proliferation by stimulation of G protein–coupled receptors. Effector systems include enzymes that hydrolyze phospholipids, such as phosphatidylinositol- and phosphatidylcholine-specific phospholipase C (PI-PLC and PC-PLC, respectively), as well as phospholipase D and phospholipase A$_2$, in addition to inhibition of adenylyl cyclase. Although protein kinase C has been shown to be involved in NHE1 activation, this route is only minimal. First, TPA or DAG have only a small effect. Second, downregulation of PKC only attenuates α-thrombin activation of NHE1. At present all the elements of the signaling pathways that are modulated by receptors with intrinsic tyrosine kinase activity and by the above-mentioned effector systems are still not known, and the most widely accepted hypothesis is that the stimulation of these different pathways activate phosphorylation cascades. At the end of these cascades, the activity of different target proteins would be modified by phosphorylation, as seems to be the case for both the ribosomal protein S6 and for several transcription factors (Fig. 5).

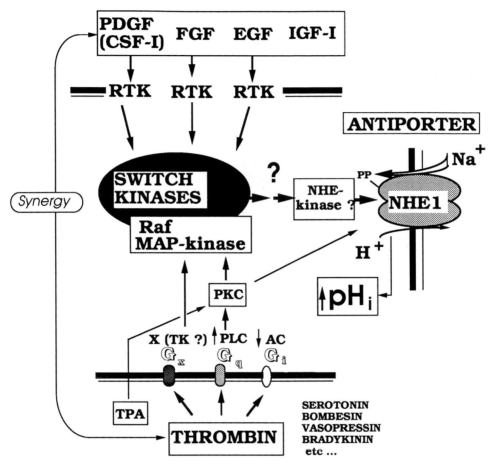

Figure 5. Signaling pathways involved in the growth factor activation of the Na$^+$/H$^+$ antiporter NHE1 isoform.

The Activation of the Na$^+$/H$^+$ Exchanger Correlates with an Increase in Its Phosphorylation

The NHE1 activation process seems to be consistent with the previous working hypothesis: immunoprecipitation experiments in ^{32}P-labeled ER22 cells (hamster fibroblasts overexpressing the EGF receptor) have shown that the phosphorylation state of the antiporter correlates with activation of the Na$^+$/H$^+$ exchange by growth factors of both categories. In G0 arrested ER22 cells, α thrombin, serum, and EGF induce a rapid and persistent cytoplasmic alkalinization, parallel with an increase in phosphorylation of the Na$^+$/H$^+$ exchanger (Sardet et al., 1990). The phosphorylation of the antiporter occurs only on serine residues, while tryptic phosphopeptide maps have shown that the activation of both tyrosine kinase and G protein–coupled receptor pathways results in the increase of phosphorylation of the exchanger at the same sites (Fafournoux, P., unpublished results). A confirmation of this hypothesis has been provided by the use of okadaic acid, an inhibitor of phosphatases which can activate the Na^+/H^+ antiporter by increasing its phosphorylation state (Sardet et al., 1991) (Fig. 6).

These experiments, in addition to others performed on human platelets (Livne et al., 1991) and A431 cells suggest that the antiporter is activated by phosphorylation, and that the integration of the signals coming from the different pathways takes place upstream from the antiporter, at the level of a common kinase cascade that might involve MAP kinase kinase and MAP kinase. It is hypothesized that this route could activate one specific NHE kinase (Fig. 5).

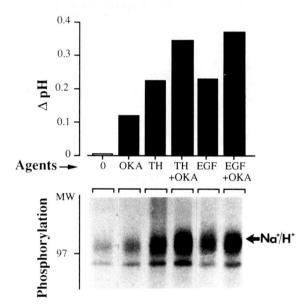

Figure 6. Correlation between activation of the NHE1 antiporter (intracellular alkalinization) and the corresponding state of phosphorylation of NHE1 in hamster fibroblasts (ER22 clone) in the presence of okadaic acid (*OKA*), thrombin (*TH*), and EGF. (Reproduced from *The Journal of Biological Chemistry*, 1991, 266: 19166–19171, by copyright permission of the American Society for Biochemistry and Molecular Biology.)

At this stage of the work, the following model of activation could be proposed: growth factors phosphorylate the antipoorter, inducing a conformational change of the molecule which will increase the affinity of its proton sensor site for intracellular protons. Therefore, phosphorylation of the antiporter modulates the set point value and makes the system more efficient in the physiological range of pH.

Identification of a Functional Domain Required for Growth Factor Activation

To investigate further the mechanisms of activation of the exchanger by growth factors, it was necessary to perform studies leading to the identification of the sites of the antiporter that are crucial for transmitting the growth factor response. To perform these studies, we were able to make use of very powerful tools: the NHE1 cDNA, which could be modified by classical molecular biology techniques, and the PS120 cell line, a derivative of CCL39 cells that are devoid of Na^+/H^+ exchanger

(Pouysségur et al., 1984). Therefore, we analyzed growth factor–induced alkalinization in PS120 cells transiently or stably expressing the wild-type or mutated forms of NHE1 cDNA.

The first step in this series of experiments was to express antiporters deleted at different points of their COOH-terminal domain; since this domain was entirely cytoplasmic, it was very unlikely that this region would be required for ionic transport. Hence, as had been shown to be the case for the Cl$^-$/HCO$_3^-$ exchanger, a regulatory role was hypothesized for this part of the protein, and the most simple way to test this was to delete this section of the protein completely. The corresponding deleted cDNA, when introduced into the PS120 cells, led to the expression of a functional antiporter. This deleted molecule possessed all the ionic transport features of the wild-type antiporter, but was no longer activatable by growth factors, and the pH activation setpoint of this deleted antiporter was shifted to more acidic values than the wild-type protein (Wakabayashi et al., 1992). This interesting result showed that the cytoplasmic COOH-terminal loop was only involved in regulatory roles and was carrying important sequence motifs for growth factor activation and pH regulation of the antiporter activity. It could therefore be hypothesized that the phosphorylation of critical sites in this part of the molecule could induce a change in conformation which would make the pKa of the proton sensor site of the antiporter (situated outside of the deleted cytoplasmic domain) more alkaline; and that reciprocally, the removal of this loop would induce a change in conformation resulting in the change of this pKa to more acidic values. Interestingly, ATP depletion by metabolic inhibitors in cells expressing the complete NHE1 was able to mimic the shift of the internal pH activation setpoint that had been observed in cells expressing the cytoplasmic domain deletion mutant. Additionally, the antiporter activation upon growth factor addition was no longer observed in ATP-depleted cells.

To identify these phosphorylation sites more precisely, different partial deletion mutants of the cytoplasmic regulatory domain were expressed in PS120 cells. These experiments have shown that the region between amino acids 566 and 635 is absolutely required for growth factor activation since the 566 deletion mutant is not activated further by growth factors, while the 635 one is still activatable (Fig. 2). Thus, it was likely that one or more serine residues present in this 566 to 635 region would be a crucial phosphorylation site for mitogen activation. Surprisingly, the mutation of each one of the 12 serines present in this domain to alanines did not impair the ability of the exchanger to be activated by growth factors (Wakabayashi et al., 1992). At least three hypotheses can explain this result: First, growth factor activation requires the multiple phosphorylation of a set of serine residues in this area. Second, the critical phosphorylation site for mitogen activation is present in the transmembrane domain, but the cytoplasmic region enclosed by amino acids 566–635 is absolutely required for the conformation of the regulatory region. Third, we cannot exclude that the growth factor–induced activation of the antiporter could involve one or many unknown proteins interacting physically with the regulatory domain mentioned above, and that phosphorylation of critical serine residues could be only one of the events of a more complex regulation process. Experiments to test these hypotheses are in progress. Additionally, consensus kinase C recognition sequences have been proven (by site-directed mutagenesis) not to be involved in the activation of the NHE1 isoform. This result is coherent with the fact that tyrosine kinases and kinase C activation induce the phosphorylation of the same sites and suggests that kinases situated downstream integrate the signals coming from these two pathways. Studies about these "integrator" kinases such as MAP and RAF kinases are actually performed in our laboratory.

The Activation of the Trout β NHE1 Antiporter

In contrast to the NHE1 antiporter, the trout red cell Na^+/H^+ exchanger is activated by catecholamines that trigger an increase in intracellular cAMP, resulting in an activation of PKA. To better understand the molecular mechanism of this activation, the human NHE1 cDNA was used as a probe to screen a library prepared from hematopoietic tissue of trout. One clone obtained by this strategy contained a complete coding sequence (467 amino acids) which exhibited a high degree of homology with the NHE1 antiporter transmembrane domain (74% amino acid identity), while the cytoplasmic domain was more divergent (48% amino acid identity) (Fig. 3). This antiporter, termed β NHE1, was able to restore the functional features of the trout red cell antiporter when expressed in antiporter-deficient hamster fibroblasts. Cells expressing β NHE1 exhibited an amiloride-sensitive antiporter activity which was activated by intracellular protons, PKC activators, and cAMP. An examination of the sequence of the cytoplasmic regulatory domain of this antiporter revealed two very close consensus sites for PKA which are not present in the human NHE1 isoform. A partial deletion of the cytoplasmic domain, removing these consensus sites, totally abolished the activation of the β NHE1 by cAMP, but did not impair its activation by PKC (Borghese et al., 1992). This interesting result suggests that the trout antiporter can be directly activated by PKA phosphorylation, and confirms the finding that the cytoplasmic domain of the Na^+/H^+ antiporters, although unnecessary for ion transport, is crucial for mediating hormonal response.

Conclusions

The cloning of a cDNA coding for the human NHE1 Na^+/H^+ exchanger isoform, revealing its sequence, provided us with the possibility of performing new experiments in order to understand the relations between the structure and function of the antiporter. Specific antibodies, obtained from a fusion protein constructed with this cDNA, allowed the characterization of the NHE1 isoform as a phosphoglycoprotein, which can exist as an oligomer and is phosphorylated in response to growth factors. Expression of deleted NHE1 and β NHE1 cDNAs revealed that the cytoplasmic domain of the exchanger was dispensable for ion transport, but was required for hormonal regulation by phosphorylation. The critical phosphorylation sites of this domain have not been identified yet, but experiments are in progress.

Additionally, amino acids involved in amiloride binding and sodium transport have been identified from the study of mutated exchangers. Sequence comparisons between the cloned isoforms also revealed highly conserved domains and amino acids in the transmembrane section of the protein. These regions of the Na^+/H^+ exchanger, which have been absolutely retained during evolution, are therefore probably required for the organization of the exchanger in the membrane, and/or for ion selectivity and transport catalysis.

Another important concept of physiological relevance which appeared from these studies is that Na^+/H^+ exchange, which was previously believed to be catalyzed by one ubiquitously expressed protein, is in fact mediated by a family of Na$^+$/H$^+$ exchangers which should exhibit different regulations. To date, five distinct antiporters have been cloned. They share the same overall structure, but are encoded by distinct genes, probably derived from a unique ancestor, distinct from the prokaryotic equivalent gene.

Acknowledgments

We thank our colleagues P. Fafournoux, C. Sardet, and S. Wakabayashi for technical advice and critical discussion, F. MacKenzie for critical reading of the manuscript, and D. Grall for technical assistance on cell culture. We are grateful to C.-M. Tse and M. Donowitz (Johns Hopkins University, Baltimore, MD) and F. Borghese and R. Motais (CEA, Villefranche/mer, France) for close and fruitful collaboration.

This work was supported by grants from the Centre National de la Recherche Scientifique, (CNRS, UMR134), the Institut National de la Santé et de la Recherche Médicale, the Association Franco-Israelienne pour la Recherche Scientifique et Technologique (AFIRST), the Foundation pour la Recherche Médicale, and the Association pour la Recherche contre le Cancer (ARC).

References

Aronson, P., J. Nee, and M. Shum. 1982. Modifier role of internal H$^+$ in activating the Na$^+$/H$^+$ exchanger in renal microvillus membrane vesicles. *Nature.* 299:161–163.

Benos, D. 1982. Amiloride: a molecular probe for sodium transport in tissues and cells. *American Journal of Physiology.* 242:C131–C145.

Borghese, F., C. Sardet, M. Cappadoro, J. Pouysségur, and R. Motais. 1992. Cloning and expression of a cAMP-activatable Na$^+$/H$^+$ exchanger:evidence that the cytoplasmic domain mediates hormonal regulation. *Proceedings of the National Academy of Sciences, USA.* 89:6765–6769.

Canessa, M., K. Morgan, and A. Semplicini. 1988. Genetic difference in lithium-sodium exchange and regulation of the sodium hydrogen exchanger in essential hypertension. *Journal of Cardiovascular Pharmacology.* 2:592–598.

Cassel, D., M. Katz, and M. Rotman. 1986. Depletion of cellular ATP inhibits Na$^+$/H$^+$ antiport in cultured human cells. *The Journal of Biological Chemistry.* 261:5460–5466.

Dixon, S., J. S. Cohen, E. J. Cragoe, and S. Grinstein. 1987. Estimation of the number and turnover rate of Na$^+$/H$^+$ exchangers in lymphocytes. Effect of phorbol esters and osmotic shrinking. *The Journal of Biological Chemistry.* 262:3626–3632.

Engelman, D., T. Steitz, and A. Goldman. 1986. Identifying nonpolar transbilayer helices in amino acid sequences of membrane proteins. *Annual Review of Biophysics and Biophysical Chemistry.* 15:321–353.

Franchi, A., E. Cragoe, and J. Pouysségur. 1986a. Isolation and properties of fibroblast mutants overexpressing an altered Na$^+$/H$^+$ antiporter. *The Journal of Biological Chemistry.* 261:14614–14620.

Franchi, A., D. Perucca-Lostanlen, and J. Pouysségur. 1986b. Functional expression of a transfected Na$^+$/H$^+$ antiporter human gene into antiporter-deficient mouse L cells. *Proceedings of the National Academy of Sciences, USA.* 83:9388–9392.

Frelin, C., P. Vigne, P. Barbry, and M. Lazdunski. 1986. Interaction of guanidinium and guanidinium derivatives with the Na⁺/H⁺ exchange system. *FEBS Letters.* 154:241–245.

Grillo, F. G., and P. S. Aronson. 1986. Inactivation of the renal microvillus membrane Na⁺/H⁺ exchanger by histidine-specific reagents. *The Journal of Biological Chemistry.* 261:1120–1125.

Grinstein, S. 1988. Na⁺/H⁺ Exchange. S. Grinstein, editor. CRC Press, Inc., Boca Raton, FL.

Grinstein, S., D. Rotin, and M. J. Mason. 1989. Na⁺/H⁺ exchange and growth factor-induced cytosolic pH changes. *Biochemistry.* 988:73–97.

Haggerty, J. G., N. Agarwal, R. F. Reilly, E. A. Adelberg, and C. W. Slayman. 1988. Pharmacologically different Na/H antiporters on the apical and basolateral surfaces of cultured porcine kidney cells. *Proceedings of the National Academy of Sciences, USA.* 85:6797–6801.

Hoffmann, E. K., and L. O. Simonsen. 1989. Membrane mechanisms in volume and pH regulation in vertebrate cells. *Physiological Reviews.* 69:315–382.

Jennings, M. L. 1989. Structure and function of the red blood cell anion transport protein. *Annual Reviews of Biophysics and Biophysical Chemistry.* 18:397–430.

Kinsella, J. L., and P. Aronson. 1981. Interaction of NH_4^+ and Li^+ with the renal microvillus membrane Na⁺/H⁺ exchanger. *American Journal of Physiology.* 241:C220–C226.

Kleyman, T. R., and E. J. Cragoe. 1988. Amiloride and its analogs as tools in the study of ion transport. *The Journal of Membrane Biology.* 105:1–21.

Kyte, J., and R. F. Doolittle. 1982. A simple method for displaying the hydropathic character of a protein. *The Journal of Molecular Biology.* 157:105–132.

L'Allemain, G., A. Franchi, E. J. Cragoe, and J. Pouysségur. 1984. Blockade of the Na⁺/H⁺ antiporter abolishes growth factor-induced DNA synthesis in fibroblasts. *The Journal of Biological Chemistry.* 259:4313–4319.

Lazdunski, M. 1972. Flip-flop mechanisms and half-site enzymes. *Current Topics in Cell Regulation.* 6:227–261.

Lazdunski, M., C. Frelin, and P. Vigne. 1985. The sodium/hydrogen exchange system in cardiac cells:its biochemical and pharmacological properties and its role in regulating internal concentrations of sodium and internal pH. *Journal of Molecular and Cellular Cardiology.* 17:1029–1042.

Livne, A., C. Sardet, and J. Pouysségur. 1991. The Na⁺/H⁺ exchanger is phosphorylated in human platelets in response to activating agents. *FEBS Letters.* 284:219–222.

Margolis, L. B., I. A. Rozovskaja, and E. J. Cragoe. 1988. Intracellular pH and cell adhesion to a solid substrate. *FEBS Letters.* 234:449–450.

Miller, R. T., L. Counillon, R. P. Lifton, G. Pages, C. Sardet, and J. Pouysségur. 1991. Structure of the 5'-flanking regulatory region and gene for the human growth factor-activatable Na/H exchanger NHE-1. *The Journal of Biological Chemistry.* 266:10813–10819.

Orlowski, J., R. A. Kandasamy, and G. E. Shull. 1992. Molecular cloning of putative members of the Na⁺/H⁺ exchanger gene family. *The Journal of Biological Chemistry.* 267:9331–9339.

Otsu, K., J. Kinsella, B. Sacktor, and J. P. Froehlich. 1989. Transient state kinetic evidence for an oligomer in the mechanism of Na⁺/H⁺ exchange. *Proceedings of the National Academy of Sciences, USA.* 86:4818–4822.

Paris, S., and J. Pouysségur. 1983. Biochemical characterization of the amiloride-sensitive Na⁺/H⁺ antiport in Chinese hamster lung fibroblasts. *The Journal of Biological Chemistry.* 258:3503–3508.

Paris, S., and J. Pouysségur. 1984. Growth factors activate the Na$^+$/H$^+$ antiporter by increasing its affinity for intracellular H$^+$. *The Journal of Biological Chemistry.* 259:10989–10994.

Pouysségur, J. 1985. The growth factor-activatable Na$^+$/H$^+$ exchange system: a genetic approach. *Trends in Biochemical Sciences.* 10:453–455.

Pouysségur, J., C. Sardet, A. Franchi, G. L'Allemain, and S. Paris. 1984. A specific mutation abolishing Na$^+$/H$^+$ antiporter activity in hamster fibroblasts precludes growth at neutral and acidic pH. *Proceedings of the National Academy of Sciences, USA.* 81:4833–4837.

Raley-Susman, K. M., E. J. Cragoe, R. M. Sapolsky, and R. R. Kopito. 1991. Regulation of intracellular pH in cultured hippocampal neurones by an amiloride-insensitive Na$^+$/H$^+$ exchanger. *The Journal of Biological Chemistry.* 266:2739–2745.

Ramamoorthy, S., C. Tiruppathi, C. N. Nair, V. B. Mahesh, F. H. Leibach, and V. Ganapathy. 1991. Relative sensitivity to inhibition by cimetidine and clonidine differentiates between the two types of Na$^+$/H$^+$ exchangers in cultured cells. *Biochemical Journal.* 280:317–322.

Reilly, R. F., F. Hildebrandt, D. Biemesderfer, C. Sardet, J. Pouysségur, P. S. Aronson, C. W. Slayman, and P. Igarashi. 1991. cDNA cloning and immunolocalization of a Na$^+$/H$^+$ exchanger in LLC-PK1 renal epithelial cells. *American Journal of Physiology.* 291:F1088–F1094.

Sardet, C., L. Counillon, A. Franchi, and J. Pouysségur. 1990. Growth factors induce phosphorylation of the Na$^+$/H$^+$ antiporter, a glycoprotein of 110 kD. *Science.* 247:723–726.

Sardet, C., P. Fafournoux, and J. Pouysségur. 1991. Alpha thrombin, EGF and okadaic acid activate the Na$^+$/H$^+$ exchanger by phosphorylating a set of common sites. *The Journal of Biological Chemistry.* 266:19166–19171.

Sardet, C., A. Franchi, and J. Pouysségur. 1989. Molecular cloning, primary structure and expression of the human growth factor-activatable Na$^+$/H$^+$ antiporter. *Cell.* 56:271–280.

Schwartz, M., G. Both, and C. Lechene. 1989. Effect of cell spreading on cytoplasmic pH in normal and transformed fibroblasts. *Proceedings of the National Academy of Sciences, USA.* 86:4525–4529.

Seuwen, K., and J. Pouysségur. 1992. G protein-controlled signal transduction pathways and the regulation of cell proliferation. *Advanced Cancer Research.* 58:75–94.

Tse, C. M., S. R. Brant, S. Walker, J. Pouysségur, and M. Donowitz. 1992. Cloning and sequencing of a rabbit cDNA encoding an intestinal and kidney specific Na$^+$/H$^+$ exchanger isoform (NHE3). *The Journal of Biological Chemistry.* 267:9340–9346.

Tse, C. M., A. I. Ma, V. W. Yang, A. Watson, S. Levine, M. Montrose, J. Potter, C. Sardet, J. Pouysségur, and M. Donowitz. 1991. Molecular cloning of a cDNA encoding the rabbit ileal villus cell basolateral membrane Na$^+$/H$^+$ exchanger. *EMBO Journal.* 10:1957–1967.

Ullrich, A., and J. Schlessinger. 1990. Signal transduction by receptors with tyrosine kinase activity. *Cell.* 61:203–212.

Wakabayashi, S., C. Sardet, P. Fafournoux, and J. Pouysségur. 1992. The Na$^+$/H$^+$ antiporter cytoplasmic domain mediated growth factors signals and controls 'H$^+$ sensing'. *Proceedings of the National Academy of Sciences, USA.* 89:2424–2428.

Chapter 14

The Cardiac Sodium–Calcium Exchanger

Kenneth D. Philipson, Debora A. Nicoll, and Zhaoping Li

Cardiovascular Research Laboratory and the Departments of Medicine and Physiology, University of California, Los Angeles, School of Medicine, Los Angeles, California 90024-1760

Molecular Biology and Function of Carrier Proteins © 1993 by The Rockefeller University Press

Introduction

Cardiac muscle contracts in response to a rise in intracellular Ca^{2+} levels. The source of this Ca^{2+} is the sarcoplasmic reticulum and Ca^{2+} influx through voltage-dependent Ca^{2+} channels. Since Ca^{2+} enters myocardial cells with each contraction, an equal amount of Ca^{2+} must be pumped from the cells to bring about relaxation and to maintain Ca^{2+} homeostasis.

Cardiac muscle has two mechanisms capable of extruding Ca^{2+} across the plasma membrane (sarcolemma): an ATP-dependent Ca^{2+} pump and the Na^{+}-Ca^{2+} exchanger. The Na^{+}-Ca^{2+} exchanger extrudes one Ca^{2+} ion in exchange for three external Na^{+} ions. Thus, the exchanger uses the energy stored in the Na^{+} gradient to maintain a low intracellular Ca^{2+}. Recent experiments from several laboratories have demonstrated that the Na^{+}-Ca^{2+} exchanger is by far the dominant Ca^{2+} efflux mechanism of cardiac muscle. It has been difficult to clearly demonstrate any role for the sarcolemmal ATP-dependent Ca^{2+} pump in cardiac muscle.

Our laboratory cloned the cardiac sarcolemmal Na^{+}-Ca^{2+} exchanger in 1990 (Nicoll et al., 1990). In this article, we will briefly review some of our recent molecular studies on this important cardiac transporter. More comprehensive background on both the physiology and biochemistry of cardiac Na^{+}-Ca^{2+} exchange can be found in Philipson (1990), Reeves (1990), Blaustein et al. (1991), and Philipson and Nicoll (1992).

Amino Acid Sequence and Topology

The open reading frame of the Na^{+}-Ca^{2+} exchanger codes for a protein of 970 amino acids (Nicoll et al., 1990). The first 32 amino acids represent a leader peptide which is removed during protein synthesis (Nicoll and Philipson, 1991; Durkin et al., 1991). The mature protein is modeled by hydropathy analysis to have 11 transmembrane segments (Fig. 1). A large (520 amino acids) hydrophilic loop between transmembrane segments five and six is modeled to be intracellular. The relationship of this model to reality is unknown.

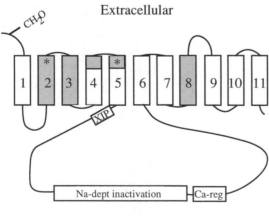

Extracellular

Intracellular

Figure 1. Model for the mature Na^{+}-Ca^{2+} exchanger. The exchanger is modeled to contain 11 transmembrane segments with a single N-linked glycosylation site near the amino terminus and a large hydrophilic, intracellular domain between transmembrane segments 5 and 6. The regions of similarity to the photoreceptor exchanger (membrane-spanning segments 2–3, and 8) and Na^{+} pump (membrane-spanning segments 4 and 5) are highlighted. The approximate location of the calmodulin binding-like region (*XIP*) and proposed Na^{+}-dependent inactivation and Ca^{2+}_{in} regulatory regions are indicated. The sites of Glu 113 and Glu 199 are indicated with asterisks.

The exchanger has six potential N-linked glycosylation sites. From site-directed mutagenesis, we know that only the asparagine at position 9 is actually glycosylated (Hryshko, L., D. A. Nicoll, J. N. Weiss, and K. D. Philipson, manuscript in preparation). Glycosylation state does not affect activity.

Homologies and Initial Mutagenesis

No known proteins have substantial amino acid similarity to the cardiac Na^+-Ca^{2+} exchanger. However, small regions of similarity exist between the Na^+-Ca^{2+} exchanger and both the Na^+, K^+-ATPase (Nicoll et al., 1990) and the Na^+-Ca^{2+}, K^+ exchanger of rod photoreceptors (Reilander et al., 1992). For the case of the Na^+, K^+-ATPase, there is a stretch of 23 amino acids within which 11 amino acids are identical to the Na^+-Ca^{2+} exchanger. This region is modeled to be intramembrane for both transporters. In addition, three of the transmembrane segments of the rod Na^+-Ca^{2+}, K^+ exchanger are 30–50% identical to analogous regions of the cardiac Na^+-Ca^{2+} exchanger.

Interestingly, these homologous transmembrane regions include two glutamic acid residues modeled to be intramembrane. It is tempting to speculate that these intramembrane anionic residues may be important in the translocation of Na^+ and Ca^{2+}. In initial site-directed mutagenesis studies, we have mutated both Glu 113 and Glu 199 to either Gln or Asp. In all cases, we obtained nonfunctioning exchangers. These regions are currently being examined in further detail.

Exchanger Inhibitory Peptide

The amino acid sequence of the Na^+-Ca^{2+} exchanger contains a potential calmodulin binding site (Nicoll et al., 1990). Although the Na^+-Ca^{2+} exchanger has never been described to be affected by calmodulin, this was of interest since calmodulin binding regions of proteins often have important auto-inhibitory roles. To test this possibility, we synthesized a peptide with the amino acid sequence of the potential calmodulin binding site (Li et al., 1991). The idea was that the potential calmodulin binding region of the exchanger might be capable of interacting with another region of the exchanger. Perhaps the synthetic peptide, designed to mimic the endogenous calmodulin binding region, would also be able to interact with this other portion of the exchanger to affect activity. Indeed, the synthetic peptide is a potent and relatively specific inhibitor of the Na^+-Ca^{2+} exchanger (Li et al., 1991). The compound has been named exchanger inhibitory peptide (XIP) and has its site of action at the intracellular surface of the exchanger.

Deletion Mutagenesis

We have constructed a deletion mutant in which most of the large intracellular loop (Fig. 1) of the Na^+-Ca^{2+} exchanger has been removed. Specifically, 440 amino acids out of the 520 amino acids comprising the loop are deleted. This removes >45% of the protein, but surprisingly, the mutant protein is still able to catalyze the Na^+-Ca^{2+} exchange reaction. A mutant with a smaller deletion (124 amino acids) in this region is fully functional.

The deletion mutants, however, do have some altered properties. Both of the deletion mutants are completely unaffected by XIP. This suggests that the binding site for XIP is located on the large intracellular loop. The exchanger also has some regulatory properties that are affected by the mutants. First, in addition to having a Ca^{2+} transport site, the exchanger has an intracellular Ca^{2+} regulatory site (Hilgemann, 1990). For example, in the presence of external Ca^{2+} and intracellular Na^+, exchange does not occur unless a low level of intracellular Ca^{2+} is also present (secondary Ca^{2+} regulation). Second, exchange activity demonstrates a phenomenon known as Na^+-dependent inactivation (Hilgemann, 1990). This is observed as a transient component of outward exchange current after application of Na^+ to the cytoplasmic surface of giant excised patches. The exchanger with the large deletion exhibits neither secondary Ca^{2+} regulation nor Na^+-dependent inactivation. Interestingly, the exchanger with the smaller deletion displays Na^+-dependent inactivation but not secondary Ca^{2+} regulation. These results allow us to tentatively localize the regions important for these regulatory properties to different portions of the large intracellular loop (Matsuoka, S., D. A. Nicoll, D. W. Hilgemann, and K. D. Philipson, manuscript submitted for publication).

Immunolocalization

Using immunogold and immunofluorescent techniques, we have examined the distribution of the Na^+-Ca^{2+} exchanger in isolated myocytes from guinea pig and rat hearts (Frank et al., 1992). Strikingly, we noted intense labeling of the sarcolemma of the transverse tubules. Labeling of the surface sarcolemma was also apparent but was less uniformly intense. Possible proximity of the exchangers in the transverse tubules to Ca^{2+} release sites on the sarcoplasmic reticulum may have important functional consequences for cardiac excitation–contraction coupling.

Acknowledgments

This work was supported by NIH grant HL-27821, the American Heart Association, Greater Los Angeles Affiliate, and the Laubisch Foundation.

References

Blaustein, M. P., R. DiPolo, and J. P. Reeves. 1991. Sodium-Calcium Exchange: Proceedings of the Second International Conference. New York Academy of Sciences, New York. 671 pp.

Durkin, J. T., D. C. Ahrens, T. C. E. Pan, and J. P. Reeves. 1991. Purification and amino terminal sequence of the bovine cardiac sodium-calcium exchanger. *Archives of Biochemistry and Biophysics.* 290:369–375.

Frank, J. S., G. Mottino, D. Reid, R. S. Molday, and K. D. Philipson. 1992. Distribution of the Na^+-Ca^{2+} exchange protein in mammalian cardiac myocytes. An immunofluorescence and immunocollodial gold-labeling study. *Journal of Cell Biology.* 117:337–345.

Hilgemann, D. W. 1990. Regulation and deregulation of cardiac Na^+-Ca^{2+} exchange in giant excised sarcolemmal membrane patches. *Nature.* 344:242–245.

Li, Z., D. A. Nicoll, A. Collins, D. W. Hilgemann, A. G. Filoteo, J. T. Penniston, J. N. Weiss, J. M. Tomich, and K. D. Philipson. 1991. Identification of a peptide inhibitor of the cardiac sarcolemmal Na^+-Ca^{2+} exchanger. *Journal of Biological Chemistry.* 266:1014–1020.

Nicoll, D. A., S. Longoni, and K. D. Philipson. 1990. Molecular cloning and functional expression of the cardiac sarcolemmal Na^+-Ca^{2+} exchanger. *Science.* 250:562–565.

Nicoll, D. A., and K. D. Philipson. 1991. Molecular studies of the cardiac sarcolemmal sodium/calcium exchanger. *Proceedings of the New York Academy of Sciences.* 639:181–188.

Philipson, K. D. 1990. The cardiac Na-Ca exchanger. *In* Calcium and the Heart. G. A. Langer, editor. Raven Press, New York. 85–108.

Philipson, K. D., and D. A. Nicoll. 1992. Na^+-Ca^{2+} exchange. *Current Opinion in Cell Biology.* 4:678–683.

Reeves, J. P. 1990. Sodium-calcium exchange. *In* Intracellular Calcium Regulation. F. Bronner, editor. Wiley-Liss, Inc., New York. 305–347.

Reilander, H., A. Achilles, U. Freidel, G. Maul, F. Lottspeich, and N. J. Cook. 1992. Primary structure and functional expression of the Na/Ca, K-exchanger from bovine rod photoreceptors. *EMBO Journal.* 11:1689–1695.

We therefore studied the kinetics of conversion of Asn-linked oligosaccharides from high-mannose to complex-type forms as a measure of transit of the proteins through the secretory pathway. HEK cells expressing AE1 or AE2 were labeled with [35S]Met and chased for various times as indicated (Fig. 2). The AE1 and AE2 polypeptides were then immunoprecipitated and subjected to digestion with endoglycosidase H (endo H). The loss of endo H sensitivity marks the conversion of Asn-linked oligosaccharides to complex forms by a process that occurs in the medial Golgi stacks. Processing of AE2 in transfected HEK cells (Fig. 2) from a 145-kD endo H–sensitive form to a 165-kD endo H–resistant form was detectable by 30 min of chase and was nearly complete by 5 h. AE1 gave rise to a characteristically broad 100-kD band which, by contrast to AE2, was never processed to a slower mobility form and never acquired endo H resistance during the 5-h chase. These data suggest that AE1, unlike AE2, is retained in a pre (or early) Golgi compartment in transfected HEK cells. This conclusion was confirmed by laser scanning confocal

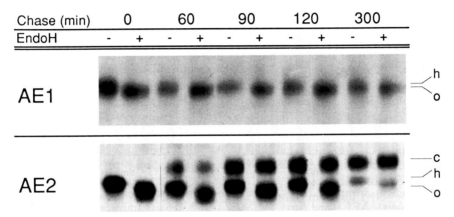

Figure 2. Endo H sensitivity of newly synthesized anion exchangers. AE transfected cells were pulse labeled with [35S]Met and chased for the indicated times. Immunoprecipitates from solubilized cells were treated (+) with endo H and resolved by 7.5% SDS-PAGE and fluorography. The mobilities of bands corresponding to protein containing no (*o*), high mannose (*h*), or complex (*c*) Asn-linked oligosaccharides are indicated.

microscopy (Fig. 3). Immunostaining of AE2 transfected cells was most intense at the cell periphery and along the fine processes that normally elaborate from HEK cells (Fig. 3 *B*). Intracellular staining was also observed, most prominently in a perinuclear reticular pattern. By contrast, AE1 transfected cells differed in both overall morphology and distribution of AE1 immunoreactivity (Fig. 3 *A*). Intense AE1 immunoreactivity was observed in a perinuclear, and occasionally juxtanuclear, reticular pattern. We conclude that AE1 and AE2 are processed differently through the secretory pathway in HEK cells and that, unlike AE2, AE1 is retained in an enzymatically active state in a pre-Golgi compartment, most probably the ER.

We used lectin affinity chromatography to separate crude microsome (CM) into plasma membrane (PM) from ER-derived (ER) vesicles. CM prepared from AE1 and AE2 transfectants were fractionated by concanavalin A Sepharose (conA) affinity chromatography into a flow-through fraction and an α-methyl mannoside eluate. The activities of the marker enzymes glucose-6-phosphatase and 5′ nucleotid-

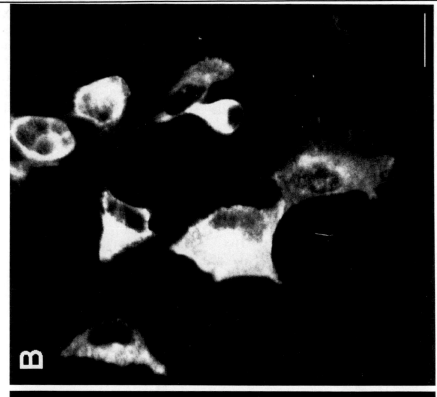

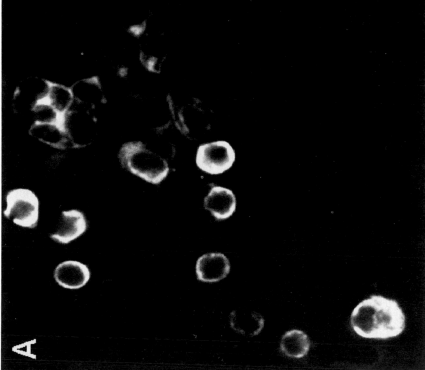

Figure 3. Confocal images of AE1 (*A*) or AE2 (*B*) transfected HEK cells. Cells were processed for indirect immunofluorescence using an antibody that reacts with the COOH termini of AE1 and AE2. *Bar*, 20 μm.

ase were used to assess the degree of enrichment of the fractions in ER and PM markers, respectively.

Immunoblots of microsomes from transfected HEK cells were used to assess the distribution of anion exchanger polypeptides in the different fractions. A 100-kD band, corresponding to immature (endo H–sensitive) AE1, was detected in immunoblots of CM from AE1 transfected cells. This band eluted exclusively with the ER markers, further supporting our conclusion that AE1 does not mature to a post-ER compartment. By contrast, the two AE2 bands, corresponding to the endo H–sensitive and –resistant species, respectively, were quantitatively resolved into ER and PM fractions by conA chromatography. We conclude that conA affinity chromatography can be used to efficiently separate CM vesicles from transfected HEK cells into pre-Golgi (ER) and post-Golgi (PM) fractions. Both fractions contain sealed membrane vesicles that are suitable for evaluating anion exchanger activity.

Processing in the Golgi Apparatus Is Not Required for Functional Activation of Anion Exchangers

To determine when, during the course of AE biogenesis, the proteins acquire the ability to catalyze anion exchange, we evaluated $Cl^-/^{35}SO_4^{2-}$ activity in the individual

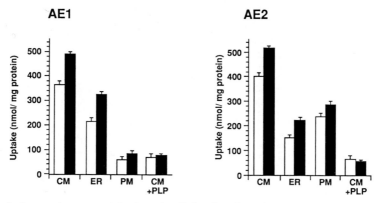

Figure 4. Anion exchange activity in subcellular fractions from AE1 and AE2 transfected HEK cells. Gradient-driven Cl^-/SO_4^{2-} exchange was determined at 60- (*open bars*) and 120-s (*filled bars*) time points on subcellular fractions as described in the legend to Fig. 1. Data are the mean ± SD of quadruplicates of three independent experiments.

microsomal subfractions (Fig. 4). As expected from the distribution of AE1 polypeptide in these fractions, AE1-mediated anion exchange activity was recovered exclusively in the ER fraction. The small amount of anion exchange activity present in the PM fraction from AE1 transfectants is similar to the background flux (CM vesicles in the presence of pyridoxal 5′-phosphate), consistent with the absence of immunodetectable AE1 in the PM fraction. By contrast, significant anion exchange activity in AE2 transfectants was recovered in both ER and PM fractions. Further, the proportion of activity in the ER (40%) and PM (60%) fractions correlated well with the distribution of the 145- and 165-kD forms of the protein in the ER and PM fractions, respectively. Since we can recover nearly all of the anion exchange activity of crude microsomes in the two conA fractions, we conclude that both AE1 and AE2 are functional as anion exchangers in their immature, high-mannose ER forms.

By contrast to the wealth of data on the kinetics of protein assembly and folding in the secretory pathway, little is known about where and when integral plasma membrane proteins become functionally activated. Such proteins, including ion channels and transporters, must transit the secretory pathway and undergo a sequence of posttranslational structural modifications before arriving at the plasma membrane. Such modifications may be necessary to activate the ion transport capacity of channels and pumps, and may be essential to maintain them in an inactive or silent state until they reach their correct destination. The nicotinic acetylcholine receptor (Smith et al., 1987; Gu et al., 1989), a ligand-gated cation channel, and the Na,K-ATPase (Caplan et al., 1990) acquire the capacity for high-affinity binding to their respective toxins early after synthesis, probably in the ER. However, the capacity of these early biosynthetic intermediates to form functional ion transporters has not been investigated. Ligand-gated transporters like the nicotinic acetylcholine receptor can be effectively silenced during intracellular transit by the absence of ligand. By contrast, the majority of transporters, including facilitated diffusion carriers, ion exchangers, and pumps must be subject to regulation by endogenous factors or contribute to the steady-state ionic composition of the lumen of secretory pathway cisternae. Conceivably, transient association of ion transport subunits with intracellular "chaperones" may be one means to preserve them in an inactive state during biosynthesis. Our data show that in highly enriched populations of ER-derived vesicles, biosynthetic intermediates of AE1 and AE2 from transfected HEK cells are catalytically active. Although our data cannot exclude the possibility that the exchangers were activated in the process of ER vesicle isolation, they suggest that modification of the AE polypeptides in post-ER compartments is not necessary for their functional activation. Further studies will be necessary to determine whether, in cells where they are endogenously expressed, AE1 or AE2 could contribute to the maintenance of steady-state anion and proton gradients across the cisternae of secretory organelles.

Perhaps overexpression of AE1 in the ER results in anionic or proton imbalance that is ultimately toxic to the cell, giving rise to the morphologic activations we observed. This may explain our consistent inability to isolate stable mammalian cell clones expressing this polypeptide. Alternatively, an additional role for the putative AE1 subunit—whose existence is suggested by the accumulation of AE1 in the ER of transfected HEK cells—is to suppress the potentially deleterious effects of premature activation of this plasma membrane anion exchanger.

Acknowledgments

This work was supported by grants from the NIH (R01 GM-8543 and R01 DK-43994), the March of Dimes Birth Defects Foundation, and the Lucille P. Markey Charitable Trust. S. Ruetz was supported by the Forschungskommission der Universität Zürich, and R. R. Kopito is a Lucille P. Markey Scholar in Biomedical Science.

References

Alper, S. L. 1991. The band 3-related anion exchanger (AE) gene family. *Annual Review of Physiology.* 53:549–564.

Barasch, J., B. Kiss, A. Prince, L. Saiman, D. Gruenert, and Q. Al-Awqati. 1991. Defective acidification of intracellular organelles in cystic fibrosis. *Nature.* 352:70–73.

Cabantchik, Z. I., M. Balshin, W. Bruer, and A. Rothstein. 1975. Pyridoxal phosphate: An anionic probe for protein amino groups exposed on the outer and inner surfaces of intact human red blood cells. *Journal of Biological Chemistry.* 250:5130–5136.

Caplan, M. J., B. Forbush, III, G. E. Palade, and J. D. Jamieson. 1990. Biosynthesis of the Na,K-ATPase in Madin-Darby canine kidney cells. *Journal of Biological Chemistry.* 265:3528–3534.

Drenckhahn, D., K. Schluter, D. P. Allen, and V. Bennett. 1985. Colocalization of band 3 with ankyrin and spectrin at the basal membrane of intercalated cells in the rat kidney. *Science.* 230:1287–1289.

Gu, Y., E. Ralston, C. Murphy-Erdosh, R. A. Black, and Z. W. Hall. 1989. Acetylcholine receptor in a C2 muscle variant is retained in the endoplasmic reticulum. *Journal of Cell Biology.* 109:729–738.

Hurtley, S. M., and A. Helenius. 1989. Protein oligomerization in the endoplasmic reticulum. *Annual Review of Cell Biology.* 5:277–307.

Kopito, R. R. 1990. Molecular biology of the anion exchanger gene family. *International Review of Cytology.* 123:177–199.

Kopito, R. R., M. A. Andersson, D. A. Herzlinger, Q. Al-Awqati, and H. F. Lodish. 1988. Structure and tissue-specific expression of the mouse anion exchanger gene in erythroid and renal cells. *In* Cell Physiology of Blood. R. B. Gunn and J. C. Parker, editors. Rockefeller University Press, New York. 151–161.

Lee, B. S., R. B. Gunn, and R. R. Kopito. 1991. Functional differences among nonerythroid anion exchangers expressed in a transfected human cell line. *Journal of Biological Chemistry.* 266:11448–11454.

Low, P. S. 1986. Structure and function of the cytoplasmic domain of band 3: center of erythrocyte membrane-peripheral protein interactions. *Biochimica et Biophysica Acta.* 864:145–167.

Smith, M. M., J. Lindstrom, and J. P. Merlie. 1987. Formation of the α-bungarotoxin binding site and assembly of the nicotinic acetylcholine receptor subunits occur in the endoplasmic retiuculum. *Journal of Biological Chemistry.* 262:4367–4376.

Steck, T. L. 1978. The band 3 protein of the human red cell membrane: a review. *Journal of Supramolecular Structure.* 8:311–324.

Chapter 16

Mitochondrial Carrier Family: ADP/ATP Carrier as a Carrier Paradigm

Martin Klingenberg

*Institute for Physical Biochemistry, University of Munich, W-8000
Munich 2, Federal Republic of Germany*

Molecular Biology and Function of Carrier Proteins © 1993 by The Rockefeller University Press

Introduction

The symbiotic nature of mitochondrial existence in eukaryotic cells is dependent on an intensive exchange of solutes between the mitochondrial matrix space and the eukaryotic cytosol. These solutes are metabolites of pathways that link mitochondria to the eukaryotic host. Therefore, they are mostly anionic in nature. The intensive traffic through the mitochondrial membrane is facilitated by a series of transport proteins or carriers in the inner mitochondrial membrane. Presently, there are an estimated 14 different carriers occurring in mitochondria (Krämer and Palmieri, 1989). They may be divided up according to their role and metabolism (Table I): (*a*)

TABLE I
Mitochondrial Solute Carriers

Carrier name	Substrates	Transport type	Control	Metabolic role
Energy transfer				
ADP/ATP	ADP^{3-}, ATP^{4-}	Exchange	$\Delta\psi$	Oxidative phos-phorylation
Phosphate	H^+/P^-	Unidirectional	ΔpH	Oxidative phos-phorylation
Uncoupling protein	H^+/OH^-	Unidirectional	ΔpH	Heat generation
Carbon transloca-tion				
Dicarboxylate	Malate, succinate (Phosphate), malinate	Exchange	—	TCC cycle Gluconeagenesis
Ketoglutarate	Ketoglutarate, malate	Exchange	—	Hydrogen import
Citrate	Citrate, isocitrate Malate	Exchange	ΔpH	Hydrogen export Fatty acid synthesis
Pyruvate	Pyruvate, lactate	Unidirectional	ΔpH	Glucose oxidation
Carnitine	Carnitine, acylcar-nitine	exchange	—	Fatty acid oxida-tion
Nitrogen transfer				
Aspartate/ glutamate	Aspartate, gluta-mate	Exchange	$\Delta\psi$	Urea cycle
Glutamate	Glutamate	Unidirectional	ΔpH	Hydrogen import
Ornithine	Ornithine, citrul-line	Exchange		Urea cycle

carriers involved in energy transfer and oxidative phosphorylation; (*b*) carriers involved in carbon and hydrogen transfer, mostly linked to the tricarboxylic acid cycle and to transaminations; and (*c*) carriers involved in nitrogen transfer linked to the urea cycle. The mitochondrial membrane also contains carriers for inorganic cations that are probably quite different from the metabolite carriers.

The evolution of these carriers is clearly linked to the emergence of mitochondria in the eukaryotes. In this way they are more unique to mitochondria than components of oxidative phosphorylation and respiration, which already occur in the prokaryotes. Therefore, it is not surprising that all these carriers appear to be encoded by nuclear genes.

The symbiotic or coordinative role of these carriers is particularly striking if one considers the large differences in two basic thermodynamic states between the matrix and cytosol, e.g., the redox potential and the phosphorylation potential. In mitochondria the redox potential is more negative than in the cytosol (Bücher and Klingenberg, 1958; Klingenberg and Slenczka, 1959), and conversely the phosphorylation potential is higher in the cytosol than in the mitochondria (Heldt et al., 1972). The differences in these major thermodynamic functions are compensated by coupling the membrane potential to some of these transport systems. With the membrane potential driving these solute transports, the thermodynamic differences are maintained across the inner mitochondrial membrane.

Among these carriers, the ADP/ATP carrier is the most abundant and occurs in all mitochondria known (Klingenberg, 1976). Also, the phosphate carrier is presumably present in all mitochondria as an obligatory partner of oxidative phosphorylation. Most of the other carriers occur in variable amounts according to the special needs of the particular cells (Krämer and Palmieri, 1989). An exception to these rules is the uncoupling protein, which specifically occurs only in mitochondria from brown adipose tissue of mammalians. This carrier transports only H^+/OH^- and surprisingly has been found to be a member of the solute carrier family (Aquila et al., 1985).

Structure

A number of these carriers have been isolated, but only the primary structure for four different carriers is known. In addition, the primary sequences of these carriers from other cells and isoforms are known. All of these carriers, those for ADP/ATP (Aquila et al., 1982), phosphate (Aquila et al., 1987), ketoglutarate-malate (Runswick et al., 1990), and the uncoupling protein (Aquila et al., 1985) have characteristic similarities in their sequence, which led to the conclusion that they belong to a mitochondrial carrier family (Fig. 1). The first common feature is that they all have a molecular mass of ~32 kD. Since other isolated carriers without known primary structure have a similar molecular mass, it can be presumed that they also belong to this carrier family.

As first noted for the ADP/ATP carrier, the primary structure can be analyzed to be composed of three similar domains of ~100 residues each (Saraste and Walker, 1982). The similarity between these repeat units is not high but nevertheless striking in terms of the conservation of certain residues and hydrophobic areas. Because of the wider distribution of polar residues, only about three transmembrane helices are clearly discernible from hydrophobicity plots. With the amphiphatic analysis four or six helices may be discerned (Aquila et al., 1985). It is only with the three-domain structure that a six-transmembrane helical arrangement can be assumed with some degree of certainty. This three-domain structure produces a threefold pseudo-symmetry (Aquila et al., 1987). As a result, each domain should have a similar folding pattern. In fact, by aligning the three domains for maximum similarity and conservation of certain residues, and by allowing insertions or additions in this manner, two transmembrane helices can be discerned within each

```
AAC   SDQALSFLKDFLAGGVAAAISKTAVAPIE..RVKLLLQVQHAS.KQISAEKQYKGIIDCVVRIPKEQGFLSFWRGNLANVIRYFPTQALNFAFKDKYKQIFLG
UCP                 IPSAGVAACVADIITFPLDTAKVRL..QIQGECLISSAIR..YKGVLGTITLAKTEGPVKLYSGLPAGLQRQISLASLRIGLYDTVQEFFIT
KMC   AATASPGASGMDGKPRTSPKSVK.FLRGGLAGMAIVPVQPFLDLVKNRM..QLSGEGAKT...RE.YKTSFHALISILRABGLRGLYTGLSAGLLRQATYTTRLGIYTTVLFERLTG
PIC   AVEEQYSCDYGSGRFFIICGLGGIISCGTTHFTALVPLDLVKCRM..QV..DPQK......YKSIFNGFSVTLKEDGFRGLAKGWAPTFIGYSLQGLCKFGFYEVFKVLYSN


AAC   GVDRHKQFWRYFAGNLASGGAAGATSLCFVYPLDFARTRLAADVG.....KGAAQREFTGLGNCITKIFKSDGLRGLYQGFNVSVQGIIIYRAAYFGVYDTAKGML
UCP   GKEASLGSKISAGLMTGGVAVFIGQ.....PTEVVKVALQAQSHLHGP.KP..RYTGIYHAY.RIIATTEGLTGLWKGTSPHLTTRVIIHCIELVTYDLMKEAL
KMC   GADGTPPGFLLKAVIGMTAGATGAFVGT.....PAEVALIRMTADGRLPVDQR.RGYKNVFNAL..FRIVQ.EEGVPLWRGCIPTMARAVVVNAAQLASYSQSKQFL
PIC   MLGEEQA.YLWRTSLYLAASASAEFFADIALA.PMEAAKVRIQTQPGYANTLRDAAPKMYK..........EEGLKAFYKGVAPLWMRQIPYTMMKFACFERTVEAL


AAC   PDPLNVHIIVSWMIAQTVTAVAGIVSIPFDTVRRRMMMQSGRKGADIMYTGTVDCWRKIAK.....DEGPKAFFKGAWSNVLRGMGGAFVLVL.YDEI......KKFV
UCP   VRMKLLADEVPCHFPVSAVVAGRCTTFVLSSPVDVVKTRFVHS.....SPGQHTSVP.HCAMMLTR....BGPSAFFKGFVPSFIRLGSWH.IMFVCFERLKQELMKCRHTMDCAI
KMC   LDSGYFSDNILCHFCASMISGLVTTAASMPVDIVKTRIQNMRMIDGKPE.YKNGLDVLIVKVV.RY...EGFFSLWKGFTPYYARLGPHTVLTFIFLEQMNKAYKRLFLSG
PIC   YKFVVPKPRSBCSKPEQLVTFVAGYIAGVFCAIVSHPADSVVSVLNKEKG...........SSASEVL.KR....LGFRGVWKGL..FARILMIGTLTALQWFIYDSVKVYFRLPRPPPP
                                                                                                            EMPESLKKKLGYTQ
```

Figure 1. The sequence of four different mitochondrial carriers from bovine tissues (heart and brown adipose tissue) arranged according to the tripartite structure. The sequences for the bovine ADP/ATP are from Aquila et al. (1982), for the bovine uncoupling protein from Casteilla et al. (1989), for the phosphate carrier from Aquila et al. (1987) and Runswick et al. (1987), and for the ketoglutarate-malate carrier from Runswick et al. (1990).

domain, amounting to a total of six transmembrane helices for the mitochondrial carrier.

Each domain has a hydrophilic central segment with 30–40 amino acids in which many polar residues are clustered. The hydrophilic connecting segments at the opposite side are only ~15 residues long. Polar residues are quite well conserved within the three central segments. Some of these residues may be involved in the gating process, as will be discussed below. Certain lysines have been shown to become accessible to pyridoxal phosphate only when the carrier is solubilized. It is proposed that in the membrane these lysines are liganded with acidic phospholipids such as cardiolipin, which are removed on solubilization with detergents (Bogner et al., 1986).

The conservation of acidic residues among all carriers and within all carriers in these three domains is striking. We consider them to be located at the termination of the transmembrane helices. Some of these acidic residues are paired with basic residues which may act in the gating mechanism (Klingenberg, 1992).

Other conspicuous polar residues are arginines within the putative second transmembrane helices of each domain. Arginine is known to occur more easily within membranes than lysine (Rees et al., 1989; Hendersen et al., 1990). In the AAC there is a lysine exclusively in the first helix that seems to be involved in binding of ADP or carboxyatractyloside as shown by the pyridoxal phosphate probe (Bogner et al., 1986). These arginines are conserved in nearly all carriers, although their position may vary. Their function will be discussed below.

Probing the Folding

The problem of the location of the NH_2 and COOH termini has been a difficult problem for several years and could only be solved by utilizing the family membership of the ADP/ATP carrier with other carriers such as the uncoupling protein. It was maintained that the COOH and NH_2 termini are opposite (Brandolin et al., 1989). This question is now settled in that both the NH_2 and COOH termini protrude to the cytosolic side of the membrane (Brandolin et al., 1989; Eckerskorn and Klingenberg, 1987). The same side of the terminals confirms the even number transmembrane helical model (Klingenberg, 1989). Thus the repeat structure and the folding pattern agree with each other (Fig. 2).

Earlier topographic investigations of the ADP/ATP carrier concerned the localization of the 22 lysine groups scattered throughout the primary structure with the membrane impermeant reagent, pyridoxalphosphate (Bogner et al., 1986). Applied from either the outside or the inside of the membrane, with this reagent a labeling pattern emerged which at first seemed to contradict the simple six-helical transmembrane folding pattern. Some lysine groups located in the matrix regions of the three domains reacted from both the matrix and cytosolic sides. The reactivity depended on the translocation states of the carrier. Since according to the single-binding, center-gated pore model there is an access route to the binding center from either the cytosolic or the matrix side, it was concluded that the double access of the lysine from either side can be explained by assuming that these lysines are in the translocation channel (Bogner et al., 1986; Klingenberg, 1989). As a consequence, portions or part of the three hydrophilic matrix sections should fold into the

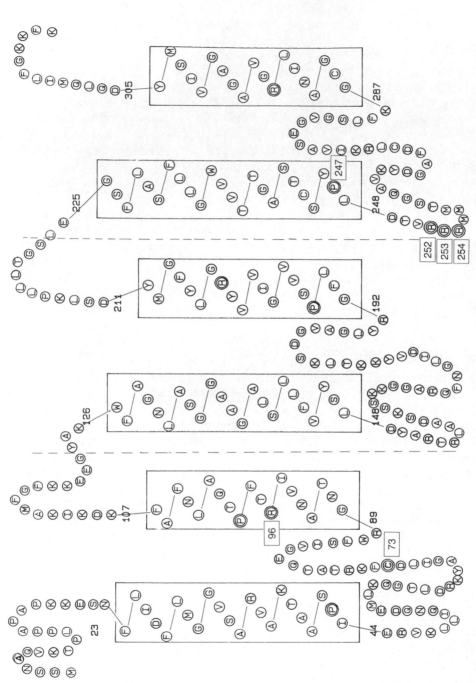

Figure 2. The folding of the yeast AAC2. The model also underlines the binding region for 8-azido-ATP. The boxed numbers show the residues changed by site-directed mutagenesis.

membrane region, where they line the translocation chain channel with the hydrophilic surface.

The Nucleotide Binding Site

This highly unconventional folding model received further support from photo-affinity labeling of the binding center in the ADP/ATP carrier. With 2-azido-ATP the ADP/ATP carrier was photo-affinity labeled in bovine mitochondria (Dalbon et al., 1988) or in the isolated state. For this purpose, the ADP/ATP carrier from yeast turned out to be more suitable because of more favorably localized CNBr cleavage sites (Mayinger et al., 1989). Both 2- and 8-azido-ATP labeling were applied to the yeast AAC in situ and in the isolated state. The binding regions both for 2- and 8-azido ATP were found to be localized at the putative intermembrane fold of the second domain. The azido-ADP must access this part from the cytosolic side, although the peptide region originates from the matrix hydrophilic central portion.

More recently, accessibility from the cytosolic side of the corresponding area in the third domain was demonstrated for the uncoupling protein (Mayinger and Klingenberg, 1992; Winkler and Klingenberg, 1992). Purine nucleotides bind to the uncoupling protein as inhibitors in a highly specific manner only from the cytosolic site. In this case the nucleotide is not transported to the matrix side. Both 2- and 8-azido-ATP incorporate into the central portion of the last domain in a section which is analogous to that labeled in the second section in the AAC. With exclusive accessibility of the nucleotide only from the cytosolic side in the uncoupling protein, it provides more stringent evidence of matrix accessibility than in the AAC, where the azido-nucleotides might be translocated to the inside, although very slowly.

Site-directed Mutagenesis

Because the ADP/ATP carrier in yeast can be easily expressed on a plasmid vector, it becomes a target for site-directed mutagenesis (O'Malley et al., 1982; Gawaz et al., 1990; Lawson et al., 1990). In this context, two types of results will be discussed that are of general significance for the mitochondrial carrier family. First, the role of cysteine: SH reagents are known to inhibit AAC transport activity in a peculiar way. Inhibition occurred only on addition of ADP, and this could be shown to be due to an unmasking in the "m" state of C 56 in the first domain. The unmasking of the SH groups played an important role in demonstrating the kinetics of the "c" to "m" state transition with fluorescent SH labels and thus the conformational change involved in the AAC (Klingenberg, 1976). All three cysteines were singly mutated to serine but failed to produce a strong inhibition of the AAC activity (Hoffmann, 1991; Klingenberg et al., 1992). Thus, the cysteine groups appeared not to be essential for transport, although the alkylation produced an inhibited carrier. This can be explained by hydrogen bonding of cysteines, which is retained by replacement with serine but interrupted on alkylation, thus inducing an inhibitory structural change in the protein. The nonessentiality of the SH group in carriers has also been noted previously for other carriers that are sensitive to SH reagents such as lac-permease (Menick et al., 1987). The inhibition by SH reagents of transport in most mitochondrial carriers can be rationalized by assuming that the hydrogen bonding capability of cysteine is essential and involved in the conformation changes during the translocation process (Fig. 2).

The intrahelical arginines 96, 204, and 295 occurring in the second helix in each domain of nearly all mitochondrial carriers have been exchanged by mutagenesis in the yeast AAC, thus eliminating the positive charge or replacing it by histidine. In all these mutations the yeast strains became dependent on fermentable sources, indicating that the AAC is largely inhibited. Most interestingly, the R96H mutation turned out to be identical with the only known natural genomic mutation of the AAC, the op1 mutant (Kolarov et al., 1990; Lawson et al., 1990). AAC was isolated from the op1 and R96H mutants. Surprisingly, the reconstituted carrier protein was ~40% as active as the wild AAC. We interpreted this finding to mean that R96H and the other interhelical arginines have an essential role in the biogenesis and folding process of the AAC (Gawaz et al., 1990). This explains their ubiquitous occurrence in the four different carriers known so far. This would also be in line with the equal folding of the three domains within each carrier, where the transmembrane segment directed toward the cytosolic side always carries an intrahelical arginine.

Interaction with Cardiolipin

Another interesting aspect involving the charge distribution in the mitochondrial carriers is the interaction with cardiolipin. In general, mitochondrial carriers are highly positively charged with an excess of up to 20 net positive charges. This renders the mitochondrial carrier proteins quite basic with a high iso-electric point. It is obvious that the high amount of negatively charged cardiolipin should interact with the positive protein charges. For the reconstitution of the wild-type ADP/ATP carrier, cardiolipins are not required. However, in reconstitution the phosphate carrier has an absolute requirement of cardiolipin (Kadenbach et al., 1982). Cardiolipin addition is also important for separating and isolating various carrier proteins, particularly the phosphate, dicarboxylate, and tricarboxylate carriers (Krämer and Palmieri, 1989). This indicates that cardiolipin interacts with these proteins in the membrane. But cardiolipin may be lost after solubilization.

The first evidence for cardiolipin binding came from the observation in NMR studies that the isolated ADP/ATP carrier contains tightly bound cardiolipin (Beyer and Klingenberg, 1985). The bound cardiolipin phosphate head groups are NMR silent and became unmasked only after denaturation by SDS and heat. Six molecules of cardiolipin are bound to the bovine ADP/ATP carrier. However, a functional role of cardiolipin could not be demonstrated because the release of cardiolipin from AAC was only possible under denaturing conditions. More recently, the C73S mutant of the AAC from yeast was found to have an absolute requirement for cardiolipin on reconstitution (Hoffmann, 1991). In this case, the cardiolipin addition was obligatory for active ADP/ATP transport. It allowed the elucidation of the high specificity of the cardiolipin requirement by using several cardiolipin analogues, lyso-forms, etc. In no other mitochondrial function the requirement of cardiolipin is found to be so essential as in the mitochondrial carriers.

The Gating Mechanism

Some of the numerous charges and charge pairs found outside of the transhelical segments in the carriers may be involved in the gating process of translocation. Gates on both sides of the translocation channel are an essential part of the single-binding center gated pore model (Klingenberg, 1976, 1991; Klingenberg et al., 1976). Either the inner or outer gate is closed in the external or internal carrier state. In the transition state, both gates are either closed or open. In line with the "induced fit transition state" mechanism, which we have described for the carrier action, the binding center has a conformation that exposes a maximum interaction with the substrates (Klingenberg, 1991). Whereas in the external or internal state the binding center is postulated not to displace substrate-like conformations, these are induced by the solute–protein interaction in the transition state. This "induced fit" is the key element of transport catalysis in carriers. In contrast, in enzymes an induced fit is catalytically counterproductive but may enhance specificity (Jencks, 1975).

For considerations of catalytic activity we assume that in the transition states both gates are open or at least partially open (Klingenberg, 1992). The gating and its control is suggested to involve alternating formation and breaking of ionic pairs. The sequence of events can be visualized as follows (Fig. 3). The anionic substrate binds

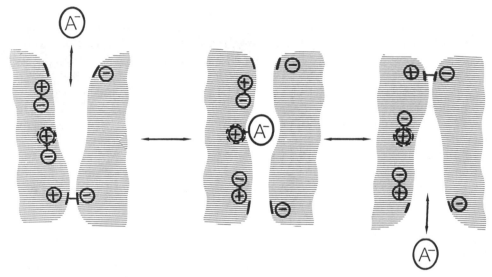

Figure 3. The charge oscillating gating model of carrier function. The role of charged residues in the translocation pathway (Klingenberg, 1992).

to one or more positive charges in the binding center. Negative residues on both sides of the binding center are postulated to be mobile and can either pair with the positive residue of the binding center or move more outside to bind with the positive residue of the gates. Without solutes present in the binding center the central positive charge is free and will pair with either the internal or external mobile negative charge. By neutralizing one mobile negative charge it releases one positive charge at the gate, which can now form an ion bond across the channel and then close the gate. At the same time, the opposite gate is opened because its positive residue is neutralized by the immobile negative charge. On binding of the anionic solute, the central positive residue is sequestered and the released mobile negative charge moves toward the outside and pairs with the positive charge of the gate. Thus, binding of the substrate is propagated to the gates by the two mobile negative residues such that both gates are open in the transition state.

The model requires the proper arrangement of charges along the translocation channel. As seen in Fig. 2, a sufficient number of charges is available in the three central domain regions that are proposed to line the translocation channel. It would be most intriguing to investigate by site-directed mutagenesis which of these charges participate in the gating process.

Conclusions

Mitochondrial carriers are distinguished from carriers of other families by their apparently simple construction. The molecular weight is small and the three repeat structures suggest a certain internal symmetry. Yet there is evidence suggesting that these carriers may act only as dimers and therefore the functional unit would have twice the molecular weight (Klingenberg, 1981; Aquila et al., 1985). All the evidence from the primary structure indicates that the mitochondrial carriers have no predecessors in prokaryotes but have emerged with the symbiosis of the mitochondria with the eukaryotic host cell. It is surprising that carriers with such widely different functions when transporting ADP, ATP, or H^+, etc. belong to one molecular family. It shows the enormous plasticity in adapting to various substrates of a similar protein structure which is also seen in other carrier and protein families.

Since the ADP/ATP carrier is highly abundant in mitochondria from aerobic eukaryotic cells, and since it adopts two specific inhibitors in a unique way, it has been a pioneer in research on the carrier structure–function relationship. Thus the ADP/ATP carrier was the first intact carrier to be isolated (Riccio et al., 1975), and was the first carrier for which a mechanism could be demonstrated on a molecular level (Klingenberg, 1976), and further it was the first for which a primary structure became known (Aquila et al., 1982). It can be expected that this protein and other members of the mitochondrial carrier family will also in the future play a decisive role in elucidating carrier mechanisms.

References

Aquila, H., D. Misra, M. Eulitz, and M. Klingenberg. 1982. Complete amino acid sequence of the ADP/ATP carrier from beef heart mitochondria. *Hoppe-Seyler's Journal of Physiological Chemistry.* 363:345–349.

Aquila, H., T. A. Link, and M. Klingenberg. 1985. The uncoupling protein from brown fat mitochondria is related to the mitochondrial ADP/ATP carrier: analysis of sequence homologies and of folding of the protein in the membrane. *EMBO Journal.* 4:2369–2376.

Aquila, H., T. A. Link, and M. Klingenberg. 1987. Solute carriers involved in energy transfer of mitochondria form a homologous protein family. *FEBS Letters.* 212:1–9.

Beyer, K., and M. Klingenberg. 1985. ADP/ATP carrier protein from beef heart mitochondria has high amounts of tightly bound cardiolipin, as revealed by ^{31}P nuclear magnetic resonance. *Biochemistry.* 24:3821–3826.

Bogner, W., H. Aquila, and M. Klingenberg. 1986. The transmembrane arrangement of the ADP/ATP carrier as elucidated by the lysine reagent pyridoxal 5-phosphate. *European Journal of Biochemistry.* 161:611–620.

Brandolin, G., F. Boulay, P. Dalbon, and P. V. Vignais. 1989. Orientation of the N-terminal region of the membrane-bound ADP/ATP carrier protein explored by antibodies and an arginine-specific endoprotease: evidence that the accessibility of the N-terminal residues depends on the conformational state of the carrier. *Biochemistry.* 28:1093–1100.

Bücher, T., and M. Klingenberg. 1958. Wege des Wasserstoffs in der lebendigen Organisation. *Angewandte Chemie.* 70:552–570.

Casteilla, F., F. Bouillaud, C. Forest, and D. Ricquier. 1989. Nucleotide sequence of a cDNA encoding bovine brown fat uncoupling protein: homology with ADP binding site of ADP/ATP carrier. *Nucleic Acids Research.* 17:2131.

Dalbon, P., G. Brandolin, F. Boulay, J. Hoppe, and P. V. Vignais. 1988. Mapping of the nucleotide-binding sites in the ADP/ATP carrier of beef heart mitochondria by photolabeling with 2-azido adenosine diphosphate. *Biochemistry.* 27:5141–5149.

Eckerskorn, C., and M. Klingenberg. 1987. In the uncoupling protein from brown adipose tissue the C-terminus protrudes to the c-side of the membrane, as shown by tryptic cleavage. *FEBS Letters.* 226:166–170.

Gawaz, M., M. G. Douglas, and M. Klingenberg. 1990. Structure-function studies of adenine nucleotide transport in mitochondria. II. Biochemical analysis of distinct AAC1 and AAC2 protein in yeast. *Journal of Biological Chemistry.* 265:14202–14208.

Heldt, H. W., M. Klingenberg, and M. Milovancev. 1972. Differences between the ATP/ADP ratios in the mitochondrial matrix and in the extramitochondrial space. *European Journal of Biochemistry.* 30:434–440.

Hendersen, R., J. W. Baldwin, and T. A. Ceska. 1990. Model for the structure of bacteriorhodopsin based on high-resolution electron cryo-microscopy. *Journal of Molecular Biology.* 213:899–929.

Hoffmann, B. 1991. Charakterisierung gezielter Arginin- und Cystein-Mutationen des ADP/ATP Carriers in Hefe. Diploma thesis. University of Munich, Munich, FRG. 70 pp.

Jencks, W. P. 1975. Binding energy, specificity and enzymatic catalysis: the circe effect. *Advances in Enzymology.* 43:219–410.

Kadenbach, B., P. Mende, H. V. J. Kolbe, I. Stipani, and F. Palmieri. 1982. The mitochondrial phosphate carrier has an essential requirement for cardiolipin. *FEBS Letters.* 139:109–112.

Klingenberg, M. 1976. The ADP/ATP carrier in mitochondrial membranes. *In* The Enzymes of Biological Membranes: Membrane Transport. A. N. Martonosi, editor. Plenum Publishing Corp., New York. 383–438.

Klingenberg, M. 1981. Membrane protein oligomeric structure and transport function. *Nature.* 290:449–454.

Klingenberg, M. 1989. Molecular aspects of the adenine nucleotide carrier from mitochondria. *Archives of Biochemistry and Biophysics.* 270:1–14.

Klingenberg, M. 1991. Mechanistic and energetic aspects of carrier catalysis exemplified with mitochondrial translocators. *In* A Study of Enzymes. Vol. II. S. A. Kuby, editor. CRC Press, Boca Raton, FL. 367–388.

Klingenberg, M. 1992. Structure-function of the ADP/ATP carrier. *Biochemical Society Transactions.* 20:547–550.

Klingenberg, M., M. Gawaz, M. G. Douglas, and J. E. Lawson. 1992. Mutagenized ADP/ATP carrier from Saccharomyces. *In* Molecular Mechanisms of Transport. E. Quagliariello and F. Palmieri, editors. Elsevier Science Publishers B.V., Amsterdam. 187–195.

Klingenberg, M., P. Riccio, H. Aquila, B. B. Buchanan, and K. Grebe. 1976. Mechanism of carrier transport and the ADP/ATP carrier. *In* The Structural Basis of Membrane Function. Y. Hatefi and L. Djavadi-Ohaniance, editors. Academic Press Inc., New York. 431–438.

Klingenberg, M., and W. Slenczka. 1959. Pyridinnukleotide in Leber-Mitochondrien. *Biochemische Zeitschrift.* 331:486–517.

Kolarov, J., N. Kolarov, and N. Nelson. 1990. A third ADP/ATP translocator gene in yeast. *Journal of Biological Chemistry.* 265:12711–12716.

Krämer, R., and F. Palmieri. 1989. Molecular aspects of isolated and reconstituted carrier proteins from animal mitochondria. *Biochimica et Biophysica Acta.* 974:1–23.

Lawson, J. E., M. Gawaz, M. Klingenberg, and M. G. Douglas. 1990. Structure-function studies of adenine nucleotide transport in mitochondria. I. Construction and genetic analysis of yeast mutants encoding the ADP/ATP carrier protein of mitochondria. *Journal of Biological Chemistry.* 265:14195–14202.

Mayinger, P., and M. Klingenberg. 1992. Labeling of two different regions of the nucleotide binding site of the uncoupling protein from brown adipose tissue mitochondria with two ATP analogs. *Biochemistry.* 31:10536–10543.

Mayinger, P., E. Winkler, and M. Klingenberg. 1989. The ADP/ATP carrier from yeast (AAC-2) is uniquely suited for the assignment of the binding center by photoaffinity labeling. *FEBS Letters.* 244:421–426.

Menick, D. R., J. A. Lee, R. J. Brooker, T. H. Wilson, and H. R. Kaback. 1987. Role of cysteine residues in the lac permease of Escherichia coli. *Biochemistry.* 26:1132–1136.

O'Malley, K., P. Pratt, J. Robertson, M. Lilly, and M. G. Douglas. 1982. Selection of the nuclear gene for the mitochondrial adenine nucleotide translocator by genetic complementation of the op$_1$ mutation in yeast. *Journal of Biological Chemistry.* 257:2097–2103.

Rees, D. C., H. Komiya, and T. O. Yeates. 1989. The bacterial photosynthetic reaction center as a model for membrane proteins. *Annual Reviews of Biochemistry.* 58:607–633.

Riccio, P., H. Aquila, and M. Klingenberg. 1975. Purification of the carboxy-atractylate binding protein from mitochondria. *FEBS Letters.* 56:133–138.

Runswick, M. J., S. J. Powell, P. Nyren, and J. E. Walker. 1987. Sequence of the bovine mitochondrial phosphate carrier protein: structural relationship to ADP/ATP translocase and the brown fat mitochondria uncoupling protein. *EMBO Journal.* 6:1367–1373.

Runswick, M. J., J. E. Walker, F. Bisaccia, V. Iacobazzi, and F. Palmieri. 1990. Sequence of bovine 2-oxoglutarate/malate carrier protein: structural relationship to other mitochondrial transport proteins. *Biochemistry.* 29:11033–11040.

Saraste, M., and J. E. Walker. 1982. Internal sequence repeats and the path of polypeptide in mitochondrial ADP/ATP translocase. *FEBS Letters.* 144:250–254.

Winkler, E., and M. Klingenberg. 1992. Photoaffinity labeling of the nucleotide-binding site of the uncoupling protein from hamster brown adipose tissue. *European Journal of Biochemistry.* 203:295–304.

Sodium-coupled and
Other Transporters

Chapter 17

The Melibiose Permease of *Escherichia coli:* Importance of the NH$_2$-terminal Domains for Cation Recognition by the Na$^+$/Sugar Cotransporter

Gerard Leblanc, Thierry Pourcher, and Marie-Louise Zani

Laboratoire J. Maetz, Département de Biologie Cellulaire et Moléculaire, Commissariat à l'Energie Atomique, Villefranche-sur-mer, 06230, France

Molecular Biology and Function of Carrier Proteins © 1993 by The Rockefeller University Press

Introduction

The melibiose (*mel*) permease of *Escherichia coli* (Prestige and Pardee, 1965) catalyzes the accumulation of the α-disaccharide melibiose (6-*O*-α-D-galactopyrano-side-D-glucopyranose) by a mechanism of symport or cation-coupled cotransport in response to the electrochemical potential gradient of the coupling ion. As with other cotransporters (or symporters), osmotic work performed by this energy transducer relies on the presence of a mechanism of compulsory coupling between downhill cation inflow and uphill solute influx into the cell (Mitchell, 1970; Skulachev, 1988). In contrast to many other bacterial sugar permeases (Henderson, 1990), however, *mel* permease uses either H$^+$, Na$^+$, or to a lesser extent Li$^+$ as the coupling cationic species depending on the ionic environment and/or the sugar transported (Tsuchiya et al., 1977, 1983; Tsuchiya and Wilson, 1978; Wilson and Wilson, 1987). Understanding of the mechanisms behind the cationic selectivity and coupling properties of the *mel* symporter has been furthered in the past by taking advantage of these unusual coupling characteristics. In addition, the recent cloning and sequencing of the *mel B* gene which encodes *mel* permease (Hanatani et al., 1984; Yazyu et al., 1984) has made the study of the *mel* permease structure–function relationship possible by utilizing molecular biology technology. The purpose of this chapter is to review recent progress in our understanding of the molecular mechanism of symport by *mel* permease. In the first part, mechanistic and biochemical properties of the *mel* symporter are briefly surveyed. Recent mutagenesis studies are described which suggest that several acidic residues, putatively clustered in the hydrophobic core of the NH$_2$-terminal domain of the permease, participate in cation recognition by the symporter.

Functional Properties of *Mel* Permease

The α-galactoside melibiose is actively accumulated by *mel* permease in *E. coli* cells or derived cytoplasmic membrane vesicles resuspended in Na$^+$-free medium (Tsuchiya and Wilson, 1978; Bassilana et al., 1985). Addition of NaCl (10 mM) stimulates sugar uptake by reducing the transport constant value K_t (Lopilato et al., 1978; Bassilana et al., 1985). The cations used for sugar cotransport were unambiguously identified by measuring the inward ionic movements that accompany downhill sugar entry in de-energized cells with H$^+$ or Na$^+$ (or Li$^+$)-sensitive electrodes (Tsuchiya and Wilson, 1978; Tsuchiya et al., 1983). In the absence of sodium or lithium salts in the medium, protons are taken up, but when these monovalent ions are added to the medium, proton influx is totally replaced by Na$^+$ or Li$^+$ uptake. Unlike α-galacto-sides, sugars with a β configuration (e.g., methyl 1-thio-β-D-galactopyranoside; TMG) are cotransported with either Na$^+$ or Li$^+$ but not with H$^+$ (Wilson and Wilson, 1987). Measurement of the melibiose transport constants (K_t, V_m) and of their variations as a function of the amplitude of the electrical component ($\Delta\Psi$) of the proton or Na$^+$ (or Li$^+$)-motive force already suggests that the contribution of the three coupling ions to the symport mechanism is variable (Bassilana et al., 1985). For example, the K_t of the H$^+$-linked melibiose transport reaction is ~ 10 times higher than the Na$^+$-or Li$^+$-coupled one, while V_{max} varies in the order $V_{[Na^+]}^{mel;} = V_{[H^+]}^{mel} > V_{[Li^+]}^{mel}$. Also, $\Delta\Psi$ has a V_{max} effect on Na$^+$- or Li$^+$-coupled transport activity, whereas it has a K_t effect on the H$^+$-coupled reaction (Leblanc et al., 1988).

The sugar-binding properties of *mel* permease are conveniently assessed by measuring binding of the high affinity ligand p-nitrophenyl-α-D-galactopyranoside (NPG) on de-energized membrane vesicles (Cohn and Kaback, 1980; Damiano-Forano et al., 1986). NPG binding is competitively inhibited by melibiose or TMG and is specifically enhanced by Na^+ or Li^+ (affinity increase), H^+, Na^+, and Li^+ competing for the same cationic site. A 1:1 Na^+/sugar stoichiometry deduced from the sugar-binding assay is in agreement with the transport stoichiometry calculated from simultaneous measurements of cosubstrate flows in de-energized membrane vesicles (Bassilana et al., 1987).

Mechanistic analyses of the Na^+-coupled sugar transport reaction in de-energized membrane vesicles (downhill sugar influx or efflux, sugar exchange at equilibrium) show that the carrier functions asymmetrically; i.e., Na^+-coupled efflux is much faster than Na^+-coupled influx (Bassilana et al., 1987; Leblanc et al., 1988). Moreover, sugar influx is strongly stimulated by internal sugar (*trans*-stimulation). Strikingly, sugar–sugar exchange by the permease occurs without concomitant translocation of Na^+ ions. These data suggest that in the absence of energy, release of Na^+ in the cytoplasmic compartment is strongly rate limiting, indicating some form of Na^+ occlusion. In active transport conditions, the membrane potential increases Na^+-coupled sugar influx by accelerating the rate of Na^+ (or Li^+) release in the internal compartment. In contrast, permease cycling during H^+ symport is apparently not limited by H^+ release.

Characterization of the Na^+-dependent sugar transport reaction suggested a kinetic model in which binding of cation and sugar to the permease on the outer surface of the membrane and their release into the cytoplasm are ordered processes, with Na^+ binding first and being released last (for review, see Leblanc et al., 1988). It is not known whether the process is statistical or steric or whether ordered binding or release of the cosubstrates is also characteristic of the H^+- or Li^+-coupled modes of symport by *mel* permease. Finally, the stability of the ternary complex cation–sugar–permease varies according to the chemical identity of the coupling ion ($H^+ < Na^+ < Li^+$). This stability may govern the rate of cosubstrate release into the cytoplasm, and hence of permease cycling, and is controlled by the electrical potential (Leblanc et al., 1988).

Identification and Secondary Structure of *Mel* Permease

Deduction of the primary sequence of *mel* permease from the *mel B* gene sequence suggests that the transport polypeptide is very hydrophobic and consists of 469 amino acid residues (52 kD; Yazyu et al., 1984). The transport protein has been identified by DNA recombinant and immunological techniques as an inner membrane protein with an apparent molecular mass of 38,000–40,000 (Botfield and Wilson, 1989; Pourcher et al., 1990b), the difference between molecular weight and relative molecular mass being typical of prokaryote permeases. A secondary structure model for *mel* permease consisting of 12 membrane-spanning segments in an α-conformation has been proposed on the basis of various criteria (Fig. 1, and Pourcher et al., 1990a, c). First, hydropathy profiling suggests that the polypeptide contains at least 10 hydrophobic stretches connected by hydrophilic charged segments (Yazyu et al., 1984). On the other hand, polyclonal antipeptide antibodies directed against the COOH-terminal domain only react with permease in inverted membrane vesicles, suggesting that the hydrophilic tail of the transport protein is exposed to the cytoplasmic compartment (Botfield and Wilson, 1989). In addition, the lack of a sequence signal suggests that the NH_2 terminus also faces the cytoplasmic space. Colocalization of the COOH and NH_2 termini on the inner surface of the membrane dictates that the polypeptide must pass the membrane an even number of times. Recent topological studies of *mel* permease based on genetic fusion of the permease with alkaline phosphatase provide experimental support for a secondary structure model with 12 membrane-spanning segments (Botfield et al., 1992).

  ·   The size, the strongly hydrophobic character, and a secondary structure model containing 12 putative transmembrane segments of *mel* permease are all properties shared by other permeases of *E. coli* (Kaback, 1989; Henderson, 1990). However, there is little homology between the primary structures of *mel* permease and most carriers of *E. coli,* including *Lac* permease, which has many sugar substrates in common with *mel* permease (Yazyu et al., 1984). In contrast to this, a significant degree of homology has been reported between *mel* permease and either the NH_2 domain of the lactose transporter of *Streptococcus thermophilus* (*LacS* carrier; Poolman et al., 1989) or the *gusB* protein that effects uptake of D-glucuronides in *E. coli* by a mechanism not entirely clarified (Henderson, P. J. F., personal communication).

Point Mutations Altering the Cationic Selectivity of *Mel* Permease

Point mutations altering the cationic selectivity of *mel* permease have been described (Niiya et al., 1982; Shiota et al., 1984, 1985; Yazyu et al., 1985; Kawakami et al., 1988). A first set of mutants resisting Li^+ toxicity groups permeases that catalyze normal Na^+-coupled transport but do not use H^+ as coupling ion and have variable capacity for cotransporting sugar with Li^+ ions. The mutated amino acids are located either in putative membrane-spanning segments (Pro122 in helix IV) or in cytoplasmic loops connecting membrane helices (Pro 142, Leu 232, or Ala 236). In another group of mutants, melibiose utilization is no longer inhibited by the nonmetabolized sugar analogue TMG (Botfield and Wilson, 1988). Permeases of this latter group display simultaneous alteration of sugar and cation specificities and the corresponding point mutations are found in four discrete clusters in the protein (helix I, II, or IV, or cytoplasmic loops connecting helices X and XI). Such a dual effect on ionic and sugar substrate specificities has been taken as evidence for the existence of a physical interaction between the sugar and cation species and/or binding sites. Unfortunately, the wide distribution of the mutations throughout the protein and the incomplete characterization of their effects make understanding of their precise molecular participation in the cationic properties of *mel* permease unclear. Finally, site-directed mutagenesis has also been used to change residues frequently involved in metal–protein interactions such as His or Cys. Substitution of serine for Cys 106, 231, and 360 residues does not modify melibiose transport activity, and change of Cys 360 into Ser may influence, in an unspecified fashion, Na^+-coupled cotransport (Botfield, 1989). Finally, it has recently been reported that replacement of any of the seven His residues of *mel* permease by Arg, and in the case of His 94 also by Asn or Gln, does not change the cationic selectivity properties of *mel* permease (Pourcher et al., 1990*c*, 1992).

Mutation of Acidic Residues Buried in the Hydrophobic Core of *Mel* Permease

Examination of the amino acid composition of the putative membrane-spanning segments of *mel* permease (Pourcher et al., 1990*c*) suggests that most of the charged residues buried in the apolar core of the transport protein are distributed in membrane segments I–IV of the transport protein. Of these, four aspartic acids are located on either helix I (Asp 31), helix II (Asp 51 and Asp 55), or helix IV (Asp 120) (Fig. 1). Helices I–IV are amphiphatic and contain additional polar amino acids, some of which are aligned with these acidic residues. From this observation, and the fact that some point mutations altering the cationic properties of *mel* permease are located in the NH₂-terminal domain (Yazyu et al., 1985; Botfield and Wilson, 1988), it was tentatively advanced that such a cluster of negatively charged residues may be involved in recognition and/or translocation of the coupling cation. To test this

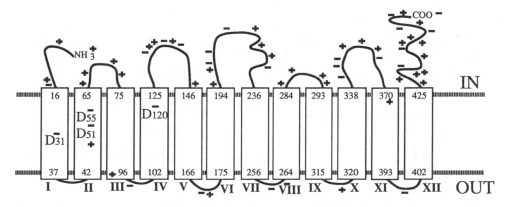

Figure 1. Secondary structure model of *mel* permease highlighting the aspartic residues located in putative membrane-spanning helices of the NH₂-terminal domain. Helices are represented as rectangles and numbered with Roman numerals. Top and bottom numbers in each rectangle correspond to the first and last residues of the helix. Aspartic residues in membrane-spanning segments are indicated by the symbol D followed by a number that indicates its position on the primary amino acid sequence.

hypothesis, site-directed mutagenesis was first used to replace individually Asp 31, Asp 51, Asp 55, and Asp 120 by uncharged residues (Pourcher et al., 1991, 1993*a*; Zani et al., 1993). Cys or Asp residue was selected as the substitute since the Cys residue is less bulky than Asp and more likely to be protonated at a physiological pH, and the Asp to Asn change corresponds to an isosteric substitution.[1] The effects of exchanging each Asp residue by a glutamic acid were then analyzed.

Permeases Carrying Asp to Cys or Asn Mutations

The consequence of these various mutations on the catalytic activity of *mel* permease was analyzed by measuring the obligatory Na⁺-coupled TMG transport reaction in

---

[1] Site-directed mutants are designated as follows: the one-letter amino acid code is used followed by a number indicating the position of the residue in wild-type *mel* permease. The sequence is followed by a second letter denoting the amino acid replacement at this position.

cells incubated in the presence of TMG (0.2 or 1 mM) and NaCl (10 mM). As shown in Fig. 2, individual replacement of Asp at position 31, 51, 55, or 120 by Cys produces complete inhibition of TMG accumulation. A similar transport defect was observed in D51N, D55N, and D120N cells. Although expression of the mutant permeases was generally reduced with respect to that of wild-type permease (from 20 to 80%), the membrane permease concentrations were too high to account by themselves for the lack of transport activity. In contrast to these mutants, D31N permease promotes TMG accumulation up to 13% of that measured in wild-type cells. This reduced transport is in proportion to the decreased quantity of permease molecules measured in the mutant cell membrane (10%).

To assess if the lack of TMG transport by the majority of the mutant permeases is the result of a defect in recognition of the sugar, of the coupling ion, or of both,

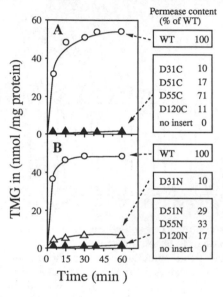

Figure 2. Na^+-dependent [^{14}C]TMG transport in cells expressing *mel* permease with Cys or Asn in place of Asp 31, 51, 55, or 120. *E. coli* cells expressing each mutant permease were equilibrated at 2 mg protein·ml^{-1} in 100 mM K phosphate (pH 6.6), 10 mM $MgSO_4$, and 10 mM NaCl at 20°C. Transport reactions were initiated by addition of [^{14}C]TMG (3 mCi·$mmol^{-1}$) to a final concentration of 0.2 mM to cell suspensions flushed with a stream of oxygen. Reactions were terminated by diluting the assay samples followed by immediate filtration. (*Top*) ▲, cells expressing D31C, D51C, D55C, or D120C permease or no permease (no insert); ○, wild-type permease (*WT*); (*bottom*) ▲, cells expressing D51N, D55N, or D120N permease or no permease (no insert); △, D31N permease; ○, wild-type permease (*WT*). Membrane permease contents were assessed by measuring binding of the high affinity ligand NPG. The results are expressed as percentage of the content found in wild-type membranes.

binding of cosubstrates on de-energized membrane vesicles was studied (Pourcher et al., 1993). Sugar binding was measured using NPG as a ligand (Damiano-Forano et al., 1986). On the other hand, owing to the poor affinity of *mel* permease for Na^+ ions (> 300 μM) and the unfavorable specific signal/background ratio, Na^+ binding had to be assessed indirectly by measuring the dependence of NPG binding activity on the presence of NaCl.

NPG binding experiments were carried out in the absence or presence of 10 mM NaCl or 1 M NaCl, the variation in medium ionic strength being compensated by addition of equimolar amounts of KCl. Table I shows that the NPG dissociation constants (K_d) measured on mutant and wild-type membranes resuspended in Na^+-free conditions are comparable (~ 1–2 μM). The only exception is that of D120N permease, which was slightly higher (6 μM). These data indicate that replacing any of these Asp residues by a neutral amino acid (Cys or Asn) has no major effect on the basic (i.e., Na+-independent) sugar recognition properties of the permease. In contrast, these substitutions greatly modify the dependence of NPG binding on NaCl. Thus, while raising the NaCl concentration to 1 M causes an eightfold reduction in the K_d for NPG binding on wild-type permease, it does not produce a significant change in the K_d value when the *mel* permease carries a Cys residue in place of either Asp residue, or when the permease has an Asn residue in place of Asp 51, Asp 55, or Asp 120. On the other hand, D31N permease retains Na+-dependent NPG binding activity. However, measurement of the Na+ activation constant (K_{Na^+}) for NPG binding on this mutant yields a K_{Na^+} value at least four times

TABLE I
Effect of NaCl on [³H]NPG Binding to Rightside-out Membrane Vesicles Carrying Permease with Cys or Asn in Place of Either Asp 31, Asp 51, Asp 55, or Asp 120

Permease*	NPG binding constant (K_d)		Relative affinity change
	Na+-free (KCl)	NaCl	
	μ*M*		
WT	1.6	0.2	8
D31C	1.5	1.5	1
D31N	1.6	0.3	5
D51C	2.4	1.3	1.8
D51N	1.3	2	0.7
D55C	2.1	1.6	1.3
D55N	2.4	3	0.8
D120C	3	3.2	0.9
D120N	6	4	1.5

Rightside-out membrane vesicles containing D31C, D51C, D55C, or D120C permease or D31N, D51N, D55N, or D120N permease were equilibrated in 0.1 M potassium phosphate solution (pH 6.6) containing 1 M KCl (or 1 M NaCl), 10 mM MgSO₄, and trace amounts of sodium salt (< 20 μM). Membrane vesicles were de-energized in the presence of FCCP (5 mM) and valinomycin (5 mM) and concentrated to ~30 mg protein · ml⁻¹). [³H]NPG binding was measured as a function of free NPG concentration (0.2–10 mM) using a flow dialysis technique (Damiano-Forano et al., 1986). Apparent K_d's for NPG binding were calculated graphically from Eadie-Hofstee representations of the data.
*Relative changes in NPG affinity were expressed as a ratio of the K_d's measured in KCl vs. NaCl media.

higher than that of wild-type permease (1.4 and 0.3 mM, respectively), suggesting that D31N permease has a decreased affinity for sodium ions. Overall, the data indicate that mutation of these acidic residues decreases or inhibits the activating effect of sodium ions on sugar binding.

Affinities of the modified permeases for the physiological sugar melibiose and the nonmetabolized analogue TMG were also estimated. Since the apparent K_d for melibiose or TMG was too high to be measured directly (0.2 and 0.6 mM, respectively), affinity changes were estimated from the competitive inhibitory effect of these sugars (K_i^{mel} K_i^{TMG}, respectively) on NPG binding (Damiano-Forano et al., 1986). In membrane expressing wild-type permease, raising the concentration of NaCl to 1 M decreases K_i^{mel} from 1.4 to 0.2 mM and K_i^{TMG} from 3.5 to 0.7 mM. Similar analysis on membranes carrying mutant permeases indicates that melibiose or TMG binds in a Na$^+$-dependent fashion on D31N permease. In contrast, affinity of the other mutant permeases for melibiose or for TMG is independent of the NaCl concentration of the medium and the observed K_i^{mel} and K_i^{TMG} values were in general comparable to those measured on wild-type permease in a Na$^+$-free medium. Strikingly, permease with Cys or Asn at position 55 displays moderate changes in the affinity for TMG but not for melibiose. As a whole, the results indicate that melibiose and TMG binding on all permeases except D31N are Na$^+$ independent.

Mel permease uses alternatively H$^+$, Na$^+$, or Li$^+$ as the coupling ion for melibiose transport (Fig. 3). Examination of the initial rate and/or maximal extent of melibiose uptake indicates that the transport efficiency of wild-type permease

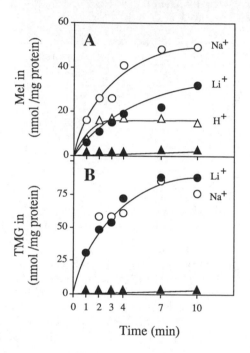

Time (min)

Figure 3. Time courses of melibiose and TMG accumulation coupled to H$^+$, Na$^+$, or Li$^+$ in cells expressing wild-type permease. [^{3}H]Melibiose (*A*) or [^{14}C]TMG (*B*) transport was measured essentially as described in the legend of Fig. 2. Sugar substrates were added into the medium at a final concentration of 0.8 mM. Na$^+$- and Li$^+$-coupled transport activities were measured in the presence of NaCl (○) or LiCl (●) at a final concentration of 10 mM. H$^+$-coupled transport (△) was measured in a medium containing < 20 μM NaCl and no lithium salts (Damiano-Forano et al., 1986). Control experiments (▲) were performed on cells previously de-energized in the presence of the uncoupler carbonyl cyanide *p*-trifluoro-methoxy-phenylhydrazone (FCCP, 50 μM).

depends on the coupling ion and decreases in the order Na$^+$ > Li$^+$ = H$^+$, Na$^+$-melibiose transport being ~2.5 times more rapid than the H$^+$- or Li$^+$-linked reactions (Fig. 3 *A*). In addition, the cationic selectivity of *mel* permease is sugar specific, as TMG is equally well transported with Na$^+$ or Li$^+$ but not with H$^+$ (Fig. 3 *B*). The cationic selectivity profiles of the different mutant permeases were analyzed. The results indicate that most of the substitutions produce dramatic inhibition of melibiose transport coupled to either Na$^+$, H$^+$, or Li$^+$, or of TMG transport coupled to Na$^+$ or Li$^+$. Fig. 4 *B* illustrates this inactivation in the particular case of permease D55C. There were, however, two exceptions. First, D31N permease retains the capacity to use the three different coupling ions for melibiose transport (data not shown). Second, D51C permease mediates a small but significant H$^+$-driven melibiose accumulation and neither the time course nor the extent of melibiose accumulation is stimulated on adding sodium or lithium salts to the medium (Fig. 4 *A*). Also,

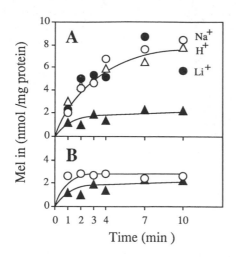

Figure 4. Time course of melibiose accumulation coupled to H^+, Na^+, or Li^+ in cells expressing D51C or D55C permease. (*A*) [^{3}H]Melibiose transport in cells expressing D51C permease was measured as described in the legend of Fig. 3 using the sugar at a final concentration of 0.8 mM. NaCl (○) or LiCl (●) was added at a final concentration of 10 mM or omitted (△). Control experiments (▲) were performed on cells previously de-energized in the presence of FCCP (50 μM). (*B*) Transport experiments in D55C cells. For clarity of the figure, time courses of H^+-, Na^+-, and Li^+-melibiose transports in cells expressing D55C permease were represented by a single symbol (○). (▲) De-energized cells.

TMG transport by D51C mutant is completely inhibited (not shown). This observation suggests that the cationic selectivity profile of *mel* permease with a Cys in place of Asp51 changes from $Na^+ > Li^+ = H^+$ to $H^+ > Na^+ = Li^+$.

Permeases Carrying Asp to Glu Mutations

The results reported above suggest that the carboxylate side chain of Asp 51, 55, or 120 may be critical for cation recognition. One can expect that changing the position of each COOH group by replacing individually these Asp residues by a glutamic acid should alter *mel* permease function and most probably its cationic selectivity. While such mutants were being engineered in the laboratory to test this hypothesis, preliminary study of the transport properties of *mel* permease carrying either a Glu residue or uncharged amino acids (Ser, Gln, Tyr, Lys, Leu, Ala, or His) at position 51 or 120 was reported (Wilson and Wilson, 1992). The results show that only permeases carrying a Glu residue in place of Asp 51 or Asp 120 still catalyze melibiose transport, albeit at a reduced rate. This strengthens the conclusion reached above that a residue carrying a negative charge at position 51 or 120 is critical for transport activity. It is, however, unclear from the transport assay conditions used in this study whether melibiose is cotransported with H^+ or Na^+ by the mutants and whether the Asp to Glu change has an effect on the cationic selectivity of *mel* permease. To answer this question, the cationic selectivity properties of *mel* permease with a Glu at position 51, 120, or 55 were analyzed using our site-directed mutants. H^+-coupled melibiose symport was measured in cells incubated in a medium containing <50 μM NaCl and no LiCl, whereas the Na^+- or Li^+-coupled melibiose (or TMG) symport activity was assessed in the presence of 10 mM NaCl or LiCl. All media contained the sugar at a final concentration of 0.8 mM (Fig. 5). The data are presented as histograms corresponding to the amount of sugar accumulated during the first 2 min in each ionic condition. The data suggest that each independent substitution modifies the cationic selectivity profile of *mel* permease. Moreover, the change observed depends on the acidic residue replaced (Fig. 5, *A* and *B*). In control experiments with wild-type membranes, the selectivity profile of the permease is $Na^+ > H^+ = Li^+$ when melibiose is the substrate and $Na^+ = Li^+ \gg H^+$ for TMG transport. The data in Fig. 5 *A* indicate that the cationic selectivity of D55E permease is probably $Na^+ = Li^+ > H^+$. Thus, Li^+ is a better coupling ion for melibiose symport by this mutant permease than by wild-type permease. In contrast, Na^+ and Li^+ have equal coupling efficiencies for TMG transport by this mutant permease (Fig. 5 *B*). On the other hand, D120E permease no longer cotransports melibiose with H^+ and, in contrast to wild-type permease, couples melibiose transport to Li^+ and Na^+ with equal efficiency. It is of interest to note that mutation of two residues very close to Asp 120 on putative helix IV, namely Pro 122 into Ser (Yazyu et al., 1985) and Y116 into Phe (unpublished observations), also produces loss of H^+ coupling and enhanced Li^+ coupling. Moreover, TMG is no longer a good substrate for D120E (Fig.

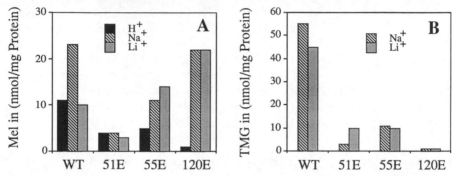

Figure 5. Variations of the rate of melibiose and TMG transport in cells expressing D51E, D55E, or D120E permease as a function of the coupling ion. (*A*) [³H]Melibiose transport was measured in the presence of the sugar at a final concentration of 0.8 mM as described in Fig. 3. Columns correspond to sugar uptake after 2 min of incubation in Na^+- and Li^+-free medium (H^+-coupled transport, *filled columns*), or media containing 10 mM NaCl (Na^+-coupled transport, *hatched columns*) or 10 mM LiCl (Li^+-coupled transport, *cross-hatched columns*). (*B*) [¹⁴C]TMG transport in the presence of 0.8 mM TMG and either 10 mM NaCl (*hatched columns*) or 10 mM LiCl (*cross-hatched columns*).

5 *B*), Pro122S (Yazyu et al., 1985), or Y116F permease (Zani, M. L., and G. Leblanc, unpublished results). This may imply cooperative interaction between the residues of this restricted domain of helix IV. Finally, it was observed that H^+ is used for coupling by D51E permease and the activity of this mutant is not stimulated by Na^+ or Li^+ ions. Strikingly, D51E permease contrasts with D51C permease in that it transports TMG at a five times higher rate in the presence of Li^+ than in the presence of Na^+ (not shown). As a whole, the transport properties of 51E permease suggest that its cationic selectivity profile is $H^+ = Li^+ \gg Na^+$ as opposed to a $Na^+ = H^+ = Li^+$ profile for a neutral residue at the same position (mutant D51C). This analysis indicates that the cationic selectivity profile of *mel* permease is modified on changing the carboxyl group's position of the residue 51, 55, or 120, with the general trend of enhancing coupling of sugar transport to Li^+ ions. Determination of the sugar-binding properties of permeases D51E, D55E, and D120E and the ionic

movements associated with downhill sugar influx in de-energized cells carrying these mutants will be useful in ascertaining that the modification of the cationic dependency of the transport rates effectively corresponds to a change in cationic selectivity of these permeases.

Putative Implication of These Asp Residues in Cation Coordination

The first important conclusion drawn from the experiments reported above is that changing Asp 51, Asp 55, or Asp 120 by a neutral residue (Cys or Asn) does not produce major modification of the affinity of the permease for the sugar substrate in Na^+-free conditions, i.e., of the basic mechanism of sugar recognition by *mel* permease. This refutes the suggestion that these substitutions produce an overall modification of the structure of *mel* permease. The second remarkable feature is that these mutations modify (D31N mutant) or prevent (all other mutants) stimulation of sugar binding by Na^+ ions. These findings suggest that an acidic residue at positions 51, 55, and 120, and to a lesser degree at position 31, is essential to confer a Na^+-dependent character to the sugar binding and transport process. Moreover, since the mutations already affect the very initial step of the transport reaction corresponding to cosubstrate binding, it is reasonable to suppose that the modified Asp residues at positions 31, 51, 55, and 120 are located at or near the cationic binding site of the permease and are involved in cation recognition. The cationic selectivity changes observed on displacing the carboxylate side chain of Asp 51, Asp 55, or Asp 120 as a result of individual Asp to Glu substitution also points to participation of these residues in recognition of the coupling ion. As a corollary, the putative distribution of these acidic amino acids in adjacent membrane-spanning segments of the NH_2-terminal domain of *mel* permease further suggests that this domain accommodates the cationic binding site, or at least is part of it.

Of interest for elucidating the molecular mechanism of cation recognition by *mel* permease is the observation that mutations directed on more than one acidic residue of this domain lead to similar transformations of the Na^+-dependent sugar binding reaction into a Na^+-independent one. This may indicate that these residues are organized as a cage in which the oxygen atoms of the carboxyl groups act as ligands for coordination of the coupling ion. Obviously, the data presented here do not exclude the possibility that one Asp residue (e.g., Asp 31), or more than one, participate in the topological organization of the cationic binding site and thus do not interact directly with the cation. That appropriately placed oxygen (or nitrogen) atoms in symporters may coordinate the coupling ion has already been envisaged by Boyer (1988). This author also suggested that modest rearrangements in the binding site organization would permit accommodation of different cations (i.e., Na^+, Li^+, and H^+ in the particular case of *mel* permease) and that H_3O^+ rather than H^+ is used as a coupling species. In this context, the well-described atomic interactions existing in the cationic coordination complexes formed by crown-ether or cryptates are relevant to speculation on the molecular basis of cation–protein interactions in the cationic binding site of symporters. This, for example, is the case of the tetracarboxylic 18-crown-6 complexes with K^+, Na^+, or H_3O^+, the structure of which reveals striking differences between K^+-, Na^+-, or H_3O^+-host atomic interactions. Crystallographic studies of these complexes indeed indicate that K^+ is coordinated to the six ether-oxygens of the compound, Na^+ to five of them, and finally, H_3O^+ to only three ether-oxygens via hydrogen bond interactions with a more exacting pyramidal geometry, and further stabilized by ion-dipolar interactions with the three remaining O atoms (Behr et al., 1982*a*, *b*; Hilgenfeld and Saenger, 1982). This could signify that only three of the four Asp residues are used for H_3O^+ coordination in the cationic site of *mel* permease, while one additional acidic residue of the site may be necessary for coordination of Na^+ or Li^+. If this is the case, mutation of Asp residues involved in coordination of both Na^+ and H_3O^+ should simultaneously modify H^+- and Na^+-linked transport reactions, whereas removing Asp residues that only participate in Na^+ or Li^+ coordination should selectively impair the Na^+-linked transport reaction. Some of the results reported here seem to be consistent with this suggestion. Thus, the observation that D51C permease catalyzes H^+-linked but not Na^+- or Li^+-linked transport reactions suggests that this acidic residue is necessary for coordination of Na^+ or Li^+ but not of H_3O^+. On the other hand, Asp 55 and Asp 120 seem to be important for coordination of either coupling ion as their replacement by neutral residues simultaneously inhibits H^+- and Na^+-linked sugar binding and transport by the symporter.

It should also be pointed out that the change in cationic selectivity of *mel* permease, and in particular enhanced Li^+-coupling, which results from individual changes of several Asp residues of the NH_2-terminal domain of the symporter (Asp 51, Asp 55, or Asp 120) into a glutamic residue, is also consistent with the proposal that they are elements of a cation coordination site. Indeed, the change in position of the carboxyl group produced by replacing the aspartic acid by glutamic acid should reduce the internal size of the site and consequently favors accommodation of alkali ions with small radii such as Li^+ ions. Finally, it is important to note that some of the changes in cationic selectivity are sugar specific, an effect which may reflect an intimate relationship between the sugar recognition site and the cation recognition site (Botfield and Wilson, 1988).

The absence of precise knowledge of the three-dimensional structure of *mel* permease makes interpretation of these individual site-directed mutations speculative. However, combination of the information obtained from independent mutations may permit identication of domains of the symporter involved in catalytic activity and suggest models stimulating further investigation. For instance, the data presented in this communication strongly suggest that several negatively charged residues distributed in putative membrane-spanning segments of the NH_2-terminal domain of *mel* permease are involved in recognition of the coupling ions by the symporter, possibly by a coordinated mechanism. It follows that the *mel* symporter has to be added to the already long list of cation or H^+ (or H_3O^+)-coupled translocating devices—among them the Na^+/K^+-ATPase (Karlish et al., 1991) and the Ca^{2+}-ATPase+ (Clarke et al., 1989), the Fo sector of *E. coli* H^+-translocating ATPase (Hope, 1982), the lactose permease (Kaback, 1989), the tetracycline antiporter of *E. coli* (Yamaguchi et al., 1992), and bacteriorhodopsin (Khorana, 1988)— for which Asp and/or Glu residues lying in membrane segments have been shown to be essential for recognition and/or translocation of the charged species across the membrane.

Acknowledgments

R. Lemonnier and P. Lahitette (Laboratoire J. Maetz, CEA, Villefranche sur mer, France) are acknowledged for excellent technical assistance.

This work was supported in part by a grant from the Centre de la Recherche Scientifique France (associated with Commissariat à l'Energie Atomique, DO 638).

References

Bassilana, M., E. Damiano-Forano, and G. Leblanc. 1985. Effect of membrane potential on the kinetic parameters of the Na⁺ or H⁺ melibiose symport in *Escherichia coli* membrane vesicles. *Biochemical and Biophysical Research Communications.* 129:626–631.

Bassilana, M., T. Pourcher, and G. Leblanc. 1987. Facilitated diffusion properties of melibiose permease in *Escherichia coli* membrane vesicles: release of co-substrates is rate limiting for permease cycling. *Journal of Biological Chemistry.* 262:16865–16870.

Behr, J. P., P. Dumas, and D. Moras. 1982a. The H₃O⁺ cation: molecular structure of an oxonium-macrocyclic polyether complex. *Journal of the American Chemical Society.* 104:4540–4543.

Behr, J. P., J. M. Lehn, A. C. Dock, and D. Moras. 1982b. Crystal structure of a polyfunctional macrocyclic K⁺ complex provides a solid-state model for a K⁺ channel. *Nature.* 295:526–527.

Botfield, M. C. 1989. Structure/Function of the Melibiose Carrier of *E. coli*. Ph. D. thesis. Harvard Medical University, Cambridge, MA. 142–181.

Botfield, M. C., K. Naguchi, T. Tsuchiya, and T. H. Wilson. 1992. Membrane topology of the melibiose carrier of *Escherichia coli*. *Journal of Biological Chemistry.* 267:1818–1822.

Botfield, M. C., and T. H. Wilson. 1988. Mutations that simultaneously alter both sugar and cation specificity in the melibiose carrier of *Escherichia coli*. *Journal of Biological Chemistry.* 263:12909–12915.

Botfield, M. C., and T. H. Wilson. 1989. Peptide-specific antibody for the melibiose carrier of *Escherichia coli* localizes the carboxyterminus to the cytoplasmic face of the membrane. *Journal of Biological Chemistry.* 264:11649–11652.

Boyer, P. D. 1988. Bioenergetic coupling to protomotive force: should we be considering hydronium ion coordination and not group protonation? *Trends in Biochemical Sciences.* 13:5–7.

Clarke, D. M., T. W. Loo, G. Inesi, and D. H. MacLennan. 1989. Location of the high affinity Ca²⁺-binding sites within the predicted transmembrane domain of the sarcoplasmic reticulum Ca²⁺-ATPase. *Nature.* 339:476–478.

Cohn, D. E., and H. R. Kaback. 1980. Mechanism of the melibiose porter in membrane vesicles of *Escherichia coli*. *Biochemistry.* 19:4237–4243.

Damiano-Forano, E., M. Bassilana, and G. Leblanc. 1986. Sugar binding properties of the melibiose permease in *Escherichia coli* membrane vesicles. *Journal of Biological Chemistry.* 261:6893–6899.

Hanatani, M., H. Yazyu, S. Shiota-Niiya, Y. Moriyama, H. Kanazawa, M. Futai, and T. Tsuchiya. 1984. Physical and genetic characterization of the melibiose operon and identification of the gene products in *Escherichia coli*. *Journal of Biological Chemistry.* 259:1807–1812.

Henderson, P. J. F. 1990. Proton-linked sugar transport in bacteria. *Journal of Bioenergetics and Biomembranes.* 22:525–569.

Hilgenfeld, R., and W. Saenger. 1982. Structural chemistry of natural and synthetic iono-phores and their complexes with cations. *Topics in Current Chemistry.* 101:3–75.

Hope, J., H. U. Schairer, P. Friedl, and W. Sebald. 1982. An Asp-Asn substitution in the proteolipid subunit of the ATP-synthase from Escherichia coli leads to a non-functional proton channel. *FEBS Letters.* 145:21–24.

Kaback, H. R. 1989. Molecular biology of active transport: from membrane to molecule to mechanism. *The Harvey Lectures.* 83:77–105.

Karlish, S. J. D., R. Goldshleger, D. M. Tal, and W. D. Stein. 1991. Structure of the cation binding sites of Na/K ATPase. *In* The Sodium Pump: Structure, Mechanism, and Regulation. P. De Weer and J. H. Kaplan, editors. Rockefeller University Press, New York. 129–141.

Kawakami, T., Y. Akizawa, T. Ishikawa, T. Shimamoto, M. Tsuda, and T. Tsuchiya. 1988. Amino acid substitutions and alteration in cation specificity in the melibiose carrier of *Escherichia coli. Journal of Biological Chemistry.* 263:14276–14280.

Khorana, H. G. 1988. Bacteriorhodopsin, a membrane protein that uses light to translocate protons. *Journal of Biological Chemistry.* 263:7439–7442.

Leblanc, G., M. Bassilana, and T. Pourcher. 1988. Na^+, H^+ or Li^+ coupled melibiose transport in *Escherichia coli. In* Molecular Basis of Biomembrane Transport. F. Palmieri and E. Quagliariello, editors. Elsevier Science Publishers B. V., Amsterdam. 53–62.

Lopilato, J., T. Tsuchiya, and T. H. Wilson. 1978. Role of Na^+ and Li^+ in thiomethylgalacto-side transport by the melibiose transport system of *Escherichia coli. Journal of Bacteriology.* 134:147–156.

Mitchell, P. 1970. Reversible coupling between transport and chemical reactions. *In* Membrane and Ion Transport. E. E. Bittar, editor. John Wiley & Sons, Inc., New York. 192–256.

Niiya, S., K. Yamasaki, T. H. Wilson, and T. Tsuchiya. 1982. Altered cation coupling to melibiose transport in mutants of *Escherichia coli. Journal of Biological Chemistry.* 257:8902–8906.

Poolman, B., T. J. Royer, S. E. Mainzer, and B. F. Schmidt. 1989. Lactose transport system of *Streptococcus thermophilus. Journal of Bacteriology.* 71:244–253.

Pourcher, T., M. Bassilana, H. K. Sarkar, H. R. Kaback, and G. Leblanc. 1990a. Na^+/melibiose symport mechanism of *Escherichia coli:* kinetic and molecular properties. *Philosophical Transactions of the Royal Society of London.* B326:411–423.

Pourcher, T., M. Bassilana, H. R. Sarkar, H. R. Kaback, and G. Leblanc. 1990b. Melibiose permease and a-galactosidase of *Escherichia coli:* identification by selective labeling using the T7 RNA polymerase/promoter expression system. *Biochemistry.* 29:690–696.

Pourcher, T., M. Bassilana, H. R. Sarkar, H. R. Kaback, and G. Leblanc. 1992. The melibiose permease of Escherichia coli: mutations of His94 alter insertion and stability into the membrane rather than catalytic activity. *Biochemistry.* 31:5225–5231.

Pourcher, T., M. Deckert, M. Bassilana, and G. Leblanc. 1991. Melibiose permease of *Escherichia coli:* mutation of Aspartic acid 55 in putative helix II abolishes activation of sugar binding by Na^+ ions. *Biochemical and Biophysical Research Communications.* 178:1176–1181.

Pourcher, T., H. K. Sarkar, M. Bassilana, H. R. Kaback, and G. Leblanc. 1990c. Histidine-94 is the only important histidine residue in the melibiose permease of *Escherichia coli. Proceedings of the National Academy of Sciences, USA.* 87:468–472.

Pourcher, T., M. L. Zani, and G. Leblanc. 1993. Mutagenesis of acidic residues in putative membrane-spanning segments of the melibiose permease of Escherichia coli: effects on Na^+-dependent transport and binding properties. *Journal of Biological Chemistry.* In press.

Prestidge, L. S., and A. B. Pardee. 1965. A second permease for methyl-thio-b-D-galactoside: *E. coli. Biochimica et Biophysica Acta.* 100:591–593.

Shiota, S., Y. Yamane, M. Futai, and T. Tsuchiya. 1985. *Escherichia coli* mutants possessing an Li+-resistant melibiose carrier. *Journal of Bacteriology.* 162:106–109.

Shiota, S., H. Yazyu, and T. Tsuchiya. 1984. Escherichia coli mutants with altered cation recognition by the melibiose carrier. *Journal of Bacteriology.* 160:445–447.

Skulachev, V. P. 1988. Membrane Bioenegetics. Springer-Verlag, Berlin, Heidelberg, New York. 301–313.

Tsuchiya, T., M. Oho, and S. Shiota-Niiya. 1983. Lithium ion-sugar cotransport via the melibiose transport in *Escherichia coli. Journal of Biological Chemistry.* 258:12765–12767.

Tsuchiya, T., J. Raven, and T. H. Wilson. 1977. Co-transport of Na+ and methyl-β-D-thiogalactopyranoside mediated by the melibiose transport system of *E. coli. Biochemical and Biophysical Research Communications.* 76:26–31.

Tsuchiya, T., and T. H. Wilson. 1978. Cation-sugar co-transport in the melibiose system of *Escherichia coli. Membrane Biochemistry.* 2:63–79.

Wilson, D. M., and T. H. Wilson. 1987. Cation specificity for sugar substrates of the melibiose carrier of *Escherichia coli. Biochimica et Biophysica Acta.* 904:191–200.

Wilson, D. M., and T. H. Wilson. 1992. Asp-51 and Asp-120 are important for the transport function of the *Escherichia coli* melibiose permease. *Journal of Bacteriology.* 174:3083–3085.

Yamaguchi, A., T. Akasaka, N. Ono, Y. Someya, M. Nakatani, and T. Sawai. 1992. Metal-tetracycline/H+ antiporter of escherichia coli encoded by transposon Tn10: roles of the aspartyl residues located in the putative transmembrane helices. *Journal of Biological Chemistry.* 267:7490–7498.

Yazyu, H., T. Shiota-Niiya, T. Shimamoto, H. Kanazawa, M. Futai, and T. Tsuchiya. 1984. Nucleotide sequence of the *melB* gene and characteristics of deduced amino acid sequence of the melibiose carrier in *Escherichia coli. Journal of Biological Chemistry.* 259:4320–4326.

Yazyu, H., S. Shiota, M. Futai, and T. Tsuchiya. 1985. Alteration in cation specificity of the melibiose transport carrier of *Escherichia coli* due to remplacement of proline 122 with serine. *Journal of Bacteriology.* 162:933–937.

Zani, M. L., T. Pourcher, and G. Leblanc. 1993. Mutagenesis of acidic residues in putative membrane-spanning segments of the melibiose permease of Escherichia coli. II. Effects on cationic selectivity and coupling properties. *Journal of Biological Chemistry.* In press.

Chapter 18

The Sodium/Glucose Cotransporter (SGLT1)

Ernest M. Wright, Bruce Hirayama, Akihiro Hazama, Donald D. F. Loo, Stephane Supplisson, Eric Turk, and Karl M. Hager

Department of Physiology, UCLA School of Medicine, Los Angeles, California 90024-1751

Molecular Biology and Function of Carrier Proteins © 1993 by The Rockefeller University Press

Introduction

The Na^+/glucose cotransporter is the prototype of a family of membrane proteins that couple the intracellular accumulation of nutrients with an ion electrochemical potential gradient across the plasma membrane. Crane and colleagues proposed in 1960 that the active transport of sugars across the intestinal brush border was coupled to the Na^+ gradient across the membrane (Crane et al., 1961). Over the past 30 years the Na^+/glucose cotransport hypothesis has been confirmed and extended to many other ion-driven transporters in both eukaryotic and prokaryotic cells, e.g., H^+ cotransporters for sugars in bacteria and plants, Na^+ cotransporters for amino acids in bacteria and animals, and Na^+ antiporters for protons and calcium in mammals. In this paper we review the recent work with the Na^+/glucose cotransporter.

It is now apparent that the mammalian Na^+/glucose cotransporter (SGLT1) is a member of a large gene family that includes the mammalian Na^+/nucleoside and Na^+/myo-inositol cotransporters, and the bacterial Na^+/proline and Na^+/pantothenate cotransporters. Another distinct family is responsible for the transport of a variety of solutes, including neurotransmitters, osmolytes, and amino acids in the brain, kidney, and intestine (Wright et al., 1992). All belong to a large, diverse group of membrane transport proteins which are predicted to form 12 membrane-spanning (M) regions (Griffith et al., 1992). In this review we will discuss the structure of SGLT1, the homology of SGLT1 to the other members of the family, and finally, the kinetics of Na^+/glucose cotransport.

Structure

The rabbit intestinal brush border Na^+/glucose cotransporter (rSGLT1) was firmly identified as a 70–75-kD protein on SDS-PAGE using covalent, fluorescent, group-specific reagents (Peerce and Wright, 1984, 1985). The transporter was cloned by a novel expression cloning technique (Hediger et al., 1987) and the cDNA sequence predicted a protein of 662 amino acids with a mass of 75 kD. As is the case with most hydrophobic, integral membrane proteins, the primary transcript runs on SDS-PAGE with a higher than expected mobility, giving an apparent size of $\sim 50,000$. Core N-linked glycosylation increases the apparent size by 6,000 (Hediger et al., 1991), and complex glycosylation increases the apparent size to $\sim 70,000$ (Hirayama and Wright, 1992), i.e., to a size consistent with the protein identified by biochemical techniques. On the basis of a Ferguson analysis of the mobility of the primary transcript, the mature protein, and the deglycosylated protein (Hirayama and Wright, 1992), we estimate that the actual size of the mature protein is 86,000. This needs to be confirmed.

An identical SGLT1 has been cloned from the rabbit kidney (Coady et al., 1990; Morrison et al., 1991), and very similar proteins have also been cloned from the human intestine (Hediger et al., 1989) and pig renal cells (Ohta et al., 1990). These proteins of 662–664 residues have 84–85% of the amino acids identical to rSGLT1, and so far only the rabbit and the human intestinal clones exhibit Na^+/glucose cotransport when expressed in *Xenopus* oocytes.

The amino acid sequence and a secondary structure model of the rSGLT1 are shown in Fig. 1. The 662-residue protein is predicted to wind through the plasma membrane 12 times. There are two N-linked glycosylation sites at Asn248 and 306, but only Asn248 is actually glycosylated (Hediger et al., 1991). There is no O-linked glycosylation and N-linked glycosylation is not required for function (Hediger et al., 1991; Hirayama and Wright, 1992). The glycosylation at Asn248 indicates that the hydrophilic loop (Glu229–Thr271) between transmembrane segments M5 and M6 is on the exterior face of the membrane. If the 12 M prediction is correct, the NH_2 and COOH termini reside on the cytosolic side of the membrane. Note that rSGLT1 does not possess a long cytoplasmic COOH terminus, a characteristic of the SGLT1 family.

We have begun to test the validity of the secondary structure model shown in Fig. 1 (Hirayama et al., 1992). Our strategy is to treat sealed, rightside-out rabbit brush border membrane vesicles with proteases, identify specific fragments of rSGLT1 using antipeptide antibodies on Western blots, and purify, analyze, and sequence these proteolytic fragments. Brush border membrane vesicles were pre-

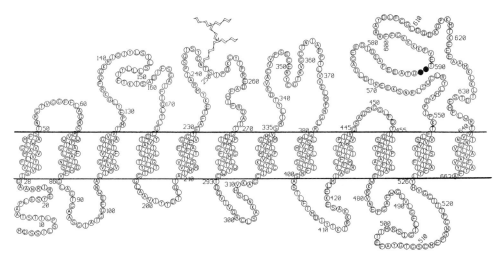

Figure 1. A two-dimensional structural model for the rabbit SGLT1 (rSGLT1) protein. The 662-residue protein is shown to possess 12 transmembrane, hydrophobic α helices. The N-linked glycosylation at Asn_{248} is indicated by the sugar residues. The positions of the two additional residues in hSGLT1 are shown (●).

pared by our Ca^{2+} precipitation method (Stevens et al., 1982), and the vesicle orientation determined by transmission electron microscopy and a DNAse I inactivation assay (Grinstein and Cohen, 1983). Preliminary estimates indicate that the brush border vesicles are >96% sealed and rightside-out, restricting proteolytic cleavage sites to the external face of the membrane. Proteolysis of the vesicles with trypsin, chymotrypsin, and papain had no significant effect on the function of rSGLT1 (Smith et al., 1992); i.e., there were no detectable changes in initial rates of uptake, the time and magnitude of the peak overshoot, or kinetics (V_{max} and K_m for Na^+ and D-glucose). However, Western blots using anti-peptide antibodies (Hirayama et al., 1992; Smith et al., 1992) showed that the transporter had been cleaved extensively. Fig. 2 shows an experiment where brush borders were cleaved with chymotrypsin and the Western blots were probed with antibodies 8792 (residues 402–419) and 8821 (residues 604–615). Both antibodies detected the 70-kD band in

Glucose-Galactose Malabsorption

One approach to defining residues important for the function of membrane transport proteins is to utilize experiments of nature; i.e., identify the residue(s) responsible for defects in protein function in patients with a hereditary disease of the transporter. In the case of the intestinal Na^+/glucose cotransporter the disease is glucose-galactose malabsorption (GGM). Over 100 patients have been described as having a defect in intestinal sugar absorption (Desjeux, 1989). It generally presents as a severe diarrhea in neonates that results in death unless fluid balance is restored, or glucose, galactose, and maltose are eliminated from the infant's diet. In these patients, there is a selective defect in glucose and galactose absorption by the Na^+/glucose cotransporter (hSGLT1). The absorption of fructose via another transporter is normal.

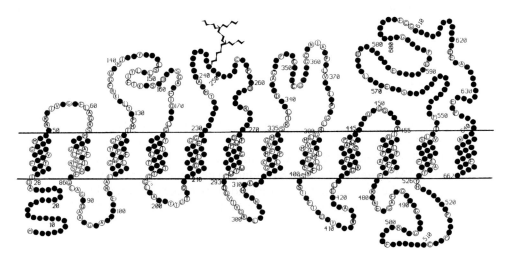

Figure 3. The 12 M secondary structure model for rSGLT1 showing the residues conserved between SGLT1 and the rabbit Na^+/nucleoside (SNST1) and dog Na^+/myoinositol (SMIT1) cotransporters. The open circles show the common residues.

In two sisters diagnosed with GGM we were able to identify the missense mutation responsible for the disease (Turk et al., 1991). The approach was to prepare cDNA from duodenal biopsies and to amplify the hSGLT1 cDNA using PCR and specific primers. Overlapping cDNA fragments were sequenced and a single base change (G → A) was found at position 92 that substituted an asparagine for an aspartic acid at position 28. This missense mutation of the residue, at the interface between the NH_2 terminus and the first transmembrane segment (Fig. 1), was sufficient to eliminate Na^+-dependent sugar transport. It is unlikely that the aspartate residue is specifically required for Na^+/glucose cotransport since an aspartate exists at this same position in at least 10 other cotransporters (Wright et al., 1992). Currently, we are screening other GGM patients to identify other mutations in the hSGLT1 gene responsible for sugar malabsorption.

Function

rSGLT1 has been expressed in a functional form in oocytes and cultured cells (COS-7, HeLa, and insect ovary cells), and, in general, the properties of rSGLT1 are independent of the expression system and similar to the native intestinal brush border transporter (Wright, 1993). This suggests that the activity of rSGLT1 is relatively independent of either posttranslational modifications or specific membrane lipids. Using the oocyte expression system, we have carried out an exhaustive kinetic study of SGLT1, mostly using electrophysiological methods (Ikeda et al., 1989; Umbach et al., 1990; Birnir et al., 1991; Parent et al., 1992*a*, *b*; Loo, D. D. F., A. Hazama, S. Supplisson, E. Turk, and E. M. Wright, manuscript submitted for

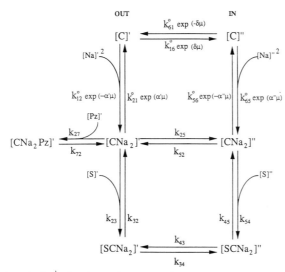

Figure 4. A transport model for the Na$^+$/glucose cotransporter. In this six-state ordered reaction model, it is assumed that the empty carrier C has a valence of -2. Thus, when two Na ions bind, the complex is neutral and subsequent reaction steps (k_{27}, k_{23}, k_{34}, k_{45}, and k_{25}) are electroneutral. Only k_{12}, k_{21}, k_{16}, and k_{61} are voltage dependent. α and δ represent the fraction of the field felt by the Na$^+$ binding and translocation of the empty carrier. $\mu = FV/RT$, where V is the membrane potential and F, R, and T have the usual meaning. For simplicity, the binding of two Na ions is described as a single reaction step. More than 2,000 computer simulations were used to fit all the electrophysiological data (Parent et al., 1992*b*) to the model. A single set of parameters (Table II) were obtained that describe the results. Note that our analysis suggest that Na$^+$ binding at the inner surface of the membrane is voltage independent ($\alpha' = 0$). (Reproduced from *Journal of Membrane Biology*, 1992, 125:63–79, by copyright permission of Springer-Verlag New York Inc.).

publication). A nonrapid equilibrium ordered model that qualitatively and quantitatively describes Na$^+$/glucose cotransport is shown in Fig. 4, and the kinetic parameters are listed in Table I.

Given that in oocytes there is an inward sodium electrochemical potential gradient, Na$_o$ = 100 mM, Na$_i$ = 10 mM, S$_i$ < 50 μM, and $E_m = -50$ mV, the addition of sugar to the external medium results in net inward transport of sugar and Na$^+$. However, the system is reversible when the oocytes are preloaded with nonmetabolized sugars (αMeDGlc). The requirement for external Na$^+$ is specific (Na$^+ \gg$ Ln$^{3+} \gg$ Li$^+$, K$^+$, Choline$^+$, NMDG$^+$), but a broad spectrum of sugar analogues are substrates (D-glucose, D-galactose, αMeDGlc > 3-*o*-methylglucoside > D-allose > D-xylose > > > L-glucose, 2-deoxyglucose). Note that αMeDGlc, a nonmetabolized substrate, is not handled by other mammalian sugar transporters (the GLUT series). Phlorizin is a competitive inhibitor of Na^+-dependent sugar transport with a K_i of ~ 5 μM. Indirect evidence indicates that the Na^+ to sugar coupling ratio is 2. The kinetics of the system are ordered; i.e., Na^+ binds first to change the conformation of the protein to allow sugar binding (that is, the affinity for sugar increases by an order of magnitude at -50 mV when the external Na^+ is increased from 2 to 100 mM). Na^+ binding at the external surface is voltage dependent; i.e., Na^+ binds in a well that extends about one-third of the way into the membrane. The apparent K_m for Na^+ at 1 mM external sugar is ~ 60 mM at 0 mV, and this decreases to ~ 2 mM at -200 mV. Because Na^+ binding is voltage dependent, the apparent K_m for sugar is sensitive to voltage at low external Na^+ concentrations. The rate of maximum transport is relatively insensitive to the external Na^+ and to membrane potential (in the range of -80 to -200 mV). These results support the ordered reaction mechanism and the voltage dependence of the transport of the fully loaded, $2Na^+$ sugar complex (k_{34}, k_{43}). Finally, our model assumes that the valence (z) of the unloaded carrier (C) is

TABLE I
Rate Constants for the Rabbit Transport Model

	$k°$	Forward	$k°$	Backward
$2\,Na_o$	12	$80{,}000\ mol^{-2}$	21	500
$Sugar_o$	23	$1 \times 10^5\ mol^{-1}$	32	20
	34	50	43	50
	45	800	54	$1.8 \times 10^7\ mol^{-1}$
	56	10	65	$50\ mol^{-2}$
	61	5	16	35
	25	0.3	52	1.4

$k°$ values are taken from Parent et al. (1992c), and the units are s^{-1}, $s^{-1}\ mol^{-1}$, or s^{-1} mol^{-2}. For the voltage-dependent Na^+ binding, k_{12} and k_{21} are described by $k°\ exp$ ($\pm 0.3\mu$) and for the voltage-dependent reorientation of the divalent, unloaded carrier, k_{16} and k_{61} are described by $k°\ exp$ ($\pm 0.7\mu$), where $\mu = FV/RT$. The phlorizin K_i for the inhibition of the sugar-independent rSGLT1 Na^+ current is 5 μM (Umbach et al., 1990).

-2, and predicts that the voltage-sensitive steps in the reaction cycle are external Na^+ binding (k_{12}, k_{21}) and transport of the unloaded carrier (k_{16}, k_{61}).

The model and the 14 rate constants quantitatively account for the major experimental observations: (*a*) the sigmoid *I-V* curve of the sugar-induced currents; (*b*) the Na^+ and voltage dependence of the apparent K_m for sugar; (*c*) the sugar and voltage dependence of the apparent K_m for Na^+; (*d*) the Na^+ leak currents in the absence of sugar; and (*e*) the pre–steady-state currents observed in the absence of sugar ($t_{1/2} \approx 13$ ms) and the Hill coefficient of >1 for Na^+ activation. The rate-limiting step for the normal sugar transport cycle under physiological conditions is the reorientation of the empty carrier from the cytoplasmic to the external surface of the membrane (k_{61}).

An important aspect of model building is the design of further experiments. There are a number of obvious experiments to test this model for Na^+/sugar cotransport. These include the measurement of transport parameters as a function of internal Na^+ and sugar concentrations. For example, (*a*) measurement of inward transport (current) as a function of internal Na⁺, sugar, and Na⁺ and sugar concentrations, and (*b*) measurement of outward transport (currents) as a function of internal Na⁺, sugar, and voltage. These will give more precise estimates of the parameters for internal sugar and Na⁺ binding, and will challenge the predicted lack of effect of voltage on internal Na⁺ binding. These experiments are currently in progress.

A more novel test of the model stems from our observation of pre–steady-state currents in the absence of sugar (Birnir et al., 1991; Parent et al., 1992*a*). This has enabled us to apply relaxation methods to Na⁺/sugar cotransport (Loo, D. D. F., A. Hazama, S. Supplisson, E. Turk, and E. M. Wright, manuscript submitted for publication). Using a fast voltage-clamp (settling time 0.5 ms), we measured the kinetics of the pre–steady-state currents (gating currents) of the hSGLT1. The approach was to overexpress hSGLT1 in oocytes, clamp the membrane potential at a holding level, rapidly step the potential to a new voltage, and record the current transient as a function of voltage, temperature, and external Na⁺ and sugar concentrations. Fig. 5 *A* shows the transient currents recorded in the absence of sugar when the membrane was held at −100 mV and then rapidly stepped to potentials ranging from −150 to +50 mV. The capacitive corrected transients decayed with a half-time (τ) of 2–10 ms for the ON response and 10 ms for the OFF response. The integral of the transient currents during the ON and OFF responses gives the carrier-mediated charge transfers (Q). The values for Q were identical for the ON and OFF responses at each voltage, and Q saturated with voltage (Fig. 5 *C*). Q as a function of voltage was described by the Boltzmann equation, where Q_{max} = 21 nC, $V_{0.5}$ = −39 mV, and Z = −1.04. Z is the effective valence of the moveable charge on hSGLT1, −1, $V_{0.5}$ is the membrane potential where the hSGLT1 charge is equally distributed between the two faces of the membrane, and Q_{max} is a measure of the number of hSGLT1 molecules in the membrane. Q_{max} in each oocyte is proportional to the maximum, steady-state, sugar-dependent current (I_{max}^S and a plot of Q_{max} vs. I_{max}^S yields a slope of 57 ± 5 s⁻¹. This is an estimate of the turnover rate of the transporter. Note that this estimate agrees well with that estimated for the rabbit clone (Table I). A Q_{max} of ∼ 25 nC is equivalent to ∼ 10¹¹ carries/oocyte, which is close to the value assumed in building the model shown in Fig. 4.

What is the origin of the charge transfer? First, it is due to the presence of hSGLT1 in the plasma membrane, because (*a*) no transients are seen in control water-injected oocytes, and (*b*) Q_{max} is proportional to the expression of hSGLT1. Second, the charge movements are intrinsic to hSGLT1 in the membrane field because (*a*) the ON and OFF responses are equal and opposite, (*b*) charge transfer saturates with voltage and is described by the Boltzmann equation, (*c*) the reversal potential is equal to the holding potential, (*d*) Q_{max} is independent of temperature, and (*e*) Q is abolished by saturating external sugar concentrations. We suggest that the charge is related to the redistribution of the Na⁺-loaded and empty carrier protein in the membrane. In the presence of high external Na⁺ (100 mM), when the membrane is held at −100 mV, SGLT1 is orientated toward the external surface where it binds Na⁺; i.e., the majority of the carrier is in the [CNa₂]′ form (Parent et al., 1992*b*). When the voltage is stepped to +50 mV, Na⁺ dissociates and the empty carrier reorientates in the field to face the cytoplasm. Since the internal Na⁺ is low (∼ 10 mM), relatively little of [C]″ is converted to the Na⁺ form [CNa₂]″. Thus the charge transfer is due to [CNa₂]′ → [C]′ → [C]″, and this reverses when the voltage is returned to -100 mV. On the addition of saturating sugar to the external surface, the transport cycle is completed to deliver sugar to the cell and most of the carrier is in the $[CNa_2]''$ form (Parent et al., 1992*b*). Therefore, when the voltage is stepped from -100 to $+50$ mV, most of the carriers are already facing the internal surface and there is virtually no charge transfer. A different situation occurs when phlorizin is added to the external surface in the absence of sugar. The inhibitor binds to the carrier to form $[CNa_2P]'$, and the charge transfer is blocked. We conclude that Na^+

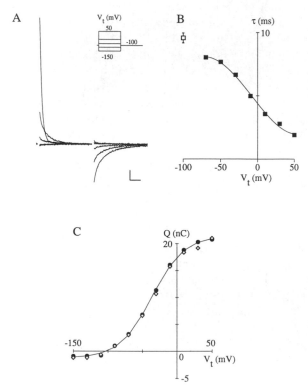

Figure 5. Charge translocation by hSGLT1 expressed in oocytes. hSGLT1 cRNA was injected into an oocyte and 6 d later a two-electrode, voltage-clamp method was used to record carrier displacement currents. In this experiment the oocyte was held at -100 mV and the currents were recorded when the membrane potential was stepped to a new voltage between -150 and $+50$ mV. Each current trace represents the average of 15 sweeps of the pulse protocol, and each was corrected for capacitive and leak currents using standard protocols. This oocyte was bathed in a solution containing 100 mM NaCl and no sugar. The capacitance of the cell was 251 nF. The settling time of the amplifier was 0.5 ms. (*A*) Carrier currents. The vertical scale is equivalent to 0.5 μA and the horizontal scale represents 10 ms. (*B*) Kinetics of current relaxation. The relaxations were described by a single time constant (τ) for the on ($\blacksquare$) and off ($\square$) responses. (*C*) The charge transfer (Q) for the on ($\bigcirc$) and off ($\bullet$) responses at each membrane potential (V). The curve is the relationship between Q and V as described by the Boltzmann relation with a Q_{max} of 21 nC, $V_{0.5} - 39$ mV, and $Z = -1.04$. Note that the current transients were not observed in H_2O injected oocytes, and that in the cRNA-injected oocyte Q_{max} was proportional to the amount of hSGLT1 expression (not shown).

cannot dissociate from $[CNa_2P]'$ and no free carrier, C', is available to reorientate in the membrane. In other words, phlorizin holds Na^+ on the carrier in an occluded state.

A detailed analysis of Q_{max} and the kinetics of charge transfer as a function of voltage and external Na^+ have enabled us to obtain estimates of k_{12}, k_{21}, k_{16}, k_{61}, α', and δ for hSGLT1. Lowering external Na^+ from 100 to 10 reduces $Q_{max} \sim 40\%$ and decreases τ, especially at small membrane depolarizations. These results suggest that

$\alpha' \approx 0.3$, $\delta \approx 0.7$, $k_{21} \approx 200$ s^{-1}, $K_{Na} \approx 0.1$ M, $k_{16}/k_{61} > 10$, and $k_{16}/k_{21} > 2$. The most obvious difference with the rabbit SGLT1 (Table I) is the ratio $k_{16}/k_{21} > 2$ (human) vs. < 0.01 (rabbit); i.e., the rate-limiting step for charge transfer in hSGLT1 is k_{21}, while in rSGLT1 it is k_{16}. This is mainly because k_{16} is an order of magnitude higher in human than in rabbit SGLT1. A detailed, direct comparison of the kinetics of the human and rabbit clones is currently under investigation. Preliminary experiments point to marked differences in the kinetics of charge transfer between the two proteins. τ increases with depolarization for the rSGLT1 (Parent et al., 1992a, Fig. 9 A), while it decreases with depolarization for hSGLT1 (Fig. 5 B). This must be due to the differences in primary amino acid sequence of the two proteins (84% identical, 96% similar).

Overall, the relaxation method provides a direct method of estimating 6 of the 14 parameters needed to fit the transport model and the agreement between the model parameters and the relaxation parameters is surprisingly good. Furthermore, the relaxation method provides a direct estimate of the number of transporters (previously estimated) and an indirect estimate of the turnover number. What about the assumed valency of the protein ($z - 2$) vs. the apparent valency (-1.04) obtained from the Boltzmann fit to the charge transfer data? The difference is largely resolved when one considers that the charge on the protein only moves through $\sim 70\%$ of the field ($\delta \approx 0.7$); i.e., the apparent valency = δZ.

In summary, the relaxation method provides direct evidence in support of the assumptions used to construct our model (i.e., provides estimates of the number of carriers in the membrane [$\sim 10^{11}$/oocyte] and the valency of the protein [-2]). In addition, the experiments provide estimates of α' (0.3), δ (0.7), k_{34}, k_{12}, k_{21}, k_{16}, and k_{61} consistent with the model parameters. We plan to use this method to examine the rSGLT1 in oocytes, and to further test the model by measuring Q_{max} and the kinetics of charge transfer as a function of internal Na$^+$ and sugar concentrations. We believe that such comparisons of the human and rabbit SGLT1 proteins and tests of our model will provide useful quantitative insights into the mechanism of Na$^+$/glucose cotransport. Ultimately, electrophysiological methods such as these can be used to investigate mutant proteins, such as the GGM mutants, to determine the precise role of individual amino acids in each step of the reaction cycle. Attempts to relate the kinetics of cotransport to the structure of the protein require firm data on the secondary and tertiary structure of SGLT1. Our testing of the secondary structure model (Fig. 1) represents our initial attempt at structural analysis of the cotransporter.

Acknowledgments

These studies were supported by grants from the USSHS (DK-19567 and NS-09666 to Ernest M. Wright and DK-41301 to Eric Turk).

References

Birnir, B., D. D. F. Loo, and E. M. Wright. 1991. Voltage clamp studies of the Na$^+$/glucose cotransporter cloned from rabbit small intestine. *Pflügers Archiv.* 418:79–85.

Coady, M. J., A. M. Pajor, and E. M. Wright. 1990. Sequence homologies among intestinal and renal Na$^+$/glucose cotransporters. *American Journal of Physiology.* 259 (*Cell Physiology* 28):C605–C610.

Crane, R. K., D. Miller, and I. Bihler. 1961. The restrictions on possible mechanisms of intestinal active transport of sugars. *In* Membrane Transport and Metabolism. A. Kleinzeller and A. Kotyk, editors. Czechoslovakian Academy of Sciences, Prague. 439–449.

Desjeux, J.-F. 1989. Congential selective Na$^+$ D-glucose cotransport defects leading to renal glycosuria and congenital selective intestinal malabsorption of glucose and galactose. *In* The Metabolic Basis of Inherited Disease. C. Scriver, A. Beaudet, W. Sly, and D. Valle, editors. McGraw-Hill Book Company, New York. 2463–2478.

Griffith, J. K., M. E. Baker, D. A. Rouch, M. G. P. Page, R. A. Skurray, I. T. Paulsen, K. F. Chater, S. A. Baldwin, and P. J. F. Henderson. 1992. Membrane transport proteins: implications of sequence comparisons. *Current Opinion in Cell Biology.* 4:684–695.

Grinstein, S., and S. Cohen. 1983. Measurement of sidedness of isolated plasma-membrane vesicles: quantitation of actin exposure by DNase I inactivation. *Analytical Biochemistry.* 130:151–157.

Hediger, M. A., M. J. Coady, T. S. Ikeda, and E. M. Wright. 1987. Expression cloning and cDNA sequencing of the Na$^+$/glucose cotransporter. *Nature.* 330:379–381.

Hediger, M. A., J. Mendlein, H.-S. Lee, and E. M. Wright. 1991. Biosynthesis of the cloned Na$^+$/glucose cotransporter. *Biochimica et Biophysica Acta.* 1064:360–364.

Hediger, M. A., E. Turk, and E. M. Wright. 1989. Homology of the human intestinal Na$^+$/glucose and *E. coli* Na$^+$/proline cotransporters. *Proceedings of the National Academy of Sciences, USA.* 86:5748–5752.

Hirayama, B. A., C. D. Smith, and E. M. Wright. 1992. Secondary structure of the Na$^+$/glucose cotransporter. *Journal of General Physiology.* 100:19*a*. (Abstr.)

Hirayama, B. A., H. C. Wong, C. D. Smith, B. A. Hagenbuch, M. A. Hediger, and E. M. Wright. 1991. Intestinal and renal Na$^+$/glucose cotransporters share common structures. *American Journal of Physiology.* 261:C296–C304.

Hirayama, B. A., and E. M. Wright. 1992. Glycosylation of the rabbit intestinal brush border Na$^+$/glucose cotransporter. *Biochimica et Biophysica Acta.* 1103:37–44.

Ikeda, T. S., E.-S. Hwang, M. J. Coady, B. A. Hirayama, M. A. Hediger, and E. M. Wright. 1989. Characterization of a Na$^+$/glucose cotransporter cloned from rabbit small intestine. *Journal of Membrane Biology.* 110:87–95.

Kwon, H. M., A. Yamauchi, S. Uchida, A. S. Preston, A. Garcia-Perez, M. D. Burg, and J. S. Handler. 1992. Expression cloning of a Na$^+$/myoinositol cotransporter, a hypertonicity stress protein. *Journal of Biological Chemistry.* 267:6297–6301.

Morrison, A. A., M. Panayotova-Heriermann, G. Feigl, B. Schölermann, and R. K. H. Kinne. 1991. Sequence comparison of the sodium-D-glucose cotransport systems in rabbit renal and intestinal epithelia. *Biochimica et Biophysica Acta.* 1089:121–123.

Ohta, T., K. J. Isselbacher, and B. D. Rhoads. 1990. Regulation of glucose transporters in LLC-PK$_1$ cells: effects of D-glucose and monosaccharides. *Molecular and Cellular Biology.* 10:6491–6499.

Pajor, A. M., and E. M. Wright. 1992. Sequence[1], tissue distribution and functional expression of a mammalian Na$^+$/nucleoside cotransporter. *Journal of Biological Chemistry.* 267:3557–3560.

Parent, L., S. Supplisson, D. D. F. Loo, and E. M. Wright. 1992*a*. Electrogenic properties of the cloned Na$^+$/glucose cotransporter. I. Voltage-clamp studies. *Journal of Membrane Biology.* 125:49–62.

Parent, L., S. Supplisson, D. D. F. Loo, and E. M. Wright. 1992*b*. Electrogenic properties of the closed Na$^+$/glucose cotransporter. II. A transport model under nonrapid equilibrium conditions. *Journal of Membrane Biology.* 125:63–79.

Parent, L., S. Supplisson, D. D. F. Loo, and E. M. Wright. 1992*c. Journal of Membrane Biology.* 130:203.

Peerce, B. E., and E. M. Wright. 1984. Sodium induced conformational changes in the glucose transporter of intestinal brush borders. *Journal of Biological Chemistry.* 259:14105–14112.

Peerce, B. E., and E. M. Wright. 1985. Evidence for tyrosyl residues at the Na$^+$ site on the intestinal Na$^+$/glucose cotransporter. *Journal of Biological Chemistry.* 260:6026–6031.

Semenza, G., M. Kessler, M. Hosang, and U. Schmidt. 1984. Biochemistry of the Na$^+$, D-glucose cotransporter of the small intestinal brush-border membrane. *Biochimica et Biophysica Acta.* 779:343–379.

Smith, C. D., B. A. Hirayama, J. Freedman, A. Klein, and E. M. Wright. 1992. Proteolysis of the Na$^+$/glucose cotransporter (SGLT1). *FASEB Journal.* 6:A1768. (Abstr.)

Stevens, B. R., H. J. Ross, and E. M. Wright. 1982. Multiple transport pathways for neutral amino acids in rabbit jejunal brush border vesicles. *Journal of Membrane Biology.* 66:213–225.

Turk, E., B. Zabel, S. Mundlos, J. Dyer, and E. M. Wright. 1991. Glucose/galactose malabsorption caused by a defect in the Na$^+$/glucose cotransporter. *Nature.* 350:354–356.

Umbach, J. A., M. J. Coady, and E. M. Wright. 1990. The intestinal Na$^+$/glucose cotransporter expressed in *Xenopus* oocytes is electrogenic. *Biophysical Journal.* 57:1217–1224.

Wright, E. M., K. M. Hager, and E. Turk. 1992. Sodium cotransport proteins. *Current Opinion in Cell Biology.* 4:696–702.

Wright, E. M. 1993. The intestinal Na$^+$/glucose transporter. *Annual Review of Physiology.* 55:575–589.

Chapter 19

Structure and Function of Sodium-coupled Neurotransmitter Transporters

Baruch I. Kanner

Department of Biochemistry, Hadassah Medical School, The Hebrew University, Jerusalem, Israel

Molecular Biology and Function of Carrier Proteins © 1993 by The Rockefeller University Press

Introduction

The reuptake of neurotransmitters from the synaptic cleft by high-affinity transporters appears to play an important role in the overall process of synaptic transmission (Iversen, 1971; Kuhar, 1973). The reuptake process is carried out by sodium-coupled neurotransmitter transport systems (Kanner, 1983, 1989; Kanner and Schuldiner, 1987) located in the plasma membrane of nerve terminals and glial cells. These transport systems have been investigated in detail by using plasma membranes obtained upon osmotic shock of synaptosomes. It appears that these transporters are coupled not only to sodium, but also to other ions like potassium or chloride.

The most abundant and well characterized of these uptake systems in rat brain are those for γ-aminobutyric acid (GABA) and L-glutamate (Kanner, 1983, 1989; Kanner and Schuldiner, 1987), two major neurotransmitters in the central nervous system. Multiple species of GABA transporters found in membrane vesicles from rat brain have been reconstituted (Mabjeesh and Kanner, 1989; Kanner and Bendahan, 1990). Two of them exhibit high affinity for their substrates and are very similar in their mechanistic properties. They are, however, inhibited selectively by different substrate analogues, namely *cis*-3-aminocyclohexane carboxylic acid and β-alanine (Kanner and Bendahan, 1990).

Mechanistic Studies

GABA is accumulated by electrogenic cotransport with sodium and chloride. We have been able to demonstrate directly that both sodium and chloride ions are cotransported with GABA by the transporter. This has been accomplished using a partly purified transporter preparation that was reconstituted into liposomes, and Dowex columns to terminate the reactions. These proteoliposomes catalyzed GABA- and chloride-dependent $^{22}[Na^+]$ transport, as well as GABA- and sodium-dependent $^{36}[Cl^-]$ translocation (Keynan and Kanner, 1988).

Using the above system, the stoichiometry for the GABA cotransporter has also been determined kinetically, i.e., by comparing the initial rate of the fluxes of $[^3H]GABA$, $^{22}[Na^+]$, and $^{36}[Cl^-]$. The results (Keynan and Kanner, 1988) are similar to those found using the thermodynamic method (Radian and Kanner, 1983), yielding an apparent stoichiometry of 2.5 Na^+:1 Cl^-:1 GABA. This is in harmony with the predicted restrictions: thus, if GABA is translocated in the zwitterionic form, which is the predominant one at physiological pH, an electrogenic cotransport of the three species requires a stoichiometry of nNa^+:mCl^-: GABA with $n > m$. The L-glutamate transporter from rat brain is also coupled to sodium (Kanner and Sharon, 1978). Although chloride is not required, the influx of L-glutamate is absolutely dependent on internal potassium (Kanner and Sharon, 1978). Interestingly, the influx of GABA is not dependent on internal potassium (Kanner, 1978). The mechanism of L-glutamate transport is such that it is cotransported with sodium, and after translocation of these two species potassium is translocated in the opposite direction, probably enabling reorientation of the binding sites for sodium and L-glutamate (Kanner and Bendahan, 1982; Pines and Kanner, 1990). The whole process is electrogenic, involving cotransport of L-glutamate with three Na^+ ions, while one K^+ is countertransported (Barbour et al., 1988).

Reconstitution and Purification

Using methodology that enables rapid and simultaneous reconstitution of many samples (Radian and Kanner, 1985), one of the subtypes of the GABA (Radian et al., 1986) and the L-glutamate transporter (Danbolt et al., 1990) have been purified to apparent homogeneity. Both transporters, like the one for glycine (Lopez-Corcuera et al., 1989) are glycoproteins and both have an apparent molecular mass of 70–80 kD. The two transporters retain all the properties observed in membrane vesicles. They are distinct not only in their different functional properties, but also in terms of their antigens. Antibodies generated against the GABA transporter (Radian et al., 1986) react (as detected by immunoblotting) only with fractions containing GABA transport activity, and not with those containing L-glutamate transport activity (Danbolt et al., 1990). The opposite is true for antibodies generated against the glutamate transporter (Danbolt et al., 1992).

Biochemical Characterization of the GABA Transporter

The GABA transporter has been subjected to deglycosylation and limited proteolysis. The treatment of the 80-kD band with endoglycosidase F results in its disappearance and reveals the presence of a polypeptide with an apparent molecular mass of ~ 60 kD, which lacks ^{125}I-labeled wheat germ agglutinin binding activity but nevertheless is recognized by the antibodies against the 80-kD band. Upon limited proteolysis with papain or pronase, the 80-kD band was degraded to one with an apparent molecular mass of ~ 60 kD. This polypeptide still contains the ^{125}I-labeled wheat germ agglutinin binding activity but is not recognized by the antibody. The effect of proteolysis on function was examined. The transporter was purified using all steps except for the lectin chromatography (Radian et al., 1986). After papain treatment and lectin chromatography, γ-aminobutyric transport activity was eluted with *N*-acetyl-glucosamine. The characteristics of transport were the same as those for the pure transporter, but the preparation contained two fragments of ~ 66 and 60 kD, instead of the 80-kD polypeptide. The ability of the anti–80-kD antibody to recognize these fragments was relatively low. These observations indicate that the transporter contains exposed domains that are not important for function (Kanner et al., 1989).

Antibodies were raised against synthetic peptides corresponding to several regions of the rat brain GABA transporter. According to our model, this glycoprotein has 12 transmembrane α-helices with amino and carboxyl termini located in the cytoplasm. The antibodies recognized the intact transporter on Western blots. Upon papain treatment, reconstitutable active transporter can be isolated upon lectin chromatography (Kanner et al., 1989). The papainized transporter runs on SDS-polyacrylamide gels as a broad band with an apparent molecular mass between 58 and 68 kD, as compared with 80 kD for the untreated transporter. The transporter fragment was recognized by all the antibodies except that raised against the amino terminus. Pronase cleaves the transporter to a relatively sharp 60-kD band, which reacts with the antibodies against the internal loops but not with either the amino or the carboxyl terminus. This pronase-treated transporter was reconstituted upon isolation by lectin chromatography. It exhibited full GABA transport activity with the same features as the intact system, including an absolute dependence on sodium and chloride as well as electrogenicity. We conclude that the amino- and carboxyl-terminal parts of the transporter, possibly including transmembrane α-helices 1, 2, and 12, are not required for the transport function (Mabjeesh and Kanner, 1992).

Membrane vesicles from rat brain were digested with pronase. The proteolytic fragments of the sodium- and chloride-coupled GABA transporter (subtype A) were analyzed with a variety of sequence-directed antibodies generated against this transporter. The major fragments detected by the various antibodies had an apparent size of ~10 kD. When protease treatment was carried out in the presence of GABA, the generation of these fragments was almost completely blocked. This was paralleled by an increase of reconstitutable GABA transport activity. The effect was specific for GABA and was not observed with a variety of other neurotransmitters and analogues. Furthermore, this protection was seen only in the presence of both sodium and chloride, the cosubstrates of the transporter. It was not observed when GABA was present only from the inside. The results indicate that the transporter can exist in at least two conformations. In the absence of one or more of the substrates, many epitopes located throughout the transporter are accessible to the protease. In the presence of all three substrates the transporter goes from a "relaxed" to a "tight" conformation and these epitopes become inaccessible to protease action (Mabjeesh, N. J., and B. I. Kanner, manuscript submitted for publication).

Immunocytochemical Localization of the GABA Transporter

Polyclonal antibodies were raised against the GABA transporter purified from rat brain tissue and used for immunocytochemical localization of the antigen in several rat brain areas, including cerebellum, hippocampus, substantia nigra, and cerebral cortex. Light microscopy studies with the peroxidase-antiperoxidase and the biotin-avidin-peroxidase techniques suggested that the GABA transporter is localized in the same axons and terminals that contain endogenous GABA, as judged by comparison with parallel sections incubated with antibodies against glutaraldehyde-conjugated GABA. However, as expected from biochemical results, neurons differed in their relative contents of GABA transporter and GABA; thus, the former was relatively low in striatonigral and Purkinje axon terminals and relatively high in the basket nerve plexus surrounding the cell bodies of cerebellar Purkinje cells, hippocampal pyramidal cells and dentate granule cells. The GABA transporter antiserum did not produce detectable labeling of nerve cell bodies. Electron microscopic studies supported the observations made by light microscopy and provided direct evidence for cellular colocalization of GABA transporter and GABA (as visualized by the peroxidase-antiperoxidase technique and postembedding immunogold labeling, respectively). The ultrastructural studies also indicated the presence of GABA transporter in glial processes but not in glial cell bodies. The relative intensity of the neuronal and glial staining varied among regions: glial staining predominated over neuronal staining in the substantia nigra, whereas the converse was true in the cerebellum and the hippocampus. The present immunocytochemical data demonstrate directly what has previously been inferred from biochemical and autoradiographic evidence: that the mechanism for high-affinity GABA uptake is selectively and differentially localized in GABAergic neurons and in glial cells (Radian et al., 1990).

Molecular Cloning of a GABA Transporter

We have recently cloned and expressed a GABA transporter from rat brain in a collaborative effort with the laboratories of H. A. Lester and N. Nelson (Guastella et al., 1990). Rat brain GABA transporter protein, purified as described (Radian et al., 1986), was subjected to cyanogen bromide degradation, and several of the resulting fragments were sequenced (Mandel et al., 1988). The sequence of the longest peptide (QPSEDIVRPENG) was used to design oligonucleotide probes. Since sucrose density RNA fractionation had shown that GABA transporter mRNA was in the 4–5-kb size range (Guastella et al., 1990), a λ-ZAPII rat brain cDNA library containing inserts of 4 kb and greater was screened with conventional plaque hybridization techniques. Two plaques screened as positives through successive platings. mRNA was synthesized in vitro from each clone and tested for its ability to express functional GABA transporters in *Xenopus* oocytes. One clone tested positive in the oocyte assay. It was selected for detailed characterization and designated GAT-1 (GABA transporter 1).

Oocytes injected with GAT-1 synthetic mRNA accumulated [^{3}H]GABA 50–100-fold over control levels. The transporter encoded by GAT-1 has a high affinity for GABA, is sodium and chloride dependent, and is pharmacologically similar to neuron-specific plasma membrane GABA transporters. GAT-1 expression in rat brain was also examined by probing polyadenylated RNA from cerebrum, cerebellum, and brain stem with nick-translated GAT-1. A single band of ~4.2 kb was visualized in each brain sample, which agrees with RNA fractionation experiments (Guastella et al., 1990); no bands were detectable in liver mRNA. The GAT-1 protein shares antigenic determinants with a native rat brain GABA transporter. The nucleotide sequence of GAT-1 predicts a protein of 599 amino acids with a molecular mass of 67 kD. Hydropathy analysis of the deduced protein suggests multiple transmembrane regions, a feature shared by several cloned transporters; however, database searches indicate that GAT-1 is not homologous to any previously identified proteins. Therefore, GAT-1 appears to be a member of a previously uncharacterized family of transport molecules.

As a matter of fact, it appears to be the first identified member of a superfamily of neurotransmitter transporters (Uhl, 1992). Transport expressed by GAT-1 is inhibited by *cis*-3-aminocyclohexane carboxylic acid but not by β-alanine (Keynan et al., 1992).

Site-directed Mutagenesis Studies with the GABA Transporter

The GAT-1 contains 599 amino acids and 12 putative membrane-spanning α-helices, and is the first described member of a neurotransmitter transporter superfamily. The membrane domain contains five charged amino acids which are basically conserved. Using site-directed mutagenesis, we show that only one of them (arginine 69) is absolutely essential for activity. It is located in a highly conserved region encompassing parts of helices 1 and 2. The three other positively charged amino acids and the only negative charged one (glutamate 467) are not critical. These results suggest that the translocation pathway of the sodium ions through the membrane does not involve charged amino acid residues, and underline the importance of the highly conserved stretch between amino acids 66 and 86 (Pantanowitz et al., 1993).

Purification and Reconstitution of the L-Glutamate Transporter

The sodium- and potassium-coupled L-glutamate transporter from rat brain has been purified to near homogeneity by reconstitution of transport as an assay. The purification steps involve lectin chromatography of the membrane proteins solubilized with 3-[(3-cholamidopropyl)dimethylammonio]-1-propanesulfonate (CHAPS), fractionation on hydroxylapatite, and ion-exchange chromatography. With this procedure the specific activity is increased 30-fold. The actual purification is higher since three- to fivefold inactivation occurs during the purification. The efficiency of reconstitution was ~20%. The properties of the pure transporter are fully preserved. They include ion dependence, electrogenicity, affinity, substrate specificity, and stereospecificity. SDS-polyacrylamide electrophoresis revealed one main band with an apparent molecular mass of ~80 kD and a few minor bands. Comparison of polypeptide composition with L-glutamate transport activity throughout the fractionation procedure reveals that only the 80-kD band can be correlated with activity. The GABA transporter, which has the same apparent molecular mass (Radian et al., 1986), is separated from it during the last two purification steps. Immunoblot experiments reveal that the antibodies against the GABA transporter only reacted with fractions exhibiting GABA transport activity and not with those containing the glutamate transporter. We conclude that the 80-kD band represents the functional sodium- and potassium-coupled L-glutamate transporter (Danbolt et al., 1990).

Polyclonal antibodies were generated against the major polypeptide (73 kD) present in a highly purified preparation of the $[Na^+ + K^+]$-coupled L-glutamate transporter from rat brain. These antibodies were able to selectively immunoprecipitate the 73-kD polypeptide as well as most of the L-glutamate transport activity, as assayed upon reconstitution, from crude detergent extracts of rat brain membranes. The immunoreactivity in the various fractions obtained during the purification procedure (Danbolt et al., 1990) closely correlated with the L-glutamate transport activity. Immunoblotting of a crude SDS brain extract, separated by two-dimensional isoelectric focusing/SDS-PAGE, showed that the antibodies recognized only one 73-kD protein species with apparent isoelectric point of about pH 6.2. Deglycosylation of the protein gave a 10-kD reduction in molecular mass, but no reduction in immunoreactivity. These findings establish that the 73-kD polypeptide represents the L-glutamate transporter or a subunit thereof. The antibodies also recognize a 73-kD polypeptide and immunoprecipitate L-glutamate transport activity in extracts of brain plasma membranes from rabbit, cat, and man (Danbolt et al., 1992).

Immunocytochemistry and Localization of the L-Glutamate Transporter

Using the antibodies raised against the glutamate transporter, the immunocytochemical localization of the transporter was studied with light and electron microscopy techniques in rat central nervous system. In all regions examined (including cerebral cortex, caudato-putamen, corpus callosum, hippocampus, cerebellum, spinal cord) it was found to be located in glial cells rather than in neurons. In particular, fine astrocytic processes were strongly stained. Putative glutamatergic axon terminals appeared nonimmunoreactive. The uptake of glutamate by such terminals has been clearly demonstrated. Hence they probably have a subtype of glutamate transporter different from the glial transporter demonstrated by us (Danbolt et al., 1992). Using a monoclonal antibody raised against the transporter, a similar glial localization of the transporter was found (Hees et al., 1992).

Molecular Cloning of the L-Glutamate Transporter

Using an antibody against the glial L-glutamate transporter from rat brain, we have isolated a complementary DNA clone (pT7-GLT-1) encoding this transporter. Expression of pT7-GLT-1 in transfected HeLa cells indicates that L-glutamate accumulation requires external sodium and internal potassium and exhibits the expected stereospecificity. The cDNA sequence predicts a protein of 573 amino acids with eight to nine putative transmembrane α-helices. Database searches indicate that this protein is not homologous to any identified protein of mammalian origin, including the recently described superfamily of neurotransmitter transporters. Therefore, GLT-1 appears to be a member of a previously uncharacterized family of transport molecules (Pines et al., 1992).

References

Barbour, B., H. Brew, and D. Attwell. 1988. Electrogenic glutamate uptake in glial cells is activated by intracellular potassium. *Nature.* 335:433–435.

Danbolt, N. C., G. Pines, and B. I. Kanner. 1990. Purification and reconstitution of the sodium- and potassium-coupled glutamate transport glycoprotein from rat brain. *Biochemistry.* 29:6734–6740.

Danbolt, N. C., J. Storm-Mathisen, and B. I. Kanner. 1992. A [$Na^+ + K^+$]-coupled L-glutamate transporter purified from rat brain is located in glial cell processes. *Neuroscience.* 51:295–310.

Guastella, J., N. Nelson, H. Nelson, L. Czyzyk, S. Keynan, M. C. Miedel, N. C. Davidson, H. A. Lester, and B. I. Kanner. 1990. Cloning and expression of a rat brain GABA transporter. *Science.* 249:1303–1306.

Hees, B., N. C. Danbolt, B. I. Kanner, W. Haase, K. Heitmann, and H. Koepsell. 1992. A monoclonal antibody against a Na^+-L-glutamate cotransporter from rat brain. *Journal of Biological Chemistry.* 267:23275–23281.

Iversen, L. L. 1971. Role of transmitter uptake mechanism in synaptic neurotransmission. *British Journal of Pharmacology.* 41:571–591.

Kanner, B. I. 1978. Active transport of γ-aminobutyric acid by membrane vesicles isolated from rat brain. *Biochemistry.* 17:1207–1211.

Kanner, B. I. 1983. Bioenergetics of neurotransmitter transport. *Biochimica et Biophysica Acta.* 726:293–316.

Kanner, B. I. 1989. Ion-coupled neurotransmitter transporter. *Current Opinion in Cell Biology.* 1:735–738.

Kanner, B. I., and A. Bendahan. 1982. Binding order of substrates in the sodium and potassium ion-coupled L-glutamate transporter from rat brain. *Biochemistry.* 21:6327–6330.

Kanner, B. I., and A. Bendahan. 1990. Two pharmacologically distinct sodium- and chloride-coupled high-affinity γ-aminobutyric acid transporters are present in plasma membrane vesicles and reconstituted preparations from rat brain. *Proceedings of the National Academy of Sciences, USA.* 87:2550–2554.

Kanner, B. I., S. Keynan, and R. Radian. 1989. Structural and functional studies on the sodium- and chloride-coupled γ-aminobutyric acid transporter: deglycosylation and limited proteolysis. *Biochemistry.* 28:3722–3727.

Kanner, B. I., and S. Schuldiner. 1987. Mechanism of transport and storage of neurotransmitters. *CRC Critical Reviews in Biochemistry.* 22:1–39.

Kanner, B. I., and I. Sharon. 1978. Active transport of L-glutamate by membrane vesicles isolated from rat brain. *Biochemistry.* 17:3949–3953.

Keynan, S., and B. I. Kanner. 1988. γ-Aminobutyric acid transport in reconstituted preparations from rat brain: coupled sodium and chloride fluxes. *Biochemistry.* 27:12–17.

Keynan, S., Y.-J. Suh, B. I. Kanner, and G. Rudnick. 1992. Expression of a cloned γ-aminobutyric acid transporter in mammalian cells. *Biochemistry.* 31:1974–1979.

Kuhar, J. M. 1973. Neurotransmitter uptake: a tool in identifying neurotransmitter-specific pathways. *Life Sciences.* 13:1623–1634.

Lopez-Corcuera, B., B. I. Kanner, and C. Aragon. 1989. Reconstitution and partial purification of the sodium and chloride-coupled glycine transporter from spinal cord. *Biochimica et Biophysica Acta.* 983:247–252.

Mabjeesh, N. J., and B. I. Kanner. 1989. Low affinity and aminobutyric acid transport in rat brain. *Biochemistry.* 28:7694–7699.

Mabjeesh, N. J., and B. I. Kanner. 1992. Neither amino nor carboxyl termini are required for function of the sodium- and chloride-coupled γ-aminobutyric acid transporter from rat brain. *Journal of Biological Chemistry.* 267:2563–2568.

Mandel, M., Y. Moriyama, J. D. Hulmes, Y. C. E. Pan, H. Nelson, and N. Nelson. 1988. cDNA sequence encoding the 16-kDa proteolipid of chromaffin granules implies gene duplication in the evolution of H$^+$-ATPases. *Proceedings of the National Academy of Sciences, USA.* 85:5521–5524.

Pantanowitz, S., A. Bendahan, and B. I. Kanner. 1993. Only one of the charged amino acids located in the transmembrane α-helices of the γ-aminobutyric acid transporter (subtype A) is essential for its activity. *Journal of Biological Chemistry.* In press.

Pines, G., N. C. Danbolt, M. Bjørås, Y. Zhang, A. Bendahan, L. Eide, H. Koepsell, J. Storm-Mathisen, E. Seeberg, and B. I. Kanner. 1992. Cloning and expression of a rat brain L-glutamate transporter. *Nature.* 360:464–467.

Pines, G., and B. I. Kanner. 1990. Counterflow of L-glutamate in plasma membrane vesicles and reconstituted preparations from rat brain. *Biochemistry.* 29:11209–11214.

Radian, R., A. Bendahan, and B. I. Kanner. 1986. Purification and identification of the functional sodium- and chloride-coupled γ-aminobutyric acid transport glycoprotein from rat brain. *Journal of Biological Chemistry.* 261:15437–15441.

Radian, R., and B. I. Kanner. 1983. Stoichiometry of sodium- and chloride-coupled γ-aminobutyric acid transport by synaptic plasma membrane vesicles isolated from rat brain. *Biochemistry.* 22:1236–1241.

Radian, R., and B. I. Kanner. 1985. Reconstitution and purification of the sodium- and chloride-coupled γ-aminobutyric acid transporter from rat brain. *Journal of Biological Chemistry.* 260:11859–11865.

Radian, R., O. L. Ottersen, J. Storm-Mathisen, M. Castel, and B. I. Kanner. 1990. Immunocytochemical localization of the GABA transporter in rat brain. *Journal of Neurosciences.* 10:1319–1330.

Uhl, G. R. 1992. Neurotransmitter transporter (plus): a promising new gene family. *Trends in Neurosciences.* 15:265–268.

Chapter 20

The Na$^+$/I$^-$ Symporter of the Thyroid Gland

Stephen M. Kaminsky, Orlie Levy, Carolina Salvador, Ge Dai, and Nancy Carrasco

Department of Molecular Pharmacology, Albert Einstein College of Medicine, Bronx, New York 10461

Molecular Biology and Function of Carrier Proteins © 1993 by The Rockefeller University Press

Introduction

The thyroid is a large and highly vascularized endocrine gland that weighs ~ 20 g in the healthy adult. However, a goitrous thyroid can multiply its size fivefold or more under extreme circumstances, proving the extraordinary capacity of the gland for growth. The thyroid hormones thyroxine (T_4) and tri-iodothyronine (T_3) are of major significance for the intermediary metabolism of virtually all tissues (DeGroot, 1989; Werner and Ingbar, 1991), and they play an essential role in the growth and maturation of the nervous system of the developing fetus and the newborn (Stubbe et al., 1986). In contrast to other endocrine glands, the thyroid contains a large store of hormones that can help maintain adequate bloodstream levels of thyroid hormones even in cases of I^- deficiency. The transport of I^- is a highly specialized process found most notably, though not exclusively, in the thyroid gland. Since iodine is an essential constituent of the thyroid hormones, the I^- concentrating mechanism of the thyroid is of considerable physiological importance: it serves as a highly specific and efficient supply route of I^- into the gland. In addition, I^- accumulation in the thyroid gland is the rate-limiting and first step in the biosynthesis of T_3 and T_4. The system is an important cellular adaptation to accumulate iodine, an environmentally scarce element (Heslop and Jones, 1976).

The basic events involved in the biosynthesis of the thyroid hormones have been largely elucidated. The following is a brief summary of these events (Fig. 1): I^- is actively transported against its electrochemical gradient across the basolateral plasma membrane of the thyroid follicular cells (Chambard et al., 1983; Nakamura et al., 1990), and passively translocated across the apical membrane into the colloid (Nilsson et al., 1990). The active uptake of I^- is mediated by the Na^+/I^- symporter, an intrinsic membrane protein. I^- transport is stimulated by thyroid-stimulating hormone (TSH) from the anterior pituitary and blocked by the well-known "classic" competitive inhibitors, the anions thiocyanate (SCN^-) and perchlorate (ClO_4^-) (Halmi, 1961; Wolff, 1964). The Na^+/I^- symporter, like other Na^+-dependent symporters, couples the energy released by the inward translocation of Na^+ down its electrochemical gradient to the driving of the simultaneous inward "uphill" translocation of I^- against its electrochemical gradient (Bagchi and Fawcett, 1973; Carrasco, 1993). The process is electrogenic (O'Neill et al., 1987; Nakamura et al., 1988). Physiologically, the Na^+ gradient acting as the driving force for I^- uptake in whole cells is generated by the ouabain-sensitive, K_{out}^+-activated Na^+/K^+ ATPase. In a complex reaction at the cell–colloid interface, often called organification of I^- and catalyzed by the thyroid peroxidase (TPO), I^- is oxidized and incorporated into some tyrosyl residues within the thyroglobulin (Tg) molecule, leading to the subsequent coupling of iodotyrosine residues (Fig. 1). The term organification refers to the incorporation of I^- into organic molecules, as opposed to nonincorporated, inorganic, or free I^-. The I^- organification reaction is inhibited by 6-n-propyl-2-thiouracil (PTU) and 1-methyl-2-mercaptoimidazole (MMI), among other compounds. Iodinated Tg is stored extracellularly in the colloid. In response to demand for thyroid hormones, phagolysosomal hydrolysis of endocytosed iodinated Tg ensues. T_3 and T_4 are secreted into the bloodstream, and nonsecreted iodotyrosines are metabolized to tyrosine and I^-, a reaction catalyzed by the microsomal enzyme iodotyrosine dehalogenase. This process facilitates reutilization of the remaining I^- (Lissitzky, 1982). All of these steps are also stimulated by TSH, the primary hormone that regulates the functions of differentiated thyroid cells (Sterling and Lazarus, 1977; Nilsen et al., 1982; Dumont et al., 1989). Thus, unlike hormone biosynthesis in other endocrine glands, hormone production in the thyroid occurs to a large extent in the colloid, an extracellular compartment.

The existence of the I$^-$ transport system in the thyroid is of major significance for the evaluation, diagnosis, and treatment of some thyroid pathological conditions (Kendal and Condon, 1969; DeGroot, 1989; Werner and Ingbar, 1991). The I$^-$

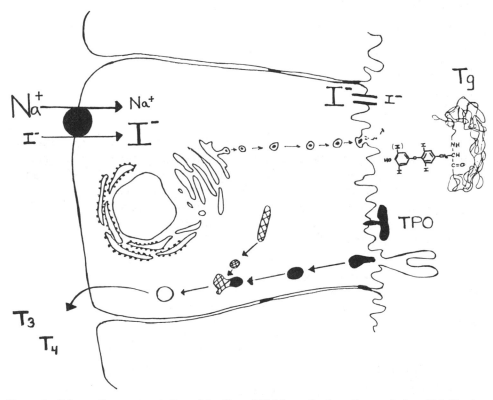

Figure 1. Schematic representation of the T$_3$ and T$_4$ biosynthetic pathway. A thyroid follicular cell is displayed with the basolateral end on the left side of the figure and the apical end on the right. (●) Active accumulation of I$^-$ mediated by the Na$^+$/I$^-$ symporter; (=) I$^-$ efflux toward the colloid; (⊙) Tg en route from the Golgi complex to the colloid; (⊣) TPO-catalyzed organification of I$^-$ into tyrosyl residues of the Tg molecule; (●) endocytosis of iodinated Tg; (⊕) lysosome; (&) phagolysosomal hydrolysis of endocytosed iodinated Tg; (⌢) release of T$_3$ and T$_4$ into the bloodstream.

uptake system of the thyroid provides the basis for radioactive iodide therapy, which is used in the treatment of toxic adenomas and thyroid cancer cases in which I$^-$ uptake is deemed sufficient to warrant such therapy (Dobyns and Maloof, 1951; DeGroot and Reilly, 1982; Gosdstein and Hart, 1983). Clearly, the special I$^-$ transport system of the thyroid ensures that the therapeutic radioisotope specifically reaches its target tissue. Endemic goiter and cretinism caused primarily by an insufficient dietary supply of iodine remain major health problems in many parts of the world, affecting millions of people (Matovinovic, 1983). Moreover, iodine deficiency still often leads to various degrees of impaired brain development, mostly in populations of children living in poor regions (Studer et al., 1974; Hetzel, 1983). The I^- transport system of the thyroid itself is rarely the primary target of thyroid disease. However, several cases of hypothyroidism caused by a specific partial or total congenital defect of the I^- transport system, a condition of considerable research interest, have been reported over the years (Leger et al., 1987; Vulsma et al., 1991). The reports indicate that the defect also involves other nonthyroid I^- uptake systems, such as those of the salivary glands, gastric mucosa, and choroid plexus.

Thyroid research in recent years has increasingly tended to address the physiology, regulation, and pathology of the gland at the molecular level. Detailed information on the physical properties, chemical composition, and primary sequence, as well as on the biosynthesis and posttranslational modifications of Tg has been obtained in the past decade (Ekholm, 1981; Lissitzky, 1984; Yoshinari and Taurog, 1985; Musti et al., 1986; Toyuyama et al., 1987). Similarly, TPO, the enzyme that catalyzes the organification of I^-, is now known to be an intrinsic, glycosylated heme-containing protein, present on the microvilli of the apical cell membrane. Human TPO has been purified and its primary amino acid sequence is known; both its catalytic mechanism and its regulation by TSH have been investigated (Magnusson et al., 1987; Magnusson, 1991; McLachlan and Rapoport, 1992). Moreover, TPO has been identified as the "microsomal antigen" in Hashimoto's thyroiditis, an autoimmune condition (Yokoyama et al., 1989). The TSH receptors from human, dog, and rat thyroids have recently been cloned and functionally expressed (Libert et al., 1989; Nagayama et al., 1989; Parmentier et al., 1989; Akamisu et al., 1990; Chazenbalk et al., 1990; Perret et al., 1990). Anti-TSH receptor antibodies have been implicated in the pathophysiology of Graves' disease (Rees Smith et al., 1988; McGregor, 1990). In contrast, the Na^+/I^- symporter has not been characterized at the molecular level, and no antibodies against it have been generated as of yet.

Expression of the Na^+/I^- Symporter in *Xenopus laevis* Oocytes

A central goal of I^- transport research is to characterize the Na^+/I^- symporter by cloning the cDNA that encodes it. Since no antibodies or oligonucleotides based on protein sequence data are yet available to screen appropriate cDNA libraries by conventional methods, it is necessary to use alternative strategies to clone the relevant cDNA. Our group has expressed the activity of the Na^+/I^- symporter in *Xenopus laevis* oocytes (Vilijn and Carrasco, 1989), thus establishing a potentially valuable system for the expression cloning of the cDNA that encodes the symporter. Poly A^+ RNA isolated from FRTL-5 cells, a highly functional line of rat thyroid cells (Ambesi-Impiombato et al., 1980, 1982; Bidey et al., 1988), was injected into oocytes, and the expression of the Na^+/I^- symporter in the plasma membrane was assayed by measuring Na^+-dependent, perchlorate-sensitive uptake of I^- by the oocytes. A five- to sevenfold increase of I^- uptake was detected in injected oocytes over the controls 5–6 d after injection. Total poly A^+ RNA was subsequently size-fractionated by sucrose gradient centrifugation. The resulting fractions were electrophoresed under denaturing conditions, pooled into five groups, and assayed for their ability to elicit I^- uptake. As shown in Fig. 2, group 2, which contained messages of 2.8–4.0 kb in length, elicited the highest transport activity. Therefore, the poly A^+ RNA that encodes the Na^+/I^- symporter is present in that fraction.

It should be pointed out that expression cloning using the oocyte system poses more difficulties than conventional cloning methods because a full-length (or at least fully functional) cDNA clone is required for the detection of a positive signal. Therefore, while efforts to find the cDNA clone that encodes the Na^+/I^- symporter continue, other studies are being performed to identify and eventually purify the symport protein itself. Included among these studies is the characterization of new inhibitors of thyroid Na^+/I^- symport activity.

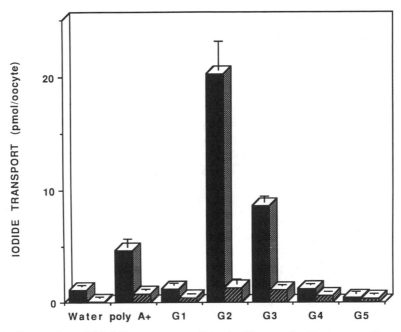

Figure 2. Expression of Na^+/I^- symport activity in *Xenopus laevis* oocytes. Oocytes were microinjected with either water, 50 ng of poly A^+ RNA, or ~30 ng of sucrose gradient-fractionated poly A^+ RNA. 7 d after injection, oocytes were assayed for Na^+-dependent I^- uptake. Oocytes (five/assay) were incubated in 100 μl of a solution containing 50 μM $Na^{125}I$ (sp act 100 mCi/mmol), 10 mM HEPES (pH 7.5), 1 mM $MgCl_2$, and 2 mM KCl with either 100 mM NaCl (*filled bars*) or 100 mM choline chloride (*hatched bars*). Reactions were terminated after 45 min by addition of 5 ml of ice-cold quenching buffer (100 mM choline chloride, 10 mM HEPES [pH 7.5]/1 mM methimazole) followed by rapid filtration through nitrocellulose filters. Filters were washed twice with an additional 5 ml each of the same quenching buffer. Radioactivity retained in the oocytes was quantitated in a gamma counter.

Identification of Novel Inhibitors of the Na^+/I^- Symporter: Harmaline and TRP-P-2

Inhibitors have often been used as probes to study and identify membrane transport proteins. However, the long-known "classic" competitive inhibitors of I^- transport (i.e., thiocyanate and perchlorate) have been of virtually no value for the characterization of the symporter. Since competition of these anions with I^- is due to their similarity in size and charge, their inhibitory activity is altered or lost when chemical modifications are performed in order to generate affinity chromatography ligands or photoaffinity labeling reagents. The only new inhibitor of I⁻ transport that has been reported in recent years is the marine toxin dysidenin, a "pseudocompetitive" inhibitor (Van Sande et al., 1990).

The hallucinogenic drug harmaline [4,9-dihydro-7-methoxy-1-methyl-3*H*-pyrido(3,4-b)indole] has been found to inhibit several Na⁺-dependent transport systems, including the intestinal Na⁺-dependent glucose carrier, the Na⁺/amino acid symporters from intestine, kidney, and erythrocytes, the Na⁺/H⁺ antiporter from placenta, and the cardiac Na⁺/Ca²⁺ antiporter (Sepulveda and Robinson, 1974; Sepulveda et al., 1976; Aronson and Bounds, 1980; Suleiman and Reeves, 1987; Young et al., 1988; Kulanthaivel et al., 1990). Inhibition by harmaline in all these cases was competitive with respect to Na⁺. We examined the possible ability of harmaline to inhibit Na⁺/I⁻ symport activity in two model systems, FRTL-5 cells and

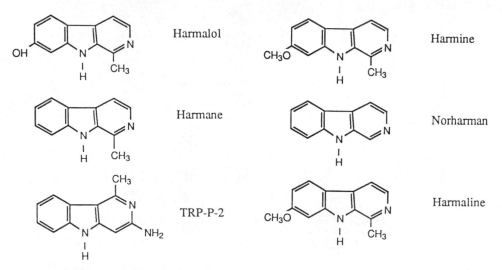

Figure 3. Chemical structure of harmaline and related compounds.

calf thyroid membrane vesicles (CTMV) (Kaminsky et al., 1991). Harmaline inhibited I⁻ uptake in a concentration-dependent fashion, displaying a relatively low affinity for the symporter molecule (K_i = 4 mM). Thus, we explored the effects of several harmaline-related compounds, the structures of which are shown in Fig. 3. Most significantly, Na⁺/I⁻ symport activity was inhibited by the chemically related convulsive agent TRP-P-2 (3-amino-1-methyl-5*H*-pyrido(4,3-b)indole acetate), which was over 10-fold more effective than harmaline. Inhibition by TRP-P-2 was competitive with respect to Na⁺ and was fully reversible. While the affinity of TRP-P-2 for the Na⁺ site of the Na⁺/I⁻ symporter was relatively low (K_i = 0.25 mM), it was > 100 times higher than that of Na⁺ (K_m = 50 mM). ⁴⁵Ca²⁺ efflux rates in CTMV were not affected by TRP-P-2, indicating that membrane integrity is not disrupted by the drug. Since TRP-P-2 contains a primary amino group (Fig. 3) that can be derivatized, the inhibitor may be a potentially useful tool for the identification of the Na⁺/I⁻ symporter and other similar symport proteins that may be inhibited as well.

Novel Observations on the Nature of TSH Stimulation of I^- Uptake

The ability of TSH to stimulate I^- uptake in the thyroid was first established in 1960, when it was observed that a single injection of TSH into rats caused a 50–100% increase in thyroid I^- accumulation after 8 h, as measured by the thyroid/serum radioiodide gradient (Halmi et al., 1960). The maximum increase was detected 24–48 h after the injection, and dissipation of the effect was complete within another 48 h. Similar experiments conducted in the 1970s in isolated bovine thyroid cells also showed 50–100% stimulation of uptake 6 h after addition of TSH, although a 2-h treatment with TSH was sufficient for the effect to occur. The addition of (Bu)2cAMP instead of TSH reproduced the same effect (Wilson et al., 1968; Knopp et al., 1970). A revealing aspect of these observations is that prolonged time lags were required for the effect of TSH to become evident after addition and to dissipate after removal of the hormone. It is also significant that the stimulation of I^- transport by TSH was prevented by the transcription inhibitor actinomycin D (ActD), and by the translation inhibitors cycloheximide and puromycin, pointing to an involvement of both transcription and translation. Still, the influence of prior TSH exposure on the observed I^- transport was impossible to evaluate in the isolated thyroid cell system. Although bound TSH was removed by gentle trypsin treatment, the effects of the hormone preceding such removal could not be assessed. In addition, the limited survival of isolated cells thus manipulated made it difficult to carry out observations beyond several hours.

The effect of TSH on I^- transport was subsequently examined using FRTL-5 cells instead of isolated cells (Weiss et al., 1984*a*, *b*), thereby confirming and extending the described observations. FRTL-5 cells exhibited I^- transport activity only when they were maintained in a medium containing TSH (TSH (+) cells). Withdrawal of TSH from the culture medium for 1 wk resulted in a decline of intracellular cAMP and complete loss of transport. Readdition of TSH to the medium fully restored I^- uptake activity within 3 d, but if TSH was added simultaneously with either actinomycin D or cycloheximide, no restoration of I^- transport activity occurred. These results have led to the proposition that TSH stimulation of I^- uptake results from increased biosynthesis of either the Na^+/I^- symporter or an activating factor. Conversely, the absence of Na^+/I^- symport activity in FRTL-5 cells maintained in TSH-free media (TSH (−) cells) has been attributed to a decrease in the biosynthesis of the Na^+/I^- symporter or an activating factor. However, we have unexpectedly observed Na^+/I^- symport activity in mixed membrane vesicles (MMV, which include all cell membranes) prepared from both TSH (+) and TSH (−) cells, even though the corresponding intact cells exhibited the established pattern (i.e., I^- uptake occurred in TSH (+) but not in TSH (−) cells) (Kaminsky, S. M., O. Levy, C. Salvador, G. Dai, and N. Carrasco, manuscript submitted for publication). These findings indicate that active Na^+/I^- symporter molecules are present in MMV prepared from TSH (−) cells, and that the proposed TSH-induced increase in the biosynthesis of the symporter fails to sufficiently explain the observed lack of I^- transport in TSH (−) cells. We have proposed the two following hypotheses to explain the phenomenon.

Redistribution Hypothesis

In the absence of TSH, Na^+/I^- symporter molecules may reside in intracellular organelles. Exposure to TSH would trigger a redistribution of these symporter molecules to the plasma membrane in an exocytosis-like fashion, much in the manner in which insulin promotes the transfer of glucose carrier molecules to the plasma membrane. This effect of insulin, however, is detected within 10 min (Simpson and Cushman, 1986), whereas the stimulation of Na^+/I^- symport activity by TSH is observed after 12–24 h (Weiss et al., 1984a). The effect may require a protein mediator. According to this model, the observed I^- transport in MMV from TSH $(-)$ cells would be due to the activity of the Na^+/I^- symporter molecules present in intracellular organelles rather than in the plasma membrane.

Activation Hypothesis

According to the activation hypothesis, Na^+/I^- symporter molecules present in the plasma membrane would be synthesized in an inactive form in the absence of TSH. The symporter would be activated by TSH by posttranslational modifications (e.g., phosphorylation, proteolytic processing), or interaction of a noncovalent activator with the Na^+/I^- symporter. TSH could also promote the removal of a posttranslational modifying group or of a noncovalent inhibitor. Therefore, addition of TSH would lead in any case to the presence of active Na^+/I^- symporter molecules in the plasma membrane. TSH could also increase biosynthesis of the symporter and/or other proteins that mediate the activation process. Preparation of MMV from TSH $(-)$ cells would mimic activation of the symporter by TSH by either exposing the symporter to an activator or removing an inhibitor.

To determine if intracellular membrane sources of Na^+/I^- symporter exist, and thus distinguish between the two proposed hypotheses, symport activity was measured in subcellular membrane fractions from TSH $(+)$ and TSH $(-)$ FRTL-5 cells. We found that maximal transport activity in both TSH $(+)$ and $(-)$ cells was associated with the plasma membrane fraction rather than with membrane fractions from intracellular organelles, a result that rules out the redistribution hypothesis and provides supporting evidence for the activation mechanism.

The precise process by which TSH ultimately stimulates I^- uptake or possibly activates Na^+/I^- symporter molecules present in the plasma membrane is unknown. However, in experiments addressing this question, we have observed (Kaminsky, S. M., O. Levy, C. Salvador, G. Dai, and N. Carrasco, manuscript submitted for publication) that I^- uptake in MMV from both TSH $(+)$ and TSH $(-)$ FRTL-5 cells was enhanced when MMV were prepared under conditions that favor proteolysis (i.e., in the absence of protease inhibitors and at room temperature) over controls prepared in the presence of protease inhibitors and at 4°C. I^- uptake in membrane vesicles prepared from intact rat thyroid glands was also increased when the vesicles were prepared under proteolysis-favorable conditions, indicating that the effect is not confined to thyroid cells in culture. These data raise the possibility that a proteolytic event may play a role in the mechanism by which TSH increases I^- uptake. Proteolysis has been reported to activate other membrane transporters, including the erythrocyte membrane Ca^{2+} ATPase (James et al., 1989) and the plasmalemma Na^+/Ca^{2+} exchanger (Hilgemann, 1990).

Concluding Remarks

In summary, (a) the activity of the Na$^+$/I$^-$ symporter has been expressed in *Xenopus laevis* oocytes, thus generating a valuable system for the functional cloning of the cDNA that encodes the symporter; (b) the harmaline-related compound TRP-P-2 has been identified as a Na$^+$ site competitive inhibitor of the Na$^+$/I$^-$ symporter; this reagent may be a potentially useful tool for the identification of the Na$^+$/I$^-$ symport protein; and (c) Na$^+$/I$^-$ symport activity has unexpectedly been observed in MMV from FRTL-5 cells maintained in the absence of TSH; since the activity is associated with the plasma membrane, this finding suggests that TSH stimulates I$^-$ uptake at least partly via activation of symporter molecules present in the plasma membrane; additional observations raise the possibility that a step in such activation may involve a proteolytic event.

Although the existence of a special I$^-$ transport system in the thyroid gland has been recognized for decades, little has been elucidated to date on the Na$^+$/I$^-$ symporter at the molecular level. Since the thyroid I$^-$ transport system is clearly of great physiological importance and considerable value in the diagnosis and treatment of thyroid conditions, it is hoped that current I$^-$ transport research will soon lead to the thorough molecular dissection of this remarkable transport process.

References

Akamisu, T., S. Ikuyama, M. Saji, S. Kosuji, C. Kozak, O. W. McBride, and L. D. Kohn. 1990. Cloning, chromosomal assignment, and regulation of the rat thyrotropin receptor: expression of the gene is regulated by thyrotropin, agents that increase cAMP levels, and thyroid autoantibodies. *Proceedings of the National Academy of Sciences, USA.* 87:5677–5681.

Ambesi-Impiombato, F. S., L. A. M. Parks, and H. G. Coons. 1980. Culture of hormone-dependent epithelial cells from rat thyroids. *Proceedings of the National Academy of Sciences, USA.* 77:3455–3459.

Ambesi-Impiombato, F. S., R. Picone, and D. Tramontano. 1982. The influence of hormones and serum on growth and differentiation of the thyroid cell strain in growth of cells in hormonally-defined media. *Cold Spring Harbor Symposia on Quantitative Biology.* 9:483.

Aronson, P. S., and S. E. Bounds. 1980. Harmaline inhibition of Na$^+$ -dependent transport in renal microvillus membrane vesicles. *American Journal of Physiology.* 238:F210–F217.

Bagchi, N., and D. M. Fawcett. 1973. Role of sodium ion in active transport of iodide by cultured thyroid cells. *Biochimica et Bophysica Acta.* 318:235–251.

Bidey, S. P., A. Lambert, and W. R. Robertson. 1988. Thyroid cell growth, differentiation and function in the FRTL-5 cell line: a survey. *Journal of Endocrinology.* 119:365–376.

Carrasco, N. 1993. Iodide transport in the thyroid gland. *Biochimica et Biophysica Acta.* In press.

Chambard, M., B. Verrier, J. Gabrion, and J. Mauchamp. 1983. Polarisation of thyroid cells in culture: evidence for the baso-lateral localization of the iodide "pump" and of the thyroid-stimulating hormone receptor-adenyl cyclase complex. *Journal of Cell Biology.* 96:1172–1177.

Chazenbalk, G., Y. Nagayama, K. D. Kaufman, and B. Rapoport. 1990. The functional expression of recombinant human thyrotropin receptors in nonthyroidal eukaryotic cells provides evidence that homologous desensitization to thyrotropin stimulation requires a cell-specific factor. *Endocrinology.* 127:1240–1244.

DeGroot, L. J. 1989. Endocrinology. L. J. DeGroot, editor. Grune and Stratton Inc., Orlando, FL. 505–804.

DeGroot, L. J., and M. Reilly. 1982. Comparison of 30 and 50 mCi doses of iodine-131 for thyroid ablation. *Annals of Internal Medicine.* 96:51–53.

Dobyns, B., and F. Maloof. 1951. Study and treatment of 119 cases of carcinoma of the thyroid with radioactive iodine. *Journal of Clinical Endocrinology.* 11:1323–1360.

Dumont, J. E., S. Vassart, and S. Refetoff. 1989. The Metabolic Basis of Inherited Diseases. C. R. Scriver, A. L. Beaudet, W. S. Sly, and D. Valle, editors. McGraw-Hill, New York. 1843–1880.

Ekholm, R. 1981. Iodination of thyroglobulin: an intracellular or extracellular mechanism. *Molecular and Cellular Endocrinology.* 24:141–163.

Gosdstein, R., and I. R. Hart. 1983. Follow-up of solitary autonomous thyroid nodules treated with [131]I. *New England Journal of Medicine.* 309:1473–1476.

Halmi, N. S. 1961. Thyroidal iodide transport. *Vitamins and Hormones.* 19:133–163.

Halmi, N. S., D. K. Granner, D. J. Doughman, B. H. Peters, and D. Muller. 1960. Biphasic effect of TSH on thyroidal iodide collection in rats. *Endocrinology.* 67:70–81.

Heslop, R. B., and K. Jones. 1976. Inorganic Chemistry: A Guide to Advanced Study. Elsevier Scientific Publishing Company, Amsterdam, The Netherlands. 511.

Hetzel, B. S. 1983. Iodine deficiency disorders (IDD) and their erradication. *Lancet.* ii:1126–1129.

Hilgemann, D. W. 1990. Regulation and deregulation of cardiac Na^+-Ca^{2+} exchange in giant excised sarcolemmal membrane patches. *Nature.* 344:242–245.

James, P., T. Vorherr, J. Krebs, A. Morelli, G. Castello, D. J. McCormick, J. T. Penniston, A. De Flora, and E. Carafoli. 1989. Modulation of erythrocyte Ca^{2+} ATPase by selective calpain cleavage of the calmodulin-binding domain. *Journal of Biological Chemistry.* 264:8289–8296.

Kaminsky, S. M., O. Levy, M. T. Garry, and N. Carrasco. 1991. Inhibition of the Na^+/I^- symporter by harmaline and TRP-P2 in thyroid cells and membrane vesicles. *European Journal of Biochemistry.* 200:203–207.

Kendal, L. W., and R. E. Condon. 1969. Prediction of malignancy in solitary thyroid nodules. *Lancet.* i:1071.

Knopp, J., V. Stolc, and W. Tong. 1970. Evidence for the induction of iodide transport in bovine thyroid cells treated with thyroid-stimulating hormone or dibutyryl cyclic adenosine 3′,5′-monophosphate. *Journal of Biological Chemistry.* 245:4403–4408.

Kulanthaivel, P., F. H. Leibach, V. B. Mahesh, and E. J. Cragoe. 1990. The Na^+/H^+ exchanger of the placental brush-border membrane is pharmacologically distinct from that of the renal brush border membrane. *Journal of Biological Chemistry.* 265:1249–1252.

Leger, F. A., R. Doumith, C. Courpotin, O. B. Helal, N. Davous, A. Aurengo, and J. C. Savoie. 1987. Complete iodide trapping defect in two cases with congenital hypothyroidism: adaptation of thyroid to huge iodide supplementation. *European Journal of Clinical Investigation.* 17:249–255.

Libert, F., A. Lefort, C. Gerard, M. Parmentier, J. Perret, M. Ludgate, J. E. Dumont, and G. Vassart. 1989. Cloning, sequencing and expression of the human thyrotropin (TSH) receptor: evidence for binding of autoantibodies. *Biochemical and Biophysical Research Communications.* 165:1250–1255.

Lissitzky, S. 1982. Deiodination of iodotyrosines. *In* Diminished Thyroid Hormone Formation: Possible Causes and Clinical Aspects. D. Reinwein and E. Klein, editors. Schattauer Verlag, Stuttgart. 49–61.

Lissitzky, S. 1984. Thyroglobulin entering into molecular biology. *Journal of Endocrinological Investigation.* 7:65–76.

Magnusson, R. P. 1991. Thyroid peroxidase. *In* Peroxidases in Chemistry and Biology. J. Everse, K. E. Everse, and M. B. Grisham, editors. CRC Press, Inc., Boca Raton, FL. 199–219.

Magnusson, R. P., J. Gestautas, A. Taurog, and B. Rapoport. 1987. Molecular cloning of the structural gene for porcine thyroid peroxidase. *Journal of Biological Chemistry.* 262:13885–13888.

Matovinovic, J. 1983. Endemic goiter and cretinism at the dawn of the third millennium. *Annual Review of Nutrition.* 3:341–412.

McGregor, A. M. 1990. Autoantibodies to the TSH receptor in patients with autoimmune disease. *Clinical Endocrinology.* 33:683–685.

McLachlan, S. M., and B. Rapoport. 1992. The molecular biology of thyroid peroxidase: cloning, expression and role as autoantigen in autoimmune thyroid disease. *Endocrine Reviews.* 13:192–206.

Musti, A. M., E. V. Avvedimento, C. Polistina, V. M. Ursini, S. Obici, L. Nitsch, S. Cocozza, and R. Di Lauro. 1986. The complete structure of the whole thyroglobulin gene. *Proceedings of the National Academy of Sciences, USA.* 83:323–327.

Nagayama, Y., K. Kaufman, P. Seto, and B. Rapoport. 1989. Molecular cloning, sequence and functional expression of the cDNA for the human thyrotropin receptor. *Biochemical and Biophysical Research Communications.* 165:1184–1190.

Nakamura, Y., T. Kotani, and S. Ohtaki. 1990. Transcellular iodide transport and iodination on the apical plasma membrane by monolayer porcine thyroid cells cultured on collagen-coated filters. *Journal of Endocrinology.* 126:275–281.

Nakamura, Y., S. Ohtaki, and I. Yamazaki. 1988. Molecular mechanism of iodide transport by thyroid plasmalemmal vesicles: cooperative sodium activation and asymmetrical affinities for the ions on the outside and inside of the vesicles. *Journal of Biochemistry.* 104:544–549.

Nilsen, T. B., Y. Totsuka, and J. B. Field. 1982. Three types of desensitization of metabolic responses to thyrotropin in thyroid tissue: a review. *Endocrinologia Experimentalis.* 16:247–257.

Nilsson, M., U. Björkman, R. Ekholm, and L. E. Ericson. 1990. Iodide transport in primary cultured thyroid follicles cells: evidence of a TSH regulated channel mediating iodide efflux selectively across the apical domain of the plasma membrane. *European Journal of Cell Biology.* 52:270–281.

O'Neill, B., D. Magnolato, and G. Semenza. 1987. The electrogenic, sodium-dependent iodide transport system in plasma membrane vesicles from thyroid glands. *Biochimica et Biophysica Acta.* 896:263–274.

Parmentier, M., F. Libert, C. Maenhaut, A. Lefort, C. Gérard, J. Perret, J. Van Sande, J. E. Dumont, and G. Vassart. 1989. Molecular cloning of the thyrotropin receptor. *Science.* 246:1620–1622.

Perret, J., M. Ludgate, F. Libert, C. Gerard, J. E. Dumont, G. Vassart, and M. Parmentier. 1990. Stable expression of the human TSH receptor in CHO cells and characterization of differentially expressing clones. *Biochemical and Biophysical Research Communications.* 171:1044–1050.

Rees Smith, B., S. M. McLachlan, and J. Furmaniak. 1988. Autoantibodies to the thyrotropin receptor. *Endocrine Reviews.* 9:106–120.

Sepulveda, F. V., M. Bulcon, and J. W. L. Robinson. 1976. Differential effects of harmaline and oubain on intestinal sodium, phenylalanine and β-methyl-glucoside transport. *Naunyn-Schmiedeberg's Archives of Pharmacology*. 295:231–236.

Sepulveda, F. V., and J. W. L. Robinson. 1974. Harmaline, a potent inhibitor of sodium-dependent transport. *Biochimica et Biophysica Acta*. 373:527–531.

Simpson, I. A., and S. W. Cushman. 1986. Hormonal regulation of mammalian glucose transport. *Annual Review of Biochemistry*. 55:1059–1089.

Sterling, L., and J. H. Lazarus. 1977. The thyroid and its control. *Annual Reviews of Physiology*. 39:349–371.

Stubbe, P., F. J. Schulte, and P. Heidemann. 1986. Iodine deficiency and brain development. *Bibliotheca Nutritio et Dieta*. 38:206–208.

Studer, H., H. Kohler, and H. Burgi. 1974. Iodine deficiency. *Handbook of Physiology*. 3:974.

Suleiman, M. S., and J. P. Reeves. 1987. Inhibition of Na^+/Ca^{2+} exchange mechanisms in cardiac sarcolemmal vesicles by harmaline. *Comparative Biochemistry and Physiology*. 88C:197–200.

Toyuyama, T., M. Yoshinari, A. B. Rawitch, and A. Taurog. 1987. Digestion of thyroglobulin with purified thyroid lysosomes: preferential release of iodoamino acids. *Endocrinology*. 121:714–721.

Van Sande, J., F. Denebourg, R. Beauwens, J. C. Brekman, D. Daloze, and J. E. Dumont. 1990. Inhibition of iodide transport in thyroid cells by dysidenin, a marine toxin, and some of its analogs. *American Society of Pharmacological Experimental Therapy*. 37:583–589.

Vilijn, F., and N. Carrasco. 1989. Expression of the thyroid sodium/iodide symporter in *Xenopus laevis* oocytes. *Journal of Biological Chemistry*. 264:11901–11903.

Vulsma, T., J. A. Rammeloo, M. H. Gons, and J. J. M. de Vijlder. 1991. The role of serum thyroglobulin concentration and thyroid ultrasound imaging in the detection of iodide transport defects in infants. *Acta Endocrinologica*. 124:405–410.

Weiss, S. J., N. J. Philp, F. S. Ambesi-Impiombato, and E. F. Grollman. 1984*a*. Thyrotropin-stimulated iodide transport mediated by adenosine 3',5'-monophosphate and dependent on protein synthesis. *Endocrinology*. 114:1099–1107.

Weiss, S. J., N. J. Philp, and E. F. Grollman. 1984*b*. Iodide transport in a continuous line of cultured cells from rat thyroid. *Endocrinology*. 114:1090–1098.

Werner, S. C., and S. Ingbar. 1991. The Thyroid: A Fundamental and Clinical Text. L. E. Braverman and R. D. Utiger, editors. J. B. Lippincott Company, Philadelphia. 1365 pp.

Wilson, B., E. Raghupathy, T. Tonoue, and W. Tong. 1968. TSH-like actions of dibutyryl-cAMP on isolated bovine thyroid cells. *Endocrinology*. 83:877–884.

Wolff, J. 1964. Transport of iodide and other anions in the thyroid gland. *Physiological Reviews*. 44:45–90.

Yokoyama, N., A. Taurog, and G. G. Klee. 1989. Thyroid peroxidase and thyroid microsomal autoantibodies. *Journal of Clinical Endocrinology and Metabolism*. 68:766–773.

Yoshinari, M., and A. Taurog. 1985. Lysosomal digestion of thyroglobulin: role of cathespin D and thiol proteases. *Endocrinology*. 117:1621–1631.

Young, J. D., D. K. Mason, and D. A. Fincham. 1988. Topographical similarities between harmaline inhibition sites on Na^+-dependent amino acid transport system ASC in human erythrocytes and Na^+-independent system asc in horse erythrocytes. *Journal of Biological Chemistry*. 263:140–143.

Chapter 21

Electrophysiology of the Na^+/Glucose Cotransporter

Lucie Parent and Ernest M. Wright

Department of Molecular Physiology and Biophysics, Baylor College of Medicine, Houston, Texas 77030; and Department of Physiology, UCLA School of Medicine, Los Angeles, California 90024-1751

Molecular Biology and Function of Carrier Proteins © 1993 by The Rockefeller University Press

Introduction

Thirty years after Crane et al. (1961, 1962) proposed that sugar transport was coupled to the Na^+ ion gradient in the small intestine, there is still no definitive kinetic model of the Na^+/glucose cotransporter (Kimmich, 1990; Centelles et al., 1991). Although we have come a long way since Crane's pioneer hypothesis (see, for instance, Crane, 1977; Semenza et al., 1984; Schultz, 1986; Hopfer, 1987; Wright et al., 1992), we are still striving to understand the complex relationship between the structure and function of the cotransporter. In 30 years, the focus of research has shifted from the physiological role of the cotransporter and its phenomenological description to the characterization of the gene product. This development is best seen through the evolution of the experimental models used for the study of the Na^+/glucose cotransporter; starting with isolated tissues (see, for instance, Schultz and Curran, 1970), followed by isolated enterocytes (Kimmich, 1970), then to isolated brush border membrane vesicles (BBMV) (Hopfer et al., 1973; Kessler et al., 1978; and reviewed by Berteloot and Semenza, 1990), and finally to the cloned protein expressed in *Xenopus* oocytes (Hediger et al., 1987; Ikeda et al., 1989). Early methods measured the unidirectional flux of radioactive substrate, while more recent electrophysiological studies have exploited the electrogenicity of the transport process. The electrophysiology of cotransporters had been studied using isolated tissues mounted in Ussing chambers (Lapointe et al., 1986) and whole-cell recording of cultured cells (Smith-Maxwell et al., 1990; Jauch et al., 1986; see Hoyer and Gögelein, 1991 for studies on the Na^+/alanine cotransporter).

Electrophysiological methods present certain advantages in studying electrogenic cotransport. Cotransporter activity can be measured with a greater time resolution (in the millisecond range) and under improved voltage-clamp conditions. Controlling the membrane potential is pivotal since stimulation of the cotransporter is known to depolarize the cell in situ, and this depolarization, decreasing the electrochemical driving force, influences in return the activity of the cotransporter (Murer and Hopfer, 1974; Frömter, 1982; Samaržija et al., 1982; Lang et al., 1986; Lapointe et al., 1990). Some key questions can therefore only be adequately addressed by electrophysiological methods. One of these crucial issues regards the site of the voltage dependence in the transport cycle. Most kinetic models have traditionally assumed that the carrier translocation is the voltage-dependent reaction step (Semenza et al., 1984), although some have proposed more recently that binding of Na^+ ions to the carrier is responsible for the observed voltage dependence (Na^+ well effect; Mitchell, 1969; Jauch and Läuger, 1986; Kimmich and Randles, 1988; Bennett and Kimmich, 1992). Unfortunately, the practical use of electrophysiological methods to record Na^+/glucose cotransport activity had long been hampered by the relatively poor access to the brush border membrane in most cells.

The emergence of molecular biological techniques has alleviated some of these difficulties by presumably increasing the density of functional cotransporters in a simple cell model. The Na^+/glucose cotransporter was cloned and successfully expressed in transient systems such as *Xenopus laevis* oocytes (Hediger et al., 1987; Ikeda et al., 1989), the insect cell line *Sf9* (Smith et al., 1992), and the mammalian cell lines COS-7 (Birnir et al., 1990) and HeLa (Blakely et al., 1991). Early experiments showed that the cloned cotransporter expressed in *Xenopus* oocytes can generate inward currents as high as 1 μA at -100 mV (Umbach et al., 1990). It then became possible to investigate the effect of potential on steady-state (Birnir et al., 1991; Parent et al., 1992*a*) and pre–steady-state (Parent et al., 1992*a*) kinetics using the two-electrode voltage-clamp method. Furthermore, owing both to the size of the cotransporter currents and to the noninvasive nature of the measurement technique, steady-state kinetic experiments can be performed on a single oocyte, hence minimizing the experimental error associated with heterogeneous populations of BBMV.

In this paper, we will review our current contribution to the investigation of the functional properties of the Na$^+$/glucose cotransporter (Parent et al., 1992*a,b*). Our intent is to demonstrate the universal appeal of an electrophysiological approach in elucidating Na$^+$-coupled cotransport. Although the experiments benefited from recent technological advances, the simultaneous kinetic model proposed is similar to classical models previously discussed for the Na$^+$/glucose cotransporter (Hopfer and Groseclose, 1980; Restrepo and Kimmich, 1985*a*; Schultz, 1986) with the important distinctions that the binding of external Na$^+$ ions and the carrier translocation step are both dependent upon the membrane potential and we do not assume rapid equilibrium (Parent et al., 1992*b*).

Description of the Na$^+$/Glucose Cotransporter Currents

The properties of the cloned Na$^+$/glucose cotransporter were assessed by radioactive uptake of the nonmetabolizable analogue α-methyl-D-glucopyranoside (αMDG) in intact oocytes (Ikeda et al., 1989) and were found to be similar to the properties of the cotransporter measured in native brush borders, namely: (1) Sugar transport is inhibited by phlorizin in a Na$^+$-dependent manner with an inhibition constant between 5 and 15 μM at saturating [Na$^+$]$_o$. (2) Sugar transport is Na$^+$ specific. Cation substitution by Li$^+$, K$^+$, or choline eliminates sugar transport. (3) Sugar transport is specific for pyranoses in the C-1 conformation with a free equatorial −OH group on carbon number 2 such as D-glucose, D-galactose, or αMDG. (4) Furthermore, sugar transport was found to be electrogenic. When expressed into *Xenopus* oocytes, the cloned Na$^+$/glucose cotransporter can generate sizable currents that are reversibly blocked by 10–100 μM phlorizin (Umbach et al., 1990; Birnir et al., 1991; Parent et al., 1992*a*).

cRNA coding for the rabbit Na$^+$/glucose cotransporter from the small intestine was injected into *Xenopus* oocytes and the electrical activity of the cotransporter was measured 5–10 d after injection. Fig. 1 shows the current traces obtained in the presence of 100 mM Na$^+$ medium before (left) and after (right) the addition of 1 mM αMDG to the bath. Current traces recorded at +50 mV in the Na$^+$ medium and in the absence of sugar relaxed with a time constant $\tau_{1/2}$ of ∼ 13 ms. This relaxation, absent at negative potential and after the addition of sugar, is referred to as pre–steady-state current in this paper. Cotransporter currents were also measured at steady state. Addition of 1 mM αMDG caused an increase in inward currents that was reversed upon washout of the sugar. The increase in inward current measured at the end of the pulse can be better appreciated in the accompanying *I-V* curves (Fig. 1, bottom panels). The addition of sugar shifted the zero current potential from −60 to −15 mV, a shift that is similar to the sugar-induced depolarization observed under non–voltage-clamped conditions. As seen in this figure, the steady-state sugar-dependent *I-V* curve was sigmoidal with no measurable reversal potential. It should

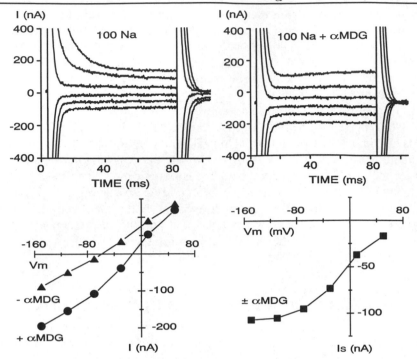

Figure 1. Current traces from cRNA-injected oocytes were recorded in a Na⁺ medium (mM):
100 NaCl, 2 KCl, 1 MgCl₂, 1 CaCl₂, 10 HEPES-Tris, pH 7.4, using the two-electrode
voltage-clamp method. Holding potential (V_h) was −50 mV. Current traces were recorded at
six voltages: −150, −110, −70, −30, 10, and 50 mV for 84 ms and records shown were not leak
subtracted. (*Top left*) Current traces recorded in the Na⁺ medium relaxed more slowly to a
steady-state current at positive potentials. This slower decay (∼13 ms) was typically observed
for voltages more positive than the holding potential in the presence of Na⁺ ions (see Fig. 3).
(*Top right*) Addition of 1 mM αMDG to the 100 mM [Na⁺] medium triggered an increase in
steady-state inward current. The developing outward current is caused by the native delayed
rectifier (Lu et al., 1990). Note the absence of pre–steady-state currents in the presence of 1
mM αMDG. (*Bottom left*) The current values recorded before (▲) and after the addition of 1
mM [αMDG]₀ to the bath (●) were averaged from 20 points taken between 79 and 84 ms and
were plotted as a function of the membrane potential. (*Bottom right*) Steady-state sugar-
dependent currents (i_s) were obtained by taking the difference between the current obtained
before and after the addition of sugar ($i_s = i_{Na} - i_{Na+s}$) and were plotted as a function of the
membrane potential. Modified from Parent et al. (1992a).

be mentioned, however, that Umbach and co-workers (1990) reported that the
cotransporter currents can reverse after overnight loading with αMDG.

   There is strong evidence that pre–steady-state and sugar-dependent currents
were generated directly by the Na⁺/glucose cotransporter. These currents were
never measured in control water-injected oocytes (Parent et al., 1992a). In addition,
pre–steady-state and sugar-dependent currents were found to be independent of the
external concentration of Cl⁻, K⁺, Mg²⁺, or Ca²⁺ ions, were insensitive to the
presence of the K⁺ channel blockers Ba²⁺ and TEA⁺, and Cl⁻ channel blockers
DIDS and SITS, and were not affected by the addition of 5 mM ouabain or 100 μM
amiloride. Finally, there was a direct correlation between the uptake of radioactive

αMDG into the oocyte, the magnitude of the sugar-dependent currents, and the magnitude of the pre–steady-state currents extrapolated to time 0. Only phlorizin (K_i = 10 μM) was found to block steady-state currents (Umbach et al., 1990; Birnir et al., 1991; Parent et al., 1992*a*) and pre–steady-state currents (Parent et al., 1992*a*). The magnitude of the sugar-dependent currents, measured at steady state, was found to depend upon the applied potential, the external Na$^+$, and sugar concentrations. The specificity of the cotransporter currents for sugars was: D-glucose, D-galactose, αMDG > 3-*O*-methyl-D-glucoside ≫ D-xylose > D-allose ≫ D-mannose (Birnir et al., 1991).

The Na$^+$/glucose cotransporter currents also exhibit some temperature dependence. The effect of temperature on transport rate has long been a very useful criterion to distinguish secondary active transport from simple diffusion processes. The temperature dependence of the sugar-dependent currents (i_s) was investigated

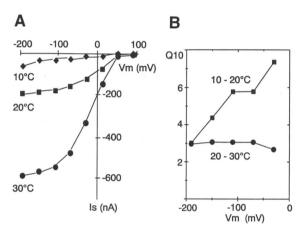

Figure 2. Temperature dependence of steady-state sugar-dependent currents. (*Left*) Steady-state sugar-dependent currents (i_s) were measured at 10 (♦), 20 (■), and 30°C (●) in the presence of 100 mM [Na$^+$]$_o$ and 1 mM [D-glucose] as a function of voltage from −190 to +90 mV. Chamber temperature was controlled by a Peltier effect unit (generously loaned from Dr. F. Bezanilla, Department of Physiology, UCLA School of Medicine, Los Angeles, CA). Results were obtained on the same oocyte. Oocyte exposure to bath temperatures lower or higher than room temperature was kept to a minimum and control *I-V* curves were periodically measured at 20°C. (*Right*) Sugar-dependent currents increased as a function of the bath temperature between 20 and 30°C with a Q_{10} of 3. The temperature dependence is more significant between 10 and 20°C. Q_{10} ranged from 7 (−30 mV) to 3 (−190 mV).

between 10 and 30°C. i_s increased with a Q_{10} of 3 (SEM = 0.1; N = 4) between 20 and 30°C (Fig. 2 *A*). The temperature dependence was, however, markedly increased at lower temperatures (10–20°C) with Q_{10} factors between 3 and 7 (Fig. 2 *B*). Such temperature dependence is much higher than that previously reported for ion channels ($Q_{10} \approx$ 1.3; Coronado et al., 1980; Barrett et al., 1982) but is within the range generally reported for Na$^+$-coupled transporter fluxes (Kippen et al., 1979; De Smedt and Kinne, 1981; Brot-Laroche et al., 1986; Malo and Berteloot, 1991) and currents (Schultz and Zalusky, 1964; Bergman and Bergman, 1985; Jauch et al., 1986).

Pre–Steady-State Currents

Pre–steady-state currents associated with the cotransporter activity were recorded in oocytes expressing the cloned cotransporter. Pre–steady-state kinetics are priceless in kinetic modeling because they give direct access to the time constant of the rate-limiting step in the transport cycle. It is widely recognized that it is practically impossible to determine individual rate constants only on the basis of a steady-state treatment (Roy et al., 1991). As observed in Fig. 1, current traces obtained from voltage-clamped RNA-injected oocytes typically reached a steady state slowly in response to a positive voltage pulse from a negative holding potential. For the rabbit Na^+/glucose cotransporter, such relaxation was only observed in response to a positive voltage pulse in the presence of high external Na^+ concentration ($[Na^+]_o > 30$ mM) from a negative holding potential. Transients were not observed in the nominal absence of Na^+ ions (choline replacement), in the presence of Na^+ ions and sugar, or in the presence of Na^+ ions and 100 μM phlorizin (Fig. 3). In addition, there appears

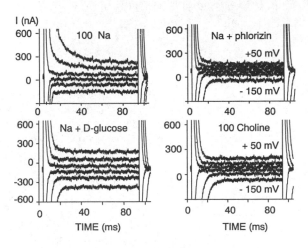

Figure 3. Na^+/glucose cotransporter pre–steady-state currents. Current traces were recorded at six potentials (-150, -110, -70, -30, 10, and 50 mV) from $V_h = -50$ mV and were not subtracted for leak current. Records were obtained during the same experiment in the presence of 100 mM $[Na^+]_o$ (*top left*); in the presence of 100 mM $[Na^+]_o$ and 100 μM phlorizin (*top right*); in the presence of 100 mM $[Na^+]_o$ and 1 mM $[\alpha MDG]_o$ (*bottom left*); and in the presence of 100 mM choline-Cl (*bottom right*). Only current traces recorded in the 100 mM Na^+ medium show current transients. Note the increase in the current scale for current traces recorded in the presence of 100 mM $[Na^+]_o$ and 1 mM $[\alpha MDG]_o$. At $V_m = -150$ mV, steady-state currents were -155 nA (Na^+), -136 nA (Na^+ + phlorizin), -373 nA (Na^+ + sugar), and -126 nA (choline). Steady-state phlorizin-sensitive currents (Na^+ ± phlorizin) were -19 nA and represent 9% of the steady-state sugar-dependent currents (Na^+ ± sugar) measured at the same potential. This steady-state phlorizin-sensitive current indicates that some Na^+ ions are translocated in the absence of sugar (Na^+ leak current). (Reproduced from *Journal of Membrane Biology,* 1992, 125:49–62, by copyright permission of Springer-Verlag New York Inc.)

to be an inverse relationship between the pre–steady-state current and the external concentration of sugar and a direct correlation between the amplitude of the pre–steady-state charge transfer and the amplitude of the sugar-dependent current (Loo et al., 1992; Parent et al., 1992*a*).

The pre–steady-state current was voltage dependent with a time constant decreasing from 13 (50 mV) to 2 ms (-30 mV). It was impossible to separate pre–steady-state currents from capacitive transients at potentials more negative than the holding potential (Parent et al., 1992*a*). These characteristics suggest that the pre–steady-state currents are associated with the cotransporter and probably result from the translocation of the empty cotransporter as we will see later.

Pre–steady-state currents were also highly dependent on the temperature. The temperature dependence of the pre–steady-state currents was also investigated between 10 and 20°C. Current traces, recorded at $V_m = +50$ mV in the presence of 100 mM $[Na^+]_o$, were fitted to two exponential functions to account for the noncompensated capacitive transient and the pre–steady-state current. The relaxation time constant τ decreased sharply as a function of the bath temperature with $Q_{10} = 3.4$ (our unpublished observations).

Voltage Dependence of External Na⁺ and Sugar Steady-State Kinetics

The single relaxation observed under pre–steady-state conditions is caused by the rate-limiting step in the transport cycle and the characteristics of the pre–steady-state currents can impose strong limitations on the individual rate constants. Steady-state kinetics are otherwise essential in characterizing the sequence of reaction steps that leads to sugar transport and in addressing questions such as the cotransporter stoichiometry and the actual voltage dependence of Na⁺ ions.

As we have seen, our experimental system is quite well suited for steady-state kinetics. The magnitude of the sugar-dependent currents (i_s) was found to depend only upon the membrane potential, the external Na⁺, and the external sugar concentrations, and to be completely reversible upon washout of the substrate. This specificity allowed us to devise a series of steady-state kinetic experiments where i_s was measured sequentially, on a single oocyte, as a function of the external Na⁺ ion concentration (Fig. 4) and as a function of the external substrate concentration (Fig. 5) between 0 and −200 mV.

To investigate the voltage dependence of Na⁺ ions binding to the transporter, steady-state sugar-dependent currents (i_s) were measured in the presence of 0.1, 1, and 20 mM $[\alpha MDG]_o$ with $[Na^+]_o$ varying from 2 to 100 mM. The increase of i_s as a function of $[Na]_o$ was best described by a Hill plot with an apparent coupling coefficient n of 2 for all sugar concentrations (see Figs. 5 and 6 of Parent et al., 1992*a*). The kinetic parameters obtained from an experiment performed in the presence of 1 mM $[\alpha MDG]_o$ are shown in Fig. 4. $K_{0.5}^{Na}$ was voltage dependent, decreasing from 53 to 3 mM from −30 to −200 mV, and i_{max}^{Na} was weakly voltage dependent. $K_{0.5}^{Na}$ was also dependent on the external sugar concentration. In a separate series of experiments, $K_{0.5}^{Na}$ was shown to decrease from 30 to 5 mM at $V_m = -70$ mV and from 14 to 1 mM at $V_m = -150$ mV when $[\alpha MDG]_o$ was increased from 0.1 to 20 mM.

The temperature dependence of $K_{0.5}^{Na}$ and i_{max}^{Na} was investigated between 22 and 32°C, as it becomes difficult to measure i_s and therefore to assess kinetics below 20°C. The results of an experiment conducted in the presence of 1 mM $[\alpha MDG]_o$ with $[Na^+]_o$ varying from 2 to 100 mM are shown in Fig. 6. The apparent Hill coupling coefficient n was not affected by the change in temperature and ranged from 1.8 (22°C) to 1.9 (32°C) (Fig. 6 *C*). At saturating external sugar concentration, there was no temperature dependence of $K_{0.5}^{Na}$. The increase in cotransporter currents is due exclusively to an increase in i_{max}^{Na}. While $K_{0.5}^{Na}$ increased from 3 to 53 mM at 32°C, $K_{0.5}^{Na}$ increased from 2.5 to 54 mM at 22°C over the same potential range. From this experiment at 32°C and four others at 22°C, $K_{0.5}^{Na}$ was found to decrease from 45

(SEM = 6; N = 5) to 17 mM (SEM = 1; N = 5) between -30 and -70 mV (Fig. 6 D). In contrast, i_{max}^{Na} was highly temperature dependent (Fig. 6 E). At -190 mV, i_{max}^{Na} increased from -195 nA at 22°C to -600 nA at 32°C (Q_{10} = 3.1). At -30 mV, i_{max}^{Na} increased from -123 nA at 22°C to -321 nA at 32°C (Q_{10} = 2.7). This is the first time, to our knowledge, that the temperature dependence of the Na$^+$/glucose cotransporter kinetics for Na$^+$ ions is being reported.

The influence of external sugar concentrations on steady-state kinetics was also

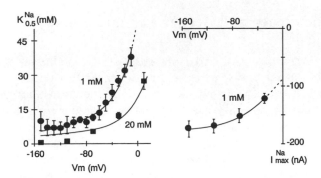

Figure 4. Voltage dependence of i_{max}^{Na} and $K_{0.5}^{Na}$. Steady-state sugar-dependent currents were measured consecutively on the same oocyte in the presence of 1 mM [αMDG]$_o$ with five external [Na$^+$]$_o$ varying from 2 to 100 mM (NaCl was replaced by choline-Cl) and were fitted to the following equation:

$$i_s = \frac{i_{max}^{Na}[Na]_o^n}{(K_{0.5}^{Na})^n + [Na]_o^n} \qquad (1)$$

where [Na$^+$]$_o$ is the external Na$^+$ ion concentration, n is the apparent coupling coefficient, i_{max}^{Na} is the maximum current at saturating external Na$^+$ concentration, and $K_{0.5}^{Na}$ is the [Na$^+$]$_o$ that gives half-maximum current. The best fit was obtained with an apparent coupling coefficient n of 1.9 (1.7–2.3) over the complete potential range. (*Left*) $K_{0.5}^{Na}$ was dependent upon the membrane potential and the external sugar concentration. As measured in two different oocytes, $K_{0.5}^{Na}$ decreased with negative potentials from 53 ± 2 mM (-30 mV) to 2.7 ± 0.5 mM (-190 mV) in the presence of 1 mM [αMDG]$_o$ and $K_{0.5}^{Na}$ decreased from 30 ± 1 mM (-30 mV) to 1 ± 0.5 mM (-190 mV) in the presence of 20 mM [αMDG]$_o$. (*Right*) In the presence of 1 mM [αMDG]$_o$, i_{max}^{Na} increased with negative potentials from -321 ± 9 nA (-30 mV) to -600 ± 47 nA (-190 mV). Error bars represent the error on the estimation of $K_{0.5}^{Na}$ and i_{max}^{Na} calculated with Eq. 1 using actual current values. Data are shown as filled symbols, and solid lines show $K_{0.5}^{Na}$ and i_{max}^{Na} simulated using the rate constants of the kinetic model (Fig. 8) with 10×10^{10} carriers oocyte^{-1}. The dotted lines show the model predictions beyond the experimental data. Modified from Parent et al. (1992b).

measured. Sugar-dependent currents (i_s) were recorded in the presence of 2, 10, 30, 50, and 100 mM [Na$^+$]$_o$ as a function of the external αMDG concentration. i_s increased as a function of [αMDG]$_o$, and the relationship between the sugar concentration and the magnitude of i_s was well described by the Michaelis-Menten equation. The results of these experiments are summarized in Fig. 5.

In contrast to $K_{0.5}^{Na}$, $K_{0.5}^{\alpha MDG}$ was not intrinsically voltage dependent. Its voltage dependence was rather a function of [Na$^+$]$_o$ (Fig 5). At 100 mM [Na$^+$]$_o$, $K_{0.5}^{\alpha MDG}$ was independent of the membrane potential with a mean value of 230 µM (SEM = 50; $N = 8$) measured between -10 and -150 mV. The $K_{0.5}^{\alpha MDG}$ is comparable to the value obtained by ^{14}[C]αMDG uptake in *Xenopus* oocytes expressing the cloned Na$^+$/glucose cotransporter (Ikeda et al., 1989). As [Na$^+$]$_o$ was gradually reduced from 100 to 2 mM, $K_{0.5}^{\alpha MDG}$ became increasingly dependent upon the membrane

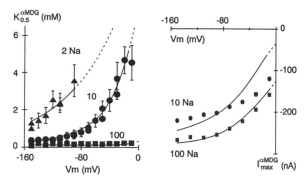

Figure 5. Voltage dependence of $i_{max}^{\alpha MDG}$ and $K_{0.5}^{\alpha MDG}$ as a function of [Na$^+$]$_o$. [Na$^+$]$_o$ was varied from 10 to 100 mM (choline-Cl substitution), and for each [Na$^+$]$_o$ steady-state sugar kinetics were measured with eight concentrations [αMDG]$_o$ from 20 µM to 10 mM. Steady-state sugar-dependent currents i_s were fitted to the following equation:

$$i_s = \frac{i_{max}^{\alpha MDG}[\alpha MDG]_o}{K_{0.5}^{\alpha MDG} + [\alpha MDG]_o} \tag{2}$$

where [αMDG]$_o$ is the external αMDG concentration, $i_{max}^{\alpha MDG}$ is the maximum current at saturating αMDG concentration, and $K_{0.5}^{\alpha MDG}$ is the [αMDG]$_o$ that gives half-maximum current. (*Left*) $K_{0.5}^{\alpha MDG}$ was measured experimentally with 2 (▲), 10 (●), and 100 (■) mM [Na$^+$]$_o$. At 100 mM [Na$^+$]$_o$, $K_{0.5}^{\alpha MDG}$ appeared independent of the membrane potential with a mean value of 230 (SEM = 50, $N = 8$) between -10 and -150 mV. The voltage dependence of $K_{0.5}^{\alpha MDG}$ increased at lower [Na$^+$]$_o$. At 10 mM [Na$^+$]$_o$, $K_{0.5}^{\alpha MDG}$ increased from 360 ± 13 to 885 ± 15 µM between -150 and -70 mV, whereas at 2 mM [Na$^+$]$_o$ $K_{0.5}^{\alpha MDG}$ increased from 1.3 to 7.2 mM over the same potential range. In the absence of membrane potential (0 mV), $K_{0.5}^{\alpha MDG}$ would be 8 mM at 10 mM [Na$^+$]$_o$ and 10 mM at 2 mM [Na$^+$]$_o$. (*Right*) $K_{0.5}^{\alpha MDG}$ and $i_{max}^{\alpha MDG}$ (*right*) were measured at 10 (●) and 100 (■) mM [Na$^+$]$_o$ on the same oocyte. $i_{max}^{\alpha MDG}$ was voltage dependent between -10 and -90 mV but showed saturation at higher potentials. $i_{max}^{\alpha MDG}$ values obtained for [Na$^+$]$_o$ between 10 and 100 mM were not found to be significantly different at each potential ($t < 95\%$). At 100 mM [Na$^+$]$_o$, $i_{max}^{\alpha MDG}$ was $-268 ± 9$ nA (-150 mV), $-222 ± 4$ nA (-50 mV), and $-158 ± 11$ nA (-10 mV). At 10 mM [Na$^+$]$_o$, $i_{max}^{\alpha MDG}$ was $-222 ± 22$ nA (-150 mV), $-176 ± 9$ nA (-50 mV), and $-120 ± 14$ nA (-10 mV). Error bars represent the error on the estimation of $K_{0.5}^{\alpha MDG}$ and $i_{max}^{\alpha MDG}$ calculated with Eq. 2 using actual current values. Solid lines show $K_{0.5}^{\alpha MDG}$ and $i_{max}^{\alpha MDG}$ simulated with the rate constants of the model (Fig. 8) and an estimated 15×10^{10} carriers oocyte^{-1}. The dotted lines show the model predictions beyond the experimental data. Modified from Parent et al. (1992*b*).

potential (Fig. 6) and $K_{0.5}^{\alpha MDG}$ decreased fivefold between -70 and -150 mV. In fact, the $K_{0.5}^{\alpha MDG}$ measured at all [Na$^+$]$_o$ approached the same limiting value of 230 µM at -150 mV, which suggests that negative membrane potential may partially compensate for the decrease in Na$^+$ ion gradient. In other words, $K_{0.5}^{\alpha MDG}$ is a function of the electrochemical gradient for Na$^+$ ions and may be decreased to its limiting value

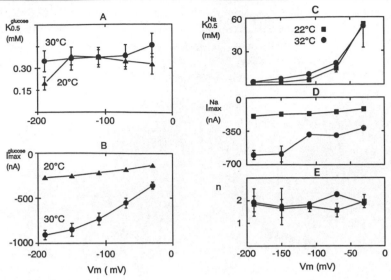

Figure 6. (*Left*) Temperature dependence of $K_{0.5}^{\text{D-glucose}}$ and $i_{\max}^{\text{D-glucose}}$. Steady-state sugar-dependent currents (i_s) were measured on the same oocyte at 20°C (▲) and 30°C (●) in the presence of the 100 mM Na^+ medium with six external D-glucose concentrations varying from 20 μM to 10 mM. Sugar-dependent currents (i_s) measured at 20°C before and after the pulse of temperature were not significantly different. (*A*) $K_{0.5}^{\text{D-glucose}}$ measured at 30°C (●) ranged from 380 ± 72 to 460 ± 161 μM and the values measured at 20°C (▲) ranged from 200 ± 44 to 390 ± 102 μM. Note that $K_{0.5}^{\text{D-glucose}}$ measured at $V_m = -190$ and $V_m = -30$ mV seemed slightly higher at 32°C than at 22°C, but more experiments are needed to determine if this is significant. (*B*) $i_{\max}^{\text{D-glucose}}$ was temperature dependent. $i_{\max}^{\text{D-glucose}}$ ranged from 275 ± 17 (-190 mV) to 134 ± 17 nA (-30 mV) at 20°C (▲) as compared with 900 ± 54 (-190 mV) and 360 ± 40 nA (-30 mV) at 30°C (●). Q_{10} factors for $i_{\max}^{\text{D-glucose}}$ were 3.3 at -190 mV and 2.7 at -30 mV. Error bars represent the error on the estimation of $K_{0.5}^{\text{D-glucose}}$ and $i_{\max}^{\text{D-glucose}}$ from actual current values using Eq. 2. (*Right*) Temperature dependence of n, $K_{0.5}^{\text{Na}}$, and $i_{\max}^{\text{Na}}$. (*C*) The apparent coupling coefficient n was independent of the bath temperature and membrane potential. At 22°C (■), $n = 1.8$ (SEM = 0.1; $N = 5$) was not significantly different than $n = 1.9$ (SEM = 0.1; $N = 5$) at 32°C (●). (*D*) $K_{0.5}^{\text{Na}}$ was not dependent upon the external temperature. From -190 to -30 mV, $K_{0.5}^{\text{Na}}$ increased from 2.7 ± 0.5 to 53 ± 2 mM at 32°C (●) and from 2.5 ± 1 to 54 ± 21 at 22°C (■). (*E*) $i_{\max}^{\text{Na}}$ increased with temperature. $i_{\max}^{\text{Na}}$ ranged from -195 ± 17 to -123 ± 9 nA at 22°C (■) and from -600 ± 47 to -321 ± 9 nA at 32°C (●) between -190 and -30 mV. Q_{10} factors were 3.1 (-190 mV) and 2.6 (-30 mV). Error bars represent the error on the estimation of $K_{0.5}^{\text{Na}}$ and $i_{\max}^{\text{Na}}$ from actual current values using Eq. 1.

either by increasing $[Na^+]_o$ to 100 mM or by increasing the membrane potential from 0 to -200 mV.

In summary, $K_{0.5}^{\alpha\text{MDG}}$ was found to be dependent upon $[Na^+]_o$ and upon the membrane potential. The effect of membrane potential was stronger at low $[Na^+]_o$ (2–30 mM). This effect is compatible with an ordered binding mechanism where the apparent voltage dependence observed at low $[Na^+]_o$ would only be secondary to the voltage dependence of the Na^+ ion binding step.

To further characterize the effect of temperature on the transport cycle, $K_{0.5}^{\text{D-glucose}}$ and $i_{\max}^{\text{D-glucose}}$ were measured on the same oocyte at both 20 and 30°C (Fig. 6, *A* and *B*). In the presence of 100 mM $[Na^+]_o$, increasing the temperature from 20 to

30°C did not affect $K_{0.5}^{\text{D-glucose}}$ but significantly enhanced $i_{\text{max}}^{\text{D-glucose}}$. The temperature dependence of $i_{\text{max}}^{\text{D-glucose}}$ between 20 and 30°C was not a function of the membrane potential. Q_{10} ranged from 3.3 to 2.7 from -190 to -30 mV. The significant temperature dependence of $i_{\text{max}}^{\text{D-glucose}}$ and the absence of temperature effect on $K_{0.5}^{\text{D-glucose}}$ are similar to results reported for transporter fluxes in guinea pig intestinal BBMV (Brot-Laroche et al., 1986) and in human intestinal BBMV (Malo and Berteloot, 1991).

Revisiting a Classical Kinetic Model for the Na⁺/Glucose Cotransporter

The next challenge was then to interpret these observations in terms of sugar transport. We chose to examine an ordered kinetic model to describe the Na⁺/glucose cotransport since the large allosteric effect of $[\text{Na}^+]_o$ observed for $K_{0.5}^{\alpha\text{MDG}}$ is compatible with an ordered binding mechanism. The steady-state results also provided an estimate of the cotransporter stoichiometry (2 Na⁺: 1 S); demonstrated that the binding of external Na⁺ ions is voltage dependent and that sugar binding is not voltage dependent; and failed to reveal large steady-state Na⁺ leak current. In parallel, the characteristics of the pre–steady-state currents are expected to significantly curtail the range of the individual rate constants valid for each reaction step. Fig. 7 *A* shows an ordered kinetic model for the Na⁺/glucose cotransporter that takes into account the effect of potential on the cotransporter steady-state and pre–steady-state currents. The sugar transport was described as a simultaneous event occurring through six distinct reaction steps. The binding of Na⁺ ions to the cotransporter was ordered rather than random and the model has mirror symmetry (Segel, 1975). External Na⁺ ions bind to the empty carrier as had been suggested for the Na⁺/glucose transport in rabbit intestinal BBMV (Hopfer and Groseclose, 1980; Kessler and Semenza, 1983; Kaunitz and Wright, 1984) and in chicken intestinal cells (Restrepo and Kimmich, 1985*a*). Rapid equilibrium was not assumed and each voltage-dependent reaction step was described by an Eyring-type barrier (Läuger and Jauch, 1986). The empty carrier was assumed to be negatively charged (Aronson, 1978; Kessler and Semenza, 1983; Restrepo and Kimmich, 1985*b*; Lapointe et al., 1986; Schultz, 1986) and the cotransporter binds two Na⁺ ions. Both Na⁺ binding and carrier translocation may be voltage dependent as suggested by the voltage dependence of $K_{0.5}^{\text{Na}}$, $i_{\text{max}}^{\text{Na}}$, and $i_{\text{max}}^{\text{sugar}}$. The absence of voltage dependence of $K_{0.5}^{\alpha\text{MDG}}$ observed in our experiments at saturating $[\text{Na}^+]_o$ strongly indicates that sugar binding is not voltage dependent per se. Translocation of the partially loaded complex CNa_2 accounted for the small cotransporter current measured in the absence of sugar (Umbach et al., 1990).

The steady-state concentrations of individual carrier states were solved using the King and Altman (1956) diagrammatic method. No analytical solution was developed for the model. Rather, we used numerical simulations to work out simultaneously the pre–steady-state time course and the steady-state kinetic parameters as they are described by two different sets of equations (see Appendix of Parent et al., 1992*b*). The following were generated simultaneously: (1) the current time course during a potential pulse in the absence and in the presence of sugar; (2) the steady-state *I-V* relationship; and (3) the variation of $K_{0.5}^{\text{sugar}}$, $i_{\text{max}}^{\text{sugar}}$, $K_{0.5}^{\text{Na}}$, and $i_{\text{max}}^{\text{Na}}$ as a function of applied voltage and of external cosubstrate concentration. In addition, simulations were performed assuming a constant internal (*trans*) Na$^+$ ion concentration of 20 mM (Dascal, 1987) and a constant internal sugar concentration of 50 μM (Umbach et al., 1990). The initial simulations were performed with a transporter turnover number of 0.1 s^{-1}. Optimization of the computer-generated simulations was accomplished by eye. Rate constants were modified gradually to improve the agreement between the model predictions and the experimental values.

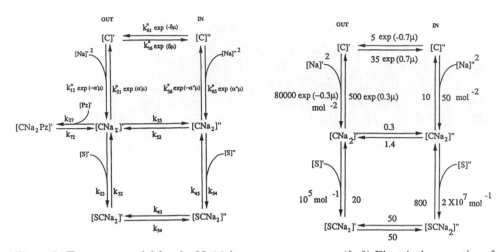

Figure 7. Transport model for the Na$^+$/glucose cotransporter. (*Left*) Electrical properties of the cotransporter were described by a six-state transport model with the empty carrier bearing two negative charges. The six states included those effectively contributing to sugar transport so the phlorizin-bound CNa$_2$Pz' complex was not accounted for. There were two binding sites for Na$^+$ ions and one binding site for sugar. Sugar transport can occur after the binding of the two Na$^+$ ions (ordered binding). The binding of the two Na$^+$ ions to the carrier was described as a single reaction step. This may imply that binding of the first Na$^+$ ion to the carrier is in rapid equilibrium (faster than any other rate) and that binding of the second Na$^+$ ion to the carrier is the rate-limiting step, or that there are two identical binding sites for Na$^+$ ions. Rate constants k_{12}, k_{21}, k_{16}, k_{61}, k_{56}, and k_{65} can be modulated by potential and/or [Na$^+$]$_o$. α' and α'' are phenomenological constants that describe the fraction of the electrical field sensed by the Na$^+$ binding to its external and internal sites, respectively. δ is the fraction of the electrical field sensed by the empty ion-binding site on the carrier during membrane translocation with $\alpha' + \alpha'' + \delta = 1$. μ is the electrochemical potential FV/RT. Membrane surface charges and unstirred layer effects at the membrane boundaries were assumed to be negligible. (*Right*) Rate constants of the Na$^+$/glucose cotransport kinetic model that were found to be compatible with our results. The numerical values were found by numerical simulations. Numerical values for rate constants k_{54} and k_{52} were mistakenly published as 4×10^7 and 0.3, respectively, in the original paper. They are shown here with the actual numbers $k_{54} = 2 \times 10^7$ and $k_{52} = 1.4$ as they were used for all simulations. An erratum was published in *Journal of Membrane Biology*, volume 130, page 203. Equations used for all simulations were detailed in the Appendix of Parent et al., 1992*b*. Units of the rate constants are either s^{-1}, s^{-1} mol^{-1}, or s^{-1} mol^{-2}. Modified from Parent et al. (1992*b*).

A single set of rate constants was found to account for all our results as shown in Fig. 7 *B*. Since we did not look for all possible numerical combinations, we cannot rule out that a different set of rate constants will give the same agreement, or that a more complex model would also reproduce our results. Nonetheless, our observations are compatible with a negatively charged carrier with two Na^+ binding sites located within 25–35% of the membrane electrical field ($\alpha' = 0.3$). The Na^+ ion binding sites on the empty carrier would traverse 60–80% of the membrane electrical field during translocation ($\delta = 0.7$). Although that would leave no voltage dependence for internal Na^+ ion binding (since by definition, $\alpha'' + \delta + \alpha' = 1$), voltage dependence at this site cannot definitively be ruled out in the absence of steady-state kinetic experiments performed as a function of $[Na^+]_i$. We expect this point to be directly investigated in the near future since new techniques are now available that allow this question to be addressed specifically. As expected, the so-called Na^+ leak

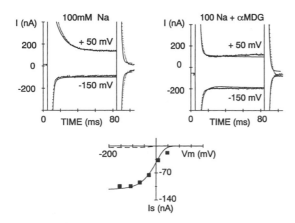

Figure 8. (*Left*) Model simulation of the current time course showing the pre–steady-state current with the oocyte capacitive current in the presence of 100 mM $[Na^+]_o$. Current traces were simulated for $V_m = -150$ and $V_m = +50$ mV ($V_h = -50$ mV) using the rate constants of Fig. 7, but for $k_{61} = 3$ s^{-1}, and are shown (*solid lines*) over the actual current traces (*dots*). Simulated pre–steady-state currents relaxed from -50 to $+50$ mV with a time constant of ~ 15 ms. No transient current was generated from -50 to -150 mV. The model also accounted for the tail (OFF) currents back at -50 mV. (*Right*) Current traces were simulated in the presence of 100 mM $[Na^+]_o$ and 1 mM $[\alpha MDG]_o$ at $V_m = -150$ and $V_m = +50$ mV and are shown (*solid lines*) superimposed to actual data (*dots*). (*Bottom*) Steady-state sugar-dependent *I-V* curve predicted by the model (*solid line*) is shown superimposed over the actual data (■). The dashed line shows the predicted carrier-mediated Na^+ leak current in the absence of external sugar. Simulations were performed assuming $[Na^+]_i = 20$ mM, $[\alpha MDG]_i = 50$ μM, and 6×10^{10} carriers oocyte^{-1}. The total membrane current was the sum of the oocyte background currents and the sugar-evoked carrier currents. Computer simulations of the current time course during a pulse of potential were carried out with a time interval $\Delta t = 10$–50 μs. The carrier pre–steady-state current was estimated as the sum of all voltage-dependent reaction steps (Läuger et al., 1981; Roy et al., 1991) with $i(t) = -F [2\alpha'(C_1 k_{12} - C_2 k_{21}) - 2\delta(C_1 k_{16} - C_6 k_{61})]$. Experimental data as shown in Fig 1. Modified from Parent et al. (1992*b*).

current (translocation of the partially loaded carrier CNa_2) was insignificant in the presence of external sugar.

Rate-limiting steps for sugar transport are dependent upon the membrane potential, the external $[Na^+]$, and the external [sugar]. In the presence of saturating 100 mM $[Na^+]_o$ and 1 mM $[S]_o$, the rate-limiting step for sugar transport is the outwardly directed translocation of the empty carrier $C' \leftarrow C''$ for membrane potential more positive than -43 mV. This is of physiological significance since the membrane resting potential for the relevant epithelial cells is around -40 mV (Parent et al., 1988). In the membrane potential range between -43 and -134 mV, the rate-limiting step becomes the dissociation of internal Na^+ ions. Because we did not assume rapid equilibrium, k_{12} (Na^+ ion-binding rate constant) can become rate limiting but only under definite conditions: membrane potentials more negative than -134 mV (-134 to $-\infty$) and external $[Na^+] < 10$ mM.

The model also accounts for the observed current time course at all membrane potentials in the presence of external Na^+ ions (Fig. 8, left) and after the addition of sugar (Fig. 8, right). Pre–steady-state currents are caused by the rapid reorientation of the cotransporter in the membrane electrical field. More specifically, they are accounted for by the reaction steps $CNa'_2 \rightarrow C' \rightarrow C''$during a positive pulse of potential in the presence of external Na^+ ions ($[Na^+]_o > 30$ mM). The relaxation is a function of both Na^+ dissociation rate and the carrier translocation, while the magnitude of the current is determined by the proportion of CNa'_2 species available at the holding potential. The absence of pre–steady-state currents after the addition of sugar to a Na^+-rich medium results from the low sugar dissociation rate constant ($k^0_{32} \approx 20$ s^{-1}), which therefore makes the dissociation of CNa_2S' into CNa'_2 very unlikely. As a result, when the potential is stepped from -50 to $+50$ mV, the clockwise reaction $CNa'_2 \rightarrow C' \rightarrow C''$ responsible for the pre–steady-state current is negligible. In fact, pre–steady-state currents are predicted to relax with the same time constant but would generate only 10% of the current observed in the Na^+-rich medium. This modest peak current together with the noncompensated capacitive transient probably account for our failure to detect experimentally pre–steady-state currents in the presence of sugar. Both ON and OFF currents are crucial in limiting rate constants k_{12}, k_{21}, k_{16}, k_{61}, and k_{32}. For instance, the values of rate constants k_{21} should be between 200 and 600 s^{-1}, k_{16} between 25 and 55 s^{-1}, k_{61} between 3 and 5 s^{-1}, and k_{32} between 1 and 30 s^{-1}.

Conclusions and Future Directions

This paper reviewed the main characteristics of the steady-state and pre–steady-state currents of the Na^+/glucose cotransporter as well as the voltage dependence of the kinetics for sugar and Na^+ ions. Pre–steady-state currents associated with charge redistribution were recorded in the presence of external Na^+ ions. According to our model, they result from the consecutive reorientation of the carrier states CNa'_2 and C' in the membrane electrical field. $K^{Na}_{0.5}$ was a function of the membrane potential at all external sugar concentrations, suggesting a true Na^+ well effect. The voltage dependence of $K^{\alpha MDG}_{0.5}$ was otherwise a function of the external Na^+ ion concentration, which suggests that it is only secondary to the voltage dependence of Na^+ ions. All these observations were accounted for by a six-state kinetic model where the external Na^+ ion binding and the carrier translocation both need to be influenced by the membrane potential to account for the cotransporter voltage dependence. The work described here was performed with the cloned rabbit Na^+/glucose cotransporter expressed in *Xenopus* oocytes, but a similar study conducted with the cloned human Na^+/glucose cotransporter (Loo et al., 1992) has yielded similar results so far.

As more cotransporters are cloned, one may reasonably expect that electrophysiology will become a prominent investigative tool in cotransport. Already, some other transporters including the Na^+/Cl^-/GABA transporter (Mager et al., 1992; Risso et al., 1992) and the y^+ amino acid transporter (Kavanaugh et al., 1992) have been shown to give rise to sizable and specific currents. Interestingly, pre–steady-state currents were also reported for the Na$^+$/Cl$^-$/GABA transporter (Mager et al., 1992), which suggests that pre–steady-state currents may be a bona fide property of Na$^+$-coupled transport proteins.

As this story unfolds, we are tempted to paraphrase a historic quote and conclude that it was "a giant step for us, but a small step overall." Indeed, what is interesting is what can be accomplished from here. Three major axes define the short-term goals: the study of the direct regulation of the Na$^+$/glucose cotransporter by kinases and G proteins; measuring the cotransporter kinetics as a function of internal Na$^+$ ions and sugar using a combination of whole-cell recordings and macro-patches (Hilgemann et al., 1991) in transfected mammalian cells; and reevaluation of the pre–steady-state current characteristics using the open-oocyte, vaseline-gap voltage-clamp technique (Taglialatela et al., 1992), developed originally for studying the gating currents of cloned channels expressed in *Xenopus* oocytes.

While the regulation of ion channels by phosphorylation-related mechanisms has been studied for quite some time, there is very little known about the possible regulation of Na$^+$/coupled transporters by phosphorylation. Although there is no evidence so far that kinases play a functional role in vivo, numerous possible phosphorylation sites were recently underlined in the secondary structure of the Na$^+$/glucose cotransporter after Kennelly and Krebs (1991) revised the canonical rules determining the consensus sequences for phosphorylation sites. Among the potentially interesting sites, the following are conserved between the rabbit and the human clones. There is one consensus sequence for phosphorylation by protein kinase A (cAMP and cGMP) at Ser415 and two consensus sequences for phosphorylation by protein kinase C at Ser418 and Ser303 (Wright et al., 1992). Phosphorylation may be involved in the fine tuning of the cotransporter activity and may up or downregulate the cotransporter activity depending on the load. For instance, one can imagine that during periods of high demand, phosphorylation could directly increase the voltage dependence of the Na$^+$ ion binding step ($\alpha' > 0.3$), therefore reducing the need for a large negative membrane potential to stimulate the cotransporter activity. On the other end of the spectrum, phosphorylation could be responsible for reducing the overall sugar transport (i_{max}) by decreasing the rate constant for the rate-limiting step without affecting $K_{0.5}^{Na}$ and $K_{0.5}^{\alpha MDG}$. Of course, these are only speculations but they illustrate the type of rigorous analysis that the innovative application of electrophysiology in the cotransport field makes possible.

However insightful they are, kinetic models simply provide a theoretical framework and cannot substitute for structural models. We still don't know the physical nature of the translocation process and the physical nature of the interaction of Na$^+$ ions and sugar at their binding sites. The translocation and the energy transduction mechanisms will ultimately be elucidated in concert with the tertiary structure of the Na$^+$/glucose cotransporter.

Acknowledgments

The authors are grateful to Dr. Luis Reuss and Dr. Ana Pajor for critical reading of the manuscript. We are indebted to Dr. Stéphane Supplisson for his invaluable collaboration in elucidating the model parameters, and we thank Dr. A. Berteloot and Dr. D. Loo for discussions.

The work was supported by a grant from the U.S. Public Health Service (DK-19567) to E. M. Wright and a postdoctoral fellowship from the Medical Research Council of Canada (to L. Parent).

References

Aronson, P. S. 1978. Energy dependence of phlorizin binding to isolated renal microvillus membranes. *Journal of Membrane Biology.* 42:81–98.

Barrett, J. N., K. L. Magleby, and B. S. Pallotta. 1982. Properties of single calcium-activated potassium channels in cultured rat muscle. *Journal of Physiology.* 331:211–230.

Bennett, E., and G. A. Kimmich. 1992. Na^+ binding to the Na^+-glucose cotransporter is potential dependent. *American Journal of Physiology.* 262 (*Cell Physiology* 31):C510–C516.

Bergman, C., and J. Bergman. 1985. Origin and voltage dependence of asparagine-induced depolarization in intestinal cells of *Xenopus* embryo. *Journal of Physiology.* 366:197–220.

Berteloot, A., and G. Semenza. 1990. Advantages and limitations of vesicles for the characterization and the kinetic analysis of transport systems. *Methods in Enzymology.* 192:409–437.

Birnir, B., H.-S. Lee, M. Hediger, and E. M. Wright. 1990. Expression and characterization of the intestinal Na^+/glucose cotransporter in COS-7 cells. *Biochimica et Biophysica Acta.* 1048:100–104.

Birnir, B., D. D. F. Loo, and E. M. Wright. 1991. Voltage-clamp studies of the Na^+/glucose cotransporter cloned from rabbit small intestine. *Pflügers Archiv.* 418:79–85.

Blakely, R. D., J. A. Clark, G. Rudnick, and S. G. Amara. 1991. Vaccinia-T7 RNA polymerase expression system: evaluation for expression cloning of plasma membrane transporters. *Analytical Biochemistry.* 194:302–308.

Brot-Laroche, E., M. A. Serrano, B. Delhomme, and F. Alvarado. 1986. Temperature sensitivity and substrate specificity of two distinct Na^+-activated D-glucose transport systems in guinea pig jejunal brush border membrane vesicles. *Journal of Biology Chemistry.* 261:6168–6176.

Centelles, J. J., R. K. H. Kinne, and E. Heinz. 1991. Energetic coupling of Na-cotransport. *Biochimica et Biophysica Acta.* 1065:239–249.

Coronado, R., R. L. Rosenberg, and C. Miller. 1980. Ionic selectivity, saturation, and block in a K^+-selective channel from sarcoplasmic reticulum. *Journal of General Physiology.* 76:425–446.

Crane, R. K. 1962. Hypothesis of mechanism of intestinal active transport of sugars. *Federation Proceedings.* 21:891–895.

Crane, R. K. 1977. The gradient hypothesis and other models of carrier-mediated active transport. *Review of Physiology and Biochemical Pharmacology.* 78:99–159.

Crane, R. K., D. Miller, and I. Bihler. 1961. The restrictions on possible mechanisms of intestinal active transport of sugars. *In* Membrane Transport and Metabolism. A. Kleinzeller and A. Kotyk, editors. Czechoslovakian Academy of Sciences, Prague. 439–449.

Dascal, N. 1987. The use of *Xenopus* oocytes for the study of ion channels. *CRC Critical Reviews in Biochemistry.* 22:317–837.

De Smedt, H., and R. Kinne. 1981. Temperature dependence of solute transport and enzyme activities in hog renal brush border membrane vesicles. *Biochimica et Biophysica Acta.* 648:247–253.

Frömter, E. 1982. Electrophysiological analysis of rat renal sugar and amino acid transport. I. Basic phenomena. *Pflügers Archiv.* 393:179–189.

Hediger, M. A., M. J. Coady, T. S. Ikeda, and E. M. Wright. 1987. Expression cloning and cDNA sequencing of the Na+/glucose co-transporter. *Nature.* 330:379–381.

Hilgemann, D. W., D. A. Nicoll, and K. Philipson. 1991. Charge movement during Na+ translocation by native and cloned cardiac Na+/Ca²+ exchanger. *Nature.* 352:715–718.

Hopfer, U. 1987. Membrane transport mechanisms for hexoses and amino acids in the small intestine. *In* Physiology of the Gastrointestinal Tract. L. R. Johnson, editor. Raven Press, New York. 1499–1526.

Hopfer, U., and R. Groseclose. 1980. The mechanism of Na+-dependent D-glucose transport. *Journal of Biological Chemistry.* 255:4453–4462.

Hopfer, U., K. Nelson, J. Perrotto, and K. J. Isselbacher. 1973. Glucose transport in isolated brush border membrane from rat small intestine. *Journal of Biological Chemistry.* 248:25–32.

Hoyer, J., and H. Gögelein. 1991. Sodium-alanine cotransport in renal proximal tubule cells investigated by whole-cell current recording. *Journal of General Physiology.* 97:1073–1094.

Ikeda, T. S., E. S. Hwang, M. J. Coady, B. A. Hirayama, M. A. Hediger, and E. M. Wright. 1989. Characterization of a Na+/glucose cotransporter cloned from rabbit small intestine. *Journal of Membrane Biology.* 110:87–95.

Jauch, P., and P. Läuger. 1986. Electrogenic properties of the sodium-alanine cotransporter in pancreatic acinar cells. II. Comparison with transport models. *Journal of Membrane Biology.* 94:117–127.

Jauch, P., O. H. Petersen, and P. Läuger. 1986. Electrogenic properties of the sodium-alanine cotransporter in pancreatic acinar cells. I. Tight-seal whole-cell recordings. *Journal of Membrane Biology.* 94:99–115.

Kaunitz, J. D., and E. M. Wright. 1984. Kinetics of sodium D-glucose cotransport in bovine intestinal brush border vesicles. *Journal of Membrane Biology.* 79:41–51.

Kavanaugh, M. P., E. Stefani, and R. A. North. 1992. Electrogenic properties of the cloned system y+ basic amino acid transporter expressed in *Xenopus* oocytes. *Journal of General Physiology.* 100:78. (Abstr.)

Kennelly, P. J., and E. G. Krebs. 1991. Consensus sequences as substrate specificity determinants for protein kinases and protein phosphates. *Journal of Biological Chemistry.* 266:15555–15558.

Kessler, M., O. Acuto, C. Storelli, H. Murer, M. Mueller, and G. Semenza. 1978. A modification procedure for the rapid preparation of efficiently transporting vesicles from small intestinal brush border membranes: their use in investigating some properties of D-glucose and choline transport systems. *Biochimica et Biophysica Acta.* 506:136–154.

Kessler, M., and G. Semenza. 1983. The small-intestinal Na+,D-glucose cotransporter: an asymmetric gated channel (or pore) responsive to "delta potential". *Journal of Membrane Biology.* 76:27–56.

Kimmich, G. A. 1970. Preparation and properties of mucosal epithelial cells isolated from small intestine of the chicken. *Biochemistry.* 9:3669–3677.

Kimmich, G. A. 1990. Membrane potentials and the mechanism of intestinal Na+-dependent sugar transport. *Journal of Membrane Biology.* 114:1–27.

Kimmich, G. A., and J. Randles. 1988. Na+-coupled sugar transport: membrane potential-dependent K_m and K_i for Na$^+$. *American Journal of Physiology.* 255 (*Cell Physiology* 24):C486–494.

King, E. L., and C. Altman. 1956. A schematic method of deriving the rate laws for enzyme-catalyzed reactions. *Journal of Physical Chemistry.* 60:1375–1378.

Kippen, I., B. Hirayama, J. R. Klinenberg, and E. M. Wright. 1979. Transport of *p*-aminohippuric acid, uric acid and glucose in highly purified rabbit renal brush border membranes. *Biochimica et Biophysica Acta.* 556:161–174.

Lang, F., G. Messner, and W. Rehwald. 1986. Electrophysiology of sodium-coupled transport in proximal renal tubules. *American Journal of Physiology.* 250 (*Renal Fluid Electrolyte Physiology* 19):F953– F962.

Lapointe, J.-Y., L. Garneau, P. D. Bell, and J. Cardinal. 1990. Membrane cross-talk in the mammalian proximal tubule during alterations in transepithelial sodium transport. *American Journal of Physiology.* 258 (*Renal Fluid Electrolyte Physiology* 27):F339–345.

Lapointe, J.-Y., R. L. Hudson, and S. G. Schultz. 1986. Current-voltage relations of sodium-coupled sugar transport across the apical membrane of Necturus small intestine. *Journal of Membrane Biology.* 93:205–219.

Läuger, P., R. Benz, G. Stark, E. Bamberg, P. C. Jordan, A. Fahr, and W. Brock. 1981. Relaxation studies of ion transport systems in lipid bilayer membranes. *Quarterly Review of Biophysics.* 14:513–598.

Läuger, P., and P. Jauch. 1986. Microscopic description of voltage effects on ion-driven cotransport systems. *Journal of Membrane Biology.* 91:275–284.

Loo, D. D. F., A. Hazama, S. Supplisson, E. Turk, and E. M. Wright. 1992. Charge translocation associated with conformational transitions of the Na$^+$/glucose cotransporter. *Journal of General Physiology.* 100:38. (Abstr.)

Lu, L., C. Montrose-Rafizadeh, T.-C. Hwang, and W. B. Guggino. 1990. A delayed rectifier potassium current in *Xenopus* oocytes. *Biophysical Journal.* 57:1117–1123.

Mager, S., J. Naeve, M. Quick, B. N. Cohen, N. Davidson, and H. A. Lester. 1992. Cloned Na/GABA cotransporter. Electrophysiological analysis yields turnover number of 5–10 per second. *Journal of General Physiology.* 100:120. (Abstr.)

Malo, C., and A. Berteloot. 1991. Analysis of kinetic data in transport studies: new insights from kinetic studies of Na$^+$/D-glucose cotransport in human intestinal brush-border membrane vesicles using a fast sampling, rapid filtration apparatus. *Journal of Membrane Biology.* 122:127–141.

Mitchell, P. 1969. Chemiosmotic coupling and energy transduction. *Theoretical and Experimental Biophysics.* 2:159–216.

Murer, H., and U. Hopfer. 1974. Demonstration of the electrogenic Na$^+$-dependent D-glucose transport in intestinal brush border membranes. *Proceedings of the National Academy of Sciences, USA.* 71:484–488.

Parent, L., J. Cardinal, and R. Sauvé. 1988. Single-channel analysis of a K channel at the basolateral membrane of rabbit proximal convoluted tubule. *American Journal of Physiology.* 254 (*Renal Fluid Electrolyte Physiology* 23):F105–F113.

Parent, L., S. Supplisson, D. D. F. Loo, and E. M. Wright. 1992a. Electrogenic properties of the cloned Na$^+$/glucose cotransporter. I. Voltage-clamp studies. *Journal of Membrane Biology.* 125:49–62.

Parent, L., S. Supplisson, D. D. F. Loo, and E. M. Wright. 1992b. Electrogenic properties of the cloned Na^+/glucose cotransporter. II. A transport model under non rapid equilibrium. *Journal of Membrane Biology.* 125:63–79.

Restrepo, D., and G. A. Kimmich. 1985a. Kinetic analysis of mechanism of intestinal Na^+-dependent sugar transport. *American Journal of Physiology.* 248 (*Cell Physiology* 17):C498–C509.

Restrepo, D., and G. A. Kimmich. 1985b. The mechanistic nature of the membrane potential dependence of sodium-sugar cotransport in small intestine. *Journal of Membrane Biology.* 87:159–172.

Risso, S., L. J. DeFelice, and R. D. Blakely. 1992. Whole-cell analysis of electrogenic neurotransmitter transport in transfected Hela cells. *Journal of General Physiology.* 100:88. (Abstr.)

Roy, G., W. Wierbicki, and R. Sauvé. 1991. Membrane transport models with fast and slow reactions: general analytical solution for a single relaxation. *Journal of Membrane Biology.* 123:105–113.

Samaržija, I., B. T. Hinton, and E. Frömter. 1982. Electrophysiological analysis of rat renal sugar and amino acid transport. II. Dependence on various transport parameters and inhibitors. *Pflügers Archiv.* 393:190–197.

Schultz, S. G. 1986. Ion-coupled transport of organic solutes across biological membranes. *In* Physiology of Membrane Disorders. T. E. Andreoli, J. F. Hoffman, D. D. Fanestil, and S. G. Schultz, editors. Plenum Publishing Corp., New York. 283–294.

Schultz, S. G., and P. F. Curran. 1970. Coupled transport of sodium and organic solutes. *Physiological Review.* 50:637–718.

Schultz, S. G., and R. Zalusky. 1964. Ion transport in isolated rabbit ileum. II. The interaction between active sodium and active sugar transport. *Journal of General Physiology.* 47:1043–1059.

Segel, I. H. 1975. Enzyme Kinetics. John Wiley & Sons, Inc., New York. 957 pp.

Semenza, G., M. Kessler, M. Hosang, and U. Schmidt. 1984. Biochemistry of the Na^+, D-glucose cotransporter of the small intestinal brush-border membrane. *Biochimica et Biophysica Acta.* 779:343–379.

Smith, C., B. A. Hirayama, and E. M. Wright. 1992. Baculovirus-mediated expression of the Na^+/glucose cotransporter in Sf9 cells. *Biochimica et Biophysica Acta.* 1104:151–159.

Smith-Maxwell, C., E. Bennett, J. Randles, and G. Kimmich. 1990. Whole-cell recording of sugar-induced currents in LLC-PK$_1$ cells. *American Journal of Physiology.* 258 (*Cell Physiology* 27):C234–C242.

Taglialatela, M., L. Toro, and E. Stefani. 1992. Novel voltage clamp to record small, fast currents from ion channels expressed in *Xenopus* oocytes. *Biophysical Journal.* 61:78–82.

Umbach, J., M. J. Coady, and E. M. Wright. 1990. Intestinal Na^+/glucose cotransporter expressed in Xenopus oocytes is electrogenic. *Biophysical Journal.* 57:1218–1224.

Wright, E. M., K. Hager, and E. Turk. 1992. Sodium cotransport proteins. *Current Opinion in Cell Biology.* 4:696–702.

Chapter 22

Tails of Serotonin and Norepinephrine Transporters: Deletions and Chimeras Retain Function

Randy D. Blakely, Kim R. Moore, and Yan Qian

Department of Anatomy and Cell Biology and Program in Neuroscience, Emory University School of Medicine, Atlanta, Georgia 30322

Molecular Biology and Function of Carrier Proteins © 1993 by The Rockefeller University Press

Introduction

Serotonin (5-hydroxytryptamine, 5HT) is transported across brain (Ross and Renyi, 1967), placental (Balkovetz et al., 1989), pulmonary (Lee and Fanburg, 1986), and platelet (Lingjarede, 1971; Rudnick, 1977) plasma membranes by a Na^+- and Cl^--dependent transporter that exhibits marked sensitivity to tricyclic and heterocyclic antidepressants, including imipramine and paroxetine (Fuller and Wong, 1990). Although the native configuration of the 5HT transporter (SERT) has yet to be elucidated, recent cDNA cloning studies (Blakely et al., 1991*a;* Hoffman et al., 1991; Ramamoorthy, S., A. L. Bauman, K. R. Moore, H. Han, T. Yang-Feng, A. S. Chang, V. Ganapathy, and R. D. Blakely, manuscript submitted for publication) have revealed that single cDNAs are sufficient to confer antidepressant- and cocaine-sensitive 5HT transport on transfected mammalian cells. Initially reported as distinct sequences, both rat brain and basophilic leukemia cell (RBL) cDNAs encode identical 630 amino acid proteins (see below), modeled with ~ 12 transmembrane (TM) domains and cytoplasmic NH_2 and COOH termini presumed to lie in the cytoplasm. The NH_2 and COOH termini are the most divergent regions in comparison of rSERT with other members of a symporter gene family (Blakely, 1992), now known to contain homologous Na^+- and Cl^--dependent transporters for γ-aminobutyric acid (GABA; Guastella et al., 1990; Clark et al., 1992), norepinephrine (NE; Pacholczyk et al., 1991), dopamine (DA; Kilty et al., 1991; Shimada et al., 1991), glycine (Smith et al., 1992; Guastella et al., 1992), proline (Fremeau et al., 1992), and betaine (Yamauchi et al., 1992).

The human norepinephrine transporter (hNET) shares with rSERT the ability to induce antidepressant- and cocaine-sensitive monoamine transport on transfected HeLa cells (Pacholczyk et al., 1991), although substrate and antagonist selectivities are distinct. Cocaine blocks both NE and 5HT transport across similar concentration ranges. On the other hand, NE transport induced by hNET is antagonized more potently by secondary amine tricyclic antidepressants such as desipramine and nortriptyline than by the tertiary amine congeners imipramine and amytriptyline (Pacholczyk et al., 1991), whereas transport of 5HT induced by rSERT exhibits a reversed antidepressant selectivity (Blakely et al., 1991*a*). Notably, tricyclic antidepressant potencies generally vary between 5HT and NE transporters by 100-fold or less, and thus many of these compounds may be viewed as only partially selective. In contrast, selective agents such as the NET transport antagonists nomifensine and mazindol and the 5HT transport selective antagonists paroxetine and citalopram differ in potencies for the two carriers by more than several orders of magnitude (Fuller and Wong, 1990; Blakely et al., 1991*a;* Pacholczyk et al., 1991). Substrate recognition may also involve shared determinants of the two carriers as structure activity studies point to the critical dependence on the protonated amine groups of both 5HT and NE (and antagonists) for transporter recognition (Koe et al., 1976; Maxwell et al., 1976; De Paulis et al., 1978; Maxwell and White, 1978). Thus, the hNET and SERT cDNAs are likely to encode domains contributing common and distinct features of substrate and antagonist recognition. In this regard, hNET, rSERT, and the DA transporter form a local cluster of more closely related carriers within the larger family of transporter molecules, suggesting their evolution from a common ancestor with pharmacologic characteristics common to all three monoamine transporters.

The sites (or domains) responsible for differential substrate specificity, shared recognition of cocaine and tricyclic antidepressants, and conserved mechanistic features (Na^+/Cl^- coupling, etc.) have yet to be determined. The large size and complex topology predicted for the 5HT carrier warrants consideration of multiple approaches to the investigation of the structural underpinnings of transporter function. One approach, the construction and analysis of chimeric proteins, has proven successful in the structural analysis of growth factor receptors (Yan et al., 1991), G protein–coupled receptors (Frielle et al., 1988; Kobilka et al., 1988), G proteins (Masters et al., 1988), and ion channel molecules (Stocker et al., 1991; Li et al., 1992). Where successful, such chimeras provide an assayable phenotype rather than a potentially uninformative loss of function mutation. Drawing from this paradigm, we have utilized recombinant DNA techniques and the polymerase chain reaction (PCR; Saiki et al., 1988) to construct various rSERT expression plasmids bearing truncated or chimeric NH_2 and COOH termini to identify or eliminate domains likely to be responsible for SERT expression, specificity, and regulation. Initial findings from these experiments suggest that regions critical for monoamine substrate and antagonist recognition lie within the regions bordered by the NH_2- and COOH-terminal substitutions and truncations, such as certain highly conserved TM domains and intervening loops. Additionally, since significant alterations of NH_2 and COOH termini can be achieved with little or no loss of function, these regions may provide targets for "epitope-tag" strategies (e.g., Campbell et al., 1992) suited to transporter purification and visualization.

Methods

Restoration of the rSERT 5′ End

Rat midbrain total RNA was purified using the guanidinium/cesium chloride centrifugation method of Chirgwin as detailed by MacDonald et al. (1987). Oligonucleotides were synthesized from nucleotide phosphoramidites on a DNA synthesizer (model 391; Applied Biosystems, Inc., Foster City, CA), manually deprotected in NH_4OH, lyophilized and resuspended in dH20, and used in PCR reactions without additional purification. First strand cDNA was synthesized from 1 μg RNA using the Superscript (GIBCO BRL, Gaithersburg, MD) preamplification system with random hexanucleotide primers. First strand reaction mix was amplified with 3 U of Hot Tub DNA polymerase (Amersham Corp., Arlington Heights, IL) for 30 cycles (94°C for 1 min, 45°C for 2 min, and 72°C for 3 min), followed by a final 12-min extension at 72°C. PCR reactions were performed in 100 μl with oligonucleotides (10 pmol/reaction) derived from the rat brain 5HT transporter 5′ end (antisense 5′-AGGTCC-GAATTCAATCCATCTTCTTGCCCCAGGTCTCCCGCTCCC-3′) and 5′ noncoding sequences of the Mayser et al. (1991) and Hoffman et al. (1991) cDNAs (sense 5′-CCGCTCGAGCTTTCCGTCTTGTCCCCATAACC-3′). Underlined bases of the sense oligonucleotide correspond to an added XhoI site for cloning. PCR product was digested with XhoI and StyI, gel purified, and ligated with XhoI/StyI digested rSERT (Blakely et al., 1991a). The resulting ligation product was sequenced by the dideoxy chain termination method (Sanger, Nicklen, and Coulson, 1977) and found to contain a sequence identical to the 5′ noncoding and NH_2 terminus of the Mayser et al. (1991) and Hoffman et al. (1991) cDNAs. The amended clone is hereafter referred to as rSERT, while the original rat brain 5HT transporter cDNA is designated rSERT(86-630) to indicate its status as a 5HT transporter cDNA partial clone with its first in-frame methionine at amino acid 86 of rSERT.

Generation of Chimeric rSERT

The general scheme for generation of transporter chimeras is given in Fig. 1. Oligonucleotides (sense 5'CCGCCTCTCTTAAGGAGAGACTGGCCTATGG-CAT-3', antisense 5'-CCGCCCCCTTGGTCGCCGTCGCGGGGCGCCAGC-3')

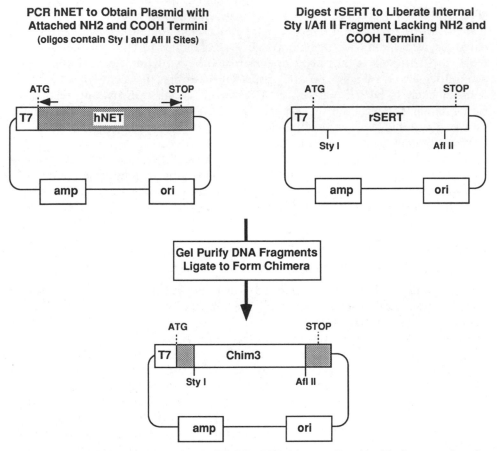

Figure 1. Scheme for construction of rSERT/hNET chimeras. Depicted is the procedure for the construction of chimera 3, which possesses NH$_2$ and COOH termini derived from hNET and the core of the 5HT carrier encoded by rSERT.

were designed to amplify the hNET plasmid (100 ng plasmid, 100 pmol primers, 100 μl reaction volume, 3 U Hot Tub polymerase, 30 cycles of PCR as above, including an additional 10-min extension on the initial cycle) so as to generate a complete linear plasmid with 5' and 3' sequences from hNET at either end, including NH$_2$-terminal (1-52) and COOH-terminal (587-617) amino acid residues, bordered by unique StyI and AflII restriction sites (underlined in oligonucleotide). Ligation of the PCR product to a 1.6-kb StyI/AflII fragment of rSERT generated chimera 3 (1-52 hNET/75-607 rSERT/587-617 hNET). Chimeras 4 and 5 were synthesized in the same basic scheme from the same starting DNAs, but in each case an internal, unique restriction site from the polylinker of pBluescript SKII- was substituted for the site encoded by one of the PCR oligonucleotides. This resulted in a linear hNET molecule with either the NH_2- or COOH-terminal sequences attached to pBluescript, permitting a complementary ligation with equivalently digested rSERT. Specifically, chimera 4 (1-52 hNET/75-630 rSERT) was formed by (*a*) digesting the hNET PCR product with StyI and NotI and (*b*) ligation of this product with StyI/NotI-digested rSERT. Chimera 5 (1-607 rSERT/587-617 hNET) was formed by (*a*) digesting the hNET PCR product with KpnI and AflII and (*b*) ligation of this product with KpnI/AflII-digested rSERT. After ligation, primary transformants were sequenced across amplification regions of hNET and adjoining sequences of rSERT to insure an absence of errors in the PCR or ligation reactions. Supercoiled plasmid DNA was prepared for transfections by alkaline lysis followed by column chromatography (QIAGEN Inc., Chatsworth, CA).

5HT Transport and Ligand Binding to Transporters Expressed in Transiently Transfected HeLa Cells

30 min before transfections, HeLa monolayers ($\sim$ 100,000/well), cultured in 24-well plates, were infected with recombinant vaccinia virus strain VTF7-3 containing T7 RNA polymerase, at 10 pfu in 100 μl as previously described (Blakely et al., 1991*a*, *b*). Plasmids with cDNAs encoding the complete rSERT cDNA, truncated coding sequences, or chimeric transporters were transfected as liposome suspensions (Lipofectin; GIBCO BRL:1 μg DNA, 3 μg lipid per well) in a total volume of 350 μl/well. pBluescript SKII- was transfected under identical conditions in separate wells on the same plate. 6 h after transfection, transfection mixes and free virus were removed from wells by aspiration and wells were preincubated for 15 min at 37°C in 400 μl Krebs-Ringers-HEPES (KRH) medium (120 mM NaCl, 4.7 mM KCl, 2.2 mM $CaCl_2$, 1.2 mM $MgSO_4$, 1.2 mM KH_2PO_4, 10 mM HEPES, 1.8 g/liter dextrose, pH 7.4). Transport assays (15 min, 37°C) were initiated by the addition of 5-[1,2-^{3}H(*N*)]hydroxytryptamine creatinine sulfate ([^{3}H]5HT, 20 nM final concentration, DuPont/NEN, Boston, MA) diluted in KRH supplemented with pargyline and L-ascorbate (100 μM final) with or without various concentrations of unlabeled 5HT and antagonists. After three 1-ml washes with ice-cold KRH medium, cells were solubilized in 500 μl of 1% SDS and released radioactivity was assessed by scintillation spectrometry. Nonspecific [^{3}H]5HT transport was determined by assays of cells transfected with the plasmid vector (pBluescript SKII-; Stratagene, La Jolla, CA) on the same plate and subtracted from the data. Rates derived from assays performed in triplicate are presented as picomoles 5HT accumulated per cell per minute $\pm$ SEM. Kinetic constants (K_m, V_{max}) were obtained by nonlinear least-squares fits (Kaleidagraph) of substrate/velocity profiles assuming a single population of noninteracting sites obeying Michaelis-Menten kinetics. To assess surface density of 5HT transporters in transfected cells, cDNA or vector-control transfected cells were washed with 2 ml KRH and then preincubated at 22°C with 450 μl KRH media for 15 min before the addition of 50 μl of various concentrations of [^{125}I]RTI-55 (3β-[4-iodophenyl]tropan-2β-carboxylic acid methyl ester tartrate; DuPont/NEN). Binding of ligand was assessed in triplicate after 1 h following three 1-ml washes with ice-cold KRH medium and solubilization of cells in 1% SDS. Competi-

tion experiments were conducted at 100 nM [^{125}I]RTI-55 ± increasing concentrations of unlabeled RTI-55, cocaine, and paroxetine. Parallel incubations were conducted ± 10 μM cocaine to assess nonspecific binding. Levels of nonspecific binding were equal to that observed with cells transfected with the plasmid vector alone and were subtracted from the data to yield specific [^{125}I]RTI-55 binding. Relative expression levels of rSERT and rSERT chimeras were estimated at a concentration of [^{125}I]RTI-55 (500 nM) above the apparent K_d of the ligand (220 nM).

Results and Discussion

Following publication of the rat brain (Blakely et al., 1991*a*) and rat basophilic leukemia (Hoffman et al., 1991) cell 5HT transporter cDNAs, differences were apparent in the NH$_2$ and COOH termini of the two carriers as predicted from conceptual translations of each open reading frame (ORF). The rat brain 5HT transporter was predicted to be encoded by a 607 amino acid protein while the RBL 5HT carrier was predicted to be encoded by a 653 amino acid protein. Both cDNAs induced saturable, Na$^+$-dependent 5HT uptake on transfected cells with similar pharmacologic properties (sensitivity to antidepressants, cocaine, etc.). After resequencing of the rat brain 5HT transporter cDNA with deoxyinosine and 7-deazaguanine nucleotides to search for overlooked sequencing compressions, a single base misread was apparent in the region of the rat brain 5HT transporter cDNA encoding the NH$_2$ terminus (base 132 of original sequence read as GG rather than G). Sequencing of the RBL cDNA (kindly provided by B. Hoffman, National Institutes of Health, Bethesda, MD) in the same manner revealed multiple differences in the RBL sequence encoding the COOH terminus and the 3′ noncoding region, also attributable to sequencing errors. Apart from these differences, both cDNAs are identical in regions encompassing transmembrane (TM) domains 1-12 and both hybridize to a single mRNA localized to rat brain and peripheral regions known to express the 5HT transporter. With the correction of sequencing errors, both brain and peripheral 5HT transporters appear to be the products of a single gene. In support of this conclusion, an orphan transporter sequence published by Mayser et al. (1991) is identical to our corrected sequence, and therefore appears to encode a 5HT transporter as well. Interestingly, the correction of the base misread in the NH$_2$ terminus of the rat brain transporter cDNA results in the utilization of a different reading frame before base 132 and surprisingly the absence of an initiating methionine. The next in-frame methionine present on the rat brain 5HT transporter cDNA lies two amino acids before TM1 but possesses a good translation initiation sequence (AAGATGG) according to Kozak (1986). Initiation of translation from this methionine would result in the synthesis of a 545 amino acid polypeptide lacking the

Figure 2. cDNA sequence of rSERT. Solid lines indicate hydrophobic stretches likely to form TM domains in the 5HT carrier. Arrows indicate the positions of oligonucleotides used to amplify the missing 5′ sequence in our original cDNA isolate. New sequences are those 5′ to the EcoRI site that bordered the original clone. A vertical line denotes the plasmid/cDNA boundary. The asterisk over base 132 indicates the position of a compression in initial sequencing reactions, read erroneously as CCAGGC rather than CCAGC in Blakely et al. (1991*a*).

```
1      CCCCTCGAG|TTTCCGTCTTGTCCCCATAACCCGAGAGAGAGATTCAAACCAAGAACCAAGAGAGCTAGCCTGGTCCTCGGCCAGATGACGGAATCCGCATCACTTACTGACCAGCAGC                    114
                 EcoRI

115    ATGGAGACCACCCCTTGAATTCACAGAAAGTGCTGTCAGAGTGTAAGGACAGAGAGGACTGTTCAGAAAATGGTGTTCCCACCACAGCGGCACGGGCA                                         228
1      M  E  T  T  P  L  N  S  Q  K  V  L  S  E  C  K  D  R  E  D  C  Q  E  N  G  V  L  Q  K  G  V  P  T  T  A  D  R  A                           38
                                                                              *
229    GAGCCTAGCCAAATATCCAATGGGTACTCTGCAGTCCCCAGCACAGAGTCAGGGGGACGAAGCTTCACACTCGATCCCAGCTGCCACCACCCTGGTGGCTGAGATTCGCCAA                           342
39     E  P  S  Q  I  S  N  G  Y  S  A  V  P  S  T  S  A  G  D  E  A  S  H  S  I  P  A  A  T  T  T  L  V  A  E  I  R  Q                           76

343    GGGGAGCGGGAGACCTGGGGCAAGAAGATGGATTTCCTCTCCGTCATTGGCTATGCCGTGGACCTGGGCAATATCTGGCGCTTCCCCTACCTGTGCTACCAGAATGGCGGA                           456
77     G  E  R  E  T  W  G  K  K  M  D  E  L  L  S  V  I  G  Y  A  V  D  L  G  N  I  W  R  F  P  Y  L  C  Y  Q  N  G  G                           114

457    GGGGCCTTCCTCCCTTATACCATCATGGCCATTTTCGGGGGGATCCCGCTCTTTTACATGGAGCTGGCACTGGGCCAGTACCACCGGAACGGCTGCATTTCCATATGAGG                           570
115    G  A  F  L  P  Y  T  I  M  A  I  F  G  G  I  P  L  F  Y  M  E  L  A  L  G  Q  Y  H  R  N  G  C  I  S  I  W  R                             152

571    AAGATCTGCCCGATTTTCAAAGGCATTGGTTACGCCCTCTGCATCATCGCCTTTTACATCGCCTCCTACTACAACACCATCATAGCCTGGGCCCTCTACTACCTCATCTCCTCC                         684
153    K  I  C  P  I  F  K  G  I  G  Y  A  L  C  I  I  A  F  Y  I  A  S  Y  Y  N  T  I  I  A  W  A  L  Y  Y  L  I  S  S                           190

685    CTCACGGACCGGCTGCCCTGGACCAGCTGCACGAACTCCTGGAACACTGGCAACTACTTCGCCCAGGACAACATCACCTGGACGCTTCACTCCACCTCCCCGCT                                   798
191    L  T  D  R  L  P  W  T  S  C  T  N  S  W  N  T  G  N  C  T  N  Y  F  A  Q  D  N  I  T  W  T  L  H  S  T  S  P  A                           228

799    GAGGAGTTCTACTTGCGCCATGTCCTGCAGATCCACCAGTCTAAGGGACTTCAGGACCTGGGCACCATCAGCTGGCAGCTCACCTTGTGCATCGTGCTCATCTTCACCGTAATC                         912
229    E  E  F  Y  L  R  H  V  L  Q  I  H  Q  S  K  G  L  Q  D  L  G  T  I  S  W  Q  L  T  L  C  I  V  L  I  F  T  V  I                           266

913    TACTTTAGCATCTGGAAAGGCGTCAAAACATCTGGCAAGGTGGTGTGGGTCACAGCCACCTTCCCACATTGTCCTCTCTCGTGGTGAGGGGCACCCTTCCTGGA                                   1026
267    Y  F  S  I  W  K  G  V  K  T  S  G  K  V  V  W  V  T  A  T  F  P  Y  I  V  L  L  S  V  L  L  V  R  G  A  T  L  P  G                        304

1027   GCCTGGAGAGGGGTCGTCTTCTACTTGAAACCCAACTGGCAGAAACTCTTGGAGACAGGGGTGTGGGTAGATGCCGCCGCTCAGATCTTCTTCTCTTTGGCCCCGGGCTTTGGG                         1140
305    A  W  R  G  V  V  F  Y  L  K  P  N  W  Q  K  L  L  E  T  G  V  W  V  D  A  A  A  Q  I  F  F  S  L  G  P  G  F  G                           342

1141   GTTCTCCTGGCTTTTGCTAGCTACAACAAGTTCAACAACAACTGTTACCAAGATGCCCTGGTCACCAGTGTGGTGAATTGCATGACAAGCTTCGTCTCTGGCTTCGTCATCTTC                         1254
343    V  L  L  A  F  A  S  Y  N  K  F  N  N  N  C  Y  Q  D  A  L  V  T  S  V  V  N  C  M  T  S  F  V  S  G  F  V  I  E                           380

1255   ACGGTGCTTGGCTACATGGCCGAGATGAGGAATGAAGATGTCAGAGAGTGTCAAAGACGAGCCCAGCCTCTTCATCACGTATGCAGAGGCAATAGCCAACATGCCA                                 1368
381    T  V  L  G  Y  M  A  E  M  R  N  E  D  V  S  E  V  A  K  D  A  G  P  S  L  L  F  I  T  Y  A  E  A  I  A  N  M  P                           418

1369   GCATCCACGTTCTTTGCCATCATCTTCTTCCTCATGTTAATCACGCTGGGATTGGACTCCACTTTTGGTGCTCTGGAAGGTGTCATCACAGCTGTGCTGGATGAGTTCCCTCAC                         1482
419    A  S  T  F  F  A  I  I  F  F  L  M  L  I  T  L  G  L  D  S  T  F  A  G  L  E  G  V  I  T  A  V  L  D  E  F  P  H                           456

1483   ATCTGGGCCAAGCGCAGGGAATGGTTCGTGCTCATCGTGGTCATCACCTGCGTCTTGGGATCCCTGCTCACACTGACTTCTGGAGGCGCATATGTGGTGACTCTGCTGGAGGAG                         1596
457    I  W  A  K  R  R  E  W  F  V  L  I  V  V  I  T  C  V  L  G  S  L  L  T  L  T  S  G  G  A  Y  V  V  T  L  L  E  E                           494

1597   TATGCCACGGGGCCAGCAGTGCTCACCGTGGCCCTCATCGAGGCCGTCGCCGTGTCTTGGTTCTATGGAATCACTCAGTTCTGCAGCGATGTGAAGGAGATGCTGGGCTTCAGC                         1710
495    Y  A  T  G  P  A  V  L  T  V  A  L  I  E  A  V  A  V  S  W  F  Y  G  I  T  Q  F  C  S  D  V  K  E  M  L  G  F  S                           532

1711   CCGGGATGGTTTTGGAGGATCTGCTGGGTGGCCATCTCTCCCCTCTTCTTGCTCTTCATCATTTGCAGTTTTCTGATGAGCCCACCAGCTACGGCTTTCCAATACAACTAT                            1824
533    P  G  W  F  W  R  I  C  W  V  A  I  S  P  L  F  L  L  F  I  I  C  S  F  L  M  S  P  P  Q  L  R  L  F  Q  Y  N  Y                           570

1825   CCCCACTGGAGTATCGTCTTTGGCTACTGCATAGGGATGTCGTCCGTCATCTGCATCCCTACCTATATCCTATATCGTTTATCCGGGACACTTAAGGAGCGCCATT                                 1938
571    P  H  W  S  I  V  L  G  Y  C  I  G  M  S  S  V  I  C  I  P  T  Y  I  L  I  Y  R  L  I  S  T  P  G  T  L  K  E  R  I                        608

1939   ATTAAAAGTATCACTCCTGAAACACCCAGAAATCCCGTGGGGACATCCCATGAATGCTGTGTAACACACCCTGGGAGAGGACACCTCTTCCAGCACCTCTCTCAGCT                                2052
609    I  K  S  I  T  P  E  T  P  T  E  I  P  C  G  D  I  R  M  N  A  V  *                                                                        630

2053   CTGAAAAGCCCACCTGGACTCCTCCCCTCTTAAGCCAAGCCTAACCATGGTGCCCAGACTCTTGTGGATTCCGACCACTTCTTCCGTGGACTCT                                             2166
2167   CAGACAGTCTACCACATTCGAATGGTGACACCACTGAGCTGGCCTCTTGGACACGTGCCGCCTTGAGGGGATGAACGCACCATGTTCAGGTTTGAAT                                          2280
2281   TAGGTCGTGGAGAGTCGTGATCATGTTTTTGTAAGATCATCATCATCACCCCGCCATCTGTTAGCTTCATAAAGCCTTCAAAGCTCTTCATGAATACATAAACCACCTAAGAGAAACAGAG              2394
2395   ATGTCTTTGCTAGCCATATAT                                                                                                                       2414
```

putative cytoplasmic NH$_2$-terminal tail sequences common to, though not conserved across, all members of the GABA/NET transporter gene family.

To restore 5' noncoding and coding sequences to the rat brain 5HT transporter cDNA, we utilized PCR on rat midbrain cDNA with oligonucleotides derived from the 5' end (antisense oligonucleotide) of our original clone and 5' noncoding elements (sense oligonucleotide) of the Mayser et al. (1991) and Hoffman et al. (1991) cDNAs. Fig. 2 presents the sequence of the restored rat brain 5HT transporter (rSERT), with the added sequence appearing upstream of the NH$_2$ terminal EcoRI site, including an initiator methionine at base 115 in agreement with the RBL transporter sequence. As differences between the COOH termini of the rat brain and RBL 5HT transporters appear to arise from sequencing errors in the report of the RBL cDNA (Blakely, R., and K. Moore, unpublished observations), the 5HT carrier from both sources is encoded by an identical 630 amino acid subunit. This conclusion is underscored by identification of a human 5HT transporter (hSERT; Ramamoorthy, S., A. L. Bauman, K. R. Moore, H. Han, T. Yang-Feng, A. S. Chang, V. Ganapathy, and R. D. Blakely, manuscript submitted for publication). Fig. 3 depicts an alignment of the rSERT and hSERT coding sequences. hSERT was also found to be encoded by a 630 amino acid polypeptide and overall exhibits 92% amino acid identity. Note that the predicted NH$_2$ termini are absolutely conserved for the first 10 amino acids, while Met86, the likely initiation site of our initial cDNA isolate, is not conserved. The greatest region of divergence between the two carriers is in the NH$_2$ terminus before TM1, where 20 of the 52 amino acid changes occur. This divergence, along with the observation that a 5HT transporter encoded by rSERT(86-630) is functional in transfected cells (Blakely et al., 1991*a*), suggests that the NH$_2$ terminus is of little importance in establishing the defining features of the 5HT carrier. Although the NH$_2$ truncated transporter induces antidepressant and cocaine-sensitive 5HT transport in transfected cells, overall expression levels are markedly elevated with the restoration of 5' sequences including the proper initiation codon (see below), revealing that this region is important for proper translation, stability, or export of the 5HT carrier. Since substrates must move across the plane of the membrane, and the TM domains of the GABA/NET gene family exhibit the greatest extent of sequence identity (Blakely, 1992), the poorly conserved NH$_2$ and COOH termini may play only a minor role in substrate recognition and translocation for members of this gene family. In support of this idea, Mabjeesh and Kanner (1992) have utilized limited proteolysis of the GABA (rGAT1) transporter reconstituted into membrane vesicles to argue that major portions of both NH$_2$ and COOH termini are not required for GABA transport.

To examine the importance of the NH$_2$ and COOH termini in defining transport activity, we swapped major portions of the presumed cytoplasmic tails of hNET for analogous amino acids of rSERT, either separately or combined. The scheme for the transfer is given in Fig. 1. An hNET expression construct was amplified to generate a linear fragment containing roughly two-thirds of the NH$_2$ (1-52) and COOH (587-617) termini of hNET, connected at either end to pBluescript SKII-. As the 5' ends of the oligonucleotides for hNET amplification encode rSERT StyI and AflII sites (absent from the plasmid vector), digestion of rSERT with both of these enzymes and ligation to the similarly digested hNET PCR product yield an rSERT chimera with major portions of NH$_2$ and COOH termini derived from hNET (chimera 3). To yield an acceptor fragment possessing only one end of hNET, the PCR fragment was

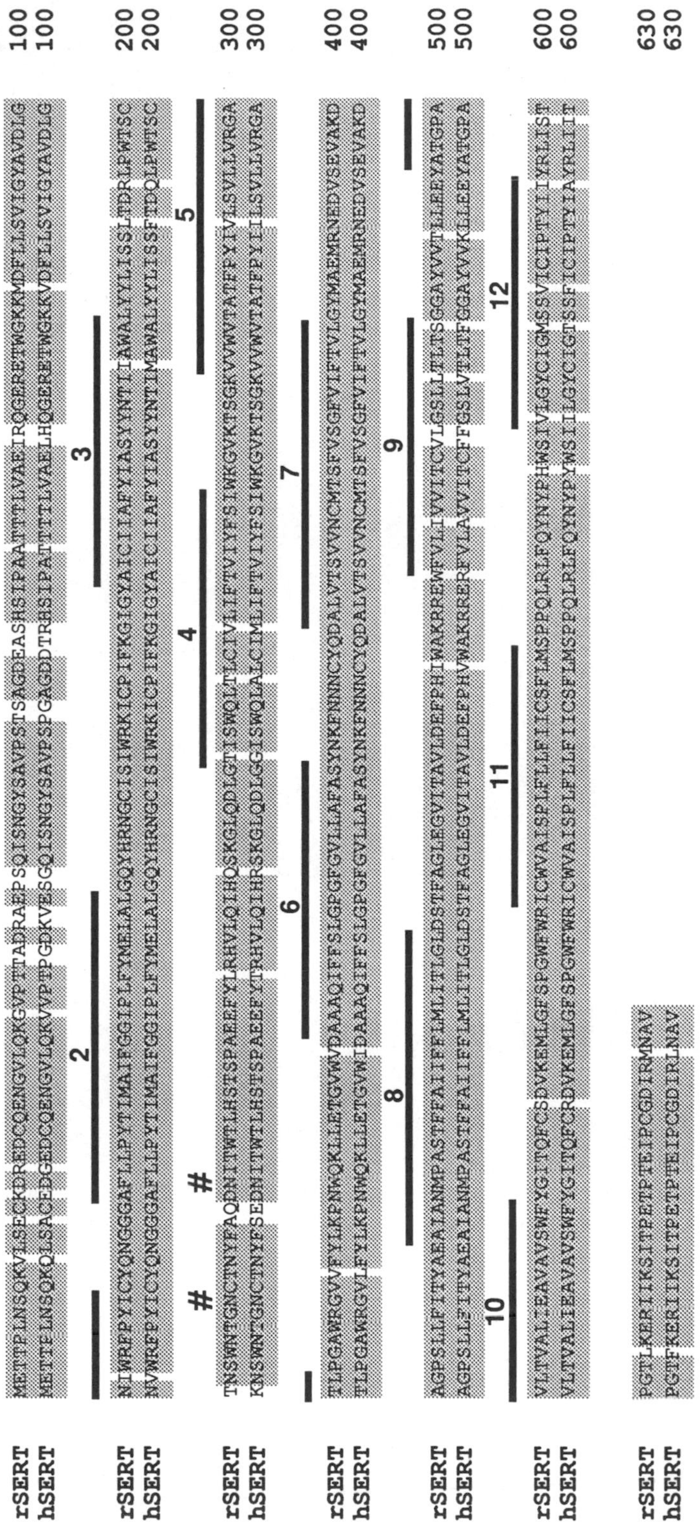

Figure 3. Comparison of rat (rSERT) and human (hSERT) 5HT transporters. Amino acid sequences are boxed to indicate absolute conservation. Solid lines indicate putative TM domains. Symbols (#) indicate the positions of canonical N-linked glycosylation sites.

digested with either StyI or AflII in combination with an enzyme in the plasmid unique to both hNET and rSERT. Ligation of identically cut rSERT into these acceptor fragments yields chimeric rSERT molecules containing only the NH_2 (chimera 4) or the COOH (chimera 5) terminus of hNET. The putative topology of chimeric and truncated rSERT molecules examined in this study is illustrated in Figs. 4 and 5.

Transport characteristics of the chimeric molecules were assessed 6 h after plasmid transfection of vaccinia-T7 infected HeLa cells. Since in our experience expression levels can vary between virus preps by approximately twofold, care was taken to make comparisons between mutant and control plasmids transfected and assayed in parallel. As demonstrated in Fig. 6, the restoration of the NH_2 terminus to

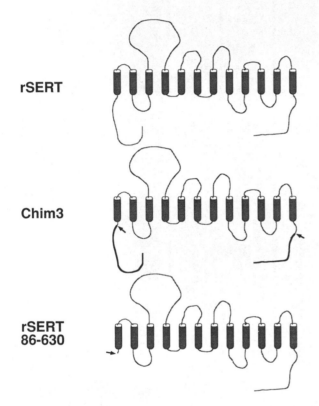

Figure 4. Putative topological representation of rSERT protein and mutants encoded by chimera 3 and rSERT(86-630). Heavy solid lines indicate domains transferred from hNET.

the rSERT cDNA boosts maximal capacity of the 5HT transporter (V_{max} = $2.64 \pm 0.18 \times 10^{-17}$ mol/cell per min) by >30-fold relative to our original cDNA, rSERT86-630 (V_{max} = $8.29 \pm 0.84 \times 10^{-19}$ mol/cell per min). In addition, the K_m for 5HT transport was found to be reduced for rSERT86-630 (100 ± 17 nM) relative to rSERT (467 ± 35 nM). In contrast to substantially reduced 5HT transport observed with an NH_2-terminal truncation, transfection of the rSERT chimera bearing NH_2- and COOH-terminal domains from hNET (chimera 3) resulted in a near-normal induction of 5HT uptake with a transport V_{max} 67% of rSERT levels. Chimera 3 transport K_m was only slightly reduced (315 ± 50 nM). NE transport was not induced by transfer of hNET domains into rSERT, nor did transfection of the plasmid vector alone yield 5HT or NE transport above nontransfected controls.

Although it is clear from these experiments that 5HT transport is induced by rSERT bearing major portions of hNET NH₂ and COOH termini, we consistently observed a 30–50% reduction in 5HT transport capacity with this chimera. To explore which of the two regions of rSERT was responsible for this effect, we constructed two additional chimeras. Chimera 4 bears sequences encoding the first 52 amino acids and 5′ noncoding elements of the hNET cDNA with the rest of the transporter formed by rSERT sequences. Chimera 5 is composed of rSERT except for residues 608-630 and 3′ noncoding sequences that are replaced by the analogous domain of hNET. 5HT transport assays conducted with these chimeras are shown in Fig. 7. Replacement of the rSERT COOH terminus and 3′ noncoding sequences in

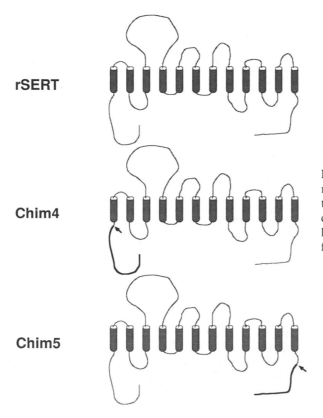

Figure 5. Putative topological representation of rSERT protein and mutants encoded by chimeras 4 and 5. Heavy solid lines indicate domains transferred from hNET.

chimera 5 yielded 5HT transport indistinguishable from rSERT (chimera 5: $V_{max} = 5.93 \pm 0.68 \times 10^{-17}$ mol/cell per min, $K_m = 513 \pm 98$ nM; rSERT: $V_{max} = 5.84 \pm 0.36 \times 10^{-17}$ mol/cell per min, $K_m = 459 \pm 53$ nM). Chimera 4, however, exhibited a twofold reduction in 5HT transport K_m (233 ± 35 nM) and a 49% reduction in V_{max}. These findings indicate that the substitutions made in the NH₂ terminus of rSERT have more pronounced effects on induced 5HT transport activity than those made in the COOH terminus. We cannot rule out at this point that the observed changes in transport activity of transfected chimeras arise from 5′ noncoding elements of hNET, although we transfected with saturating concentrations of plasmid DNA to attempt to mitigate against this possibility. Nonetheless, experiments with chimeras 3, 4, and 5 indicate that significant substitutions within NH_2 and COOH termini of rSERT can be achieved using analogous domains of a distinct transporter without altering the carrier's ability to recognize and translocate 5HT.

To determine if alterations in the NH_2 and COOH termini altered antagonist specificity of rSERT, we examined 5HT transport induced by either rSERT or chimera 3 for sensitivity to selective 5HT and NE uptake inhibitors. As shown in Fig. 8, 5HT transport of both rSERT and chimera 3 was sensitive to low concentrations of the 5HT-selective antagonist paroxetine (rSERT K_i = 678 ± 35 pM; chimera 3 K_i = 495 ± 33 pM). Similarly, neither construct induced 5HT transport sensitive to nomifensine, a potent and selective NE transport antagonist (hNET K_i = 7.68 nM; Pacholczyk et al. [1991]). Thus, substitutions within NH_2 and COOH termini of

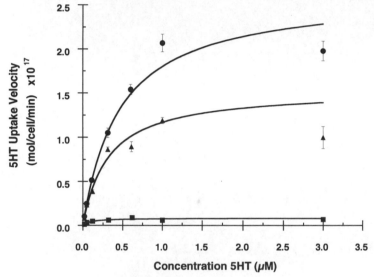

Figure 6. Transport of 5HT in HeLa cells transfected with rSERT, rSERT86-630, and chimera 3 plasmids. Uptake assays on HeLa cells transfected with SERT (*filled circles*), rSERT86-630 (*filled squares*), and chimera 3 (*filled triangles*) were performed at different 5HT concentrations in triplicate as described in Methods. Curves are nonlinear, least-squares curve fits of data, assuming single, noninteracting sites exhibiting Michaelis-Menten kinetics. Transfections with pBluescript performed in parallel were used to assess nonspecific 5HT uptake in HeLa cells and were subtracted from the data.

rSERT not only fail to alter substrate specificity and translocation, but also yield no discernible effects on antagonist selectivity and potency. These data focus attention on the TM domains (particularly 1-2 and 5-8; Blakely et al., 1991*a;* Blakely, 1992) and intervening loops as key determinants of defining features of monoamine transport antagonist recognition. In support of these conclusions, Kitayama et al. (1992) have identified residues within TM domains 1, 7, and 8 of the dopamine transporter which alter cocaine ligand recognition and/or dopamine transport. Recently, we have also identified similar TM residues in hNET and rSERT which, when mutated, abrogate NE and 5HT transport, respectively (Blakely, R. D., unpublished observations).

As outlined above, consistent reductions in maximal capacity were noted in expression experiments with chimeras 3 and 4, but not with chimera 5. These changes could arise from differences in the amount of cell surface expression of mutant transporters or in alterations in the intrinsic activity of individual transporter molecules. To address this issue, one requires a probe for cell surface expression of wild-type and mutant 5HT transporters in small numbers of transfected cells. Recently, the cocaine analogue [^{125}I]RTI-55 has been introduced as a high-affinity probe for the 5HT transporter in brain (Boja et al., 1992) and platelet (Wall et al., 1993) membranes. We observed that [^{125}I]RTI-55 binds reversibly to cells transfected with rSERT at levels markedly above those observed for mock-transfected HeLa cells, with effective inhibition of binding achieved when assays were conducted in the presence of the competitive 5HT transport antagonists paroxetine ($IC_{50} = 1.8$ nM)

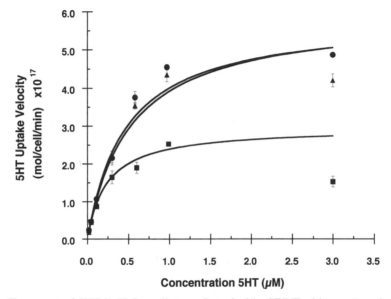

Figure 7. Transport of 5HT in HeLa cells transfected with rSERT, chimera 4, and chimera 5 plasmids. Uptake assays on HeLa cells transfected with SERT (*filled circles*), chimera 4 (*filled squares*), and chimera 5 (*filled triangles*) were performed at different 5HT concentrations in triplicate as described in Methods. Curves are nonlinear, least-squares curve fits of data, assuming single, noninteracting sites exhibiting Michaelis-Menten kinetics. Transfections with pBluescript performed in parallel were used to assess nonspecific 5HT uptake in HeLa cells and were subtracted from the data.

and cocaine ($IC_{50} = 560$ nM). As demonstrated in Fig. 9, [^{125}I]RTI-55 exhibits saturable, high-affinity specific binding to transfected HeLa cells when 10 μM cocaine (or nontransfected cells) is used to define nonspecific binding. A Scatchard transformation of specific [^{125}I]RTI-55 binding indicates that the ligand detects a single population of noninteracting sites of high affinity, $K_d = 220$ nM. The quantity V_{max}/B_{max} also gives an estimate of the maximal turnover rate of individual 5HT transporters in transfected HeLa cells, which for these particular experiments yields a value of approximately four per second. This turnover is in good agreement with values obtained from human platelet (eight per second; Talvenheimo et al., 1979) membranes. Thus, [^{125}I]RTI-55 appears to be a suitable, high specific activity ligand for the detection of surface pools of 5HT transporters in transfected cells. To determine if the reductions in 5HT transport observed with chimeras resulted from changes in surface density, we next examined the binding of [^{125}I]RTI-55 to HeLa cells transfected with rSERT, chimera 3, or chimera 5 (depicted in Fig. 10). Compared with rSERT, a small reduction (28%) in [^{125}I]RTI-55 binding was observed for chimera 5, which exhibits wild-type levels of 5HT transport activity. In contrast, cells transfected with chimera 3 exhibited a marked 70% reduction in [^{125}I]RTI-55 binding levels. These data indicate that the reductions seen in chimeric transporter uptake assays are probably attributable to changes in surface density of functionally equivalent 5HT carriers.

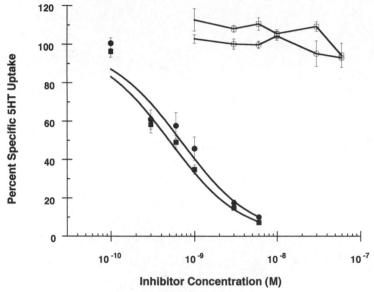

Figure 8. Antagonist sensitivity of 5HT transport induced by rSERT and chimera 3. 5HT uptake assays were conducted as described in Methods with rSERT (*filled or open circles*) and chimera 3 (*filled or open squares*) transfected HeLa cells in the presence or absence of increasing concentrations of the 5HT transport-selective antagonist paroxetine (*filled circles and squares*) or the NE transport-selective antagonist nomifensine. Assays were performed in triplicate and inhibition was expressed as a percentage of assays performed in the absence of antagonists. Transfections with pBluescript performed in parallel were used to assess nonspecific 5HT uptake in HeLa cells and were subtracted from the data.

In summary, single, highly homologous 630 amino amino acid polypeptides encode rat and human 5HT transporters, members of the GABA/NET gene family. Hydrophobicity analysis and topological considerations indicate the presence of 12 TM domains with cytoplasmic NH$_2$ and COOH termini. The greatest identity across members of the GABA/NET gene family and between SERT species homologues occurs within the TM domains and specific intervening loops, whereas NH$_2$ and COOH termini display the least conservation. These latter regions, as explored with chimeras synthesized from rSERT and hNET, can tolerate significant substitutions from analogous regions of other members of the GABA/NET transporter gene family without major alterations in substrate recognition and translocation or

A

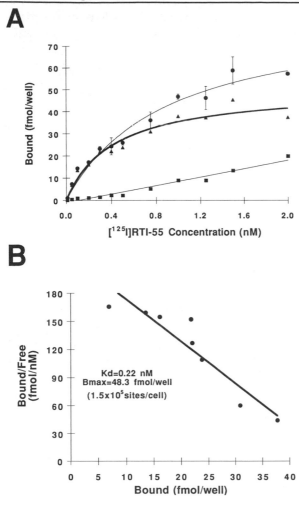

Figure 9. Binding of [^{125}I]RTI-55 to 5HT transporters in rSERT transfected HeLa cells. (*A*) Binding assays were performed in triplicate with increasing concentrations of [^{125}I]RTI-55 in the presence (*filled squares*) or absence (*filled circles*) of 10 μM cocaine to assess nonspecific binding as described in Methods. Specific binding (*filled triangles, heavy line*) is plotted as a nonlinear, least-squares fit to a single-site binding isotherm. Nonspecific binding as assessed by incubations in the presence of 10 μM cocaine yielded data equivalent to cells transfected with the plasmid vector alone. (*B*) Scatchard transformation of the saturation data presented in *A*.

B

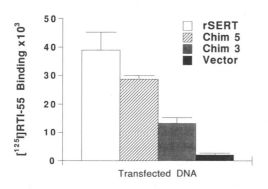

Figure 10. Binding of [^{125}I]RTI-55 to intact HeLa cells transfected with rSERT and chimeras. Incubations with transfected cells were conducted as described in Methods at 0.5 nM final ligand concentration. Assays were performed in triplicate and values are plotted ± SEM. No subtraction has been performed to indicate specific binding, although as defined by 10 μM cocaine, it is equivalent to data obtained with vector (pBluescript SKII⁻) transfected cells.

antagonist selectivity and potency. Even the apparently complete loss of the NH_2 terminus, as assessed in rSERT86-630, our original rat brain 5HT transporter isolate, induces significant Na^+-dependent 5HT transport in transfected HeLa cells with retained sensitivity to antidepressants and cocaine (Blakely et al., 1991*a*). Thus, it is likely that the intervening TM domains (particularly TM 1-2 and 5-8; Blakely, 1992) and/or connecting loops possess residues responsible for defining features of 5HT transport. Finally, these findings suggest the possibility for substitution of novel, useful epitopes within NH_2 and/or COOH termini for protein purification, transporter visualization, and subcellular trafficking studies.

Acknowledgments

We thank Walailux Nethin for excellent assistance in the construction of chimeras, F. Ivy Carrol for consultation on the use of [^{125}I]RTI-55 and the donation of unlabeled ligand, and Sammanda Ramamoorthy and Haley Berson for helpful suggestions.

This work was supported in part by NIH grant DA-07390 and a Mallinckrodt Foundation Award to R. D. Blakely, and by the Graduate Neuroscience Program of Emory University (Y. Qian).

References

Balkovetz, D. F., C. Tiruppathi, F. H. Leibach, V. B. Mahesh, and V. Ganapathy. 1989. Evidence for an imipramine-sensitive serotonin transporter in human placental brush-border membranes. *Journal of Biological Chemistry.* 264:2195–2198.

Blakely, R. D. 1992. Advances in molecular biology of neurotransmitter transporters. *Current Opinion in Psychiatry.* 5:69–73.

Blakely, R. D., H. E. Berson, R. T. Fremeau, Jr., M. G. Caron, M. M. Peek, H. K. Prince, and C. C. Bradley. 1991*a*. Cloning and expression of a functional serotonin transporter from rat brain. *Nature.* 354:66–70.

Blakely, R. D., J. A. Clark, G. Rudnick, and S. G. Amara. 1991*b*. Vaccinia-T7 RNA polymerase expression system: evaluation for the expression cloning of plasma membrane transporters. *Analytical Biochemistry.* 194:302–308.

Boja, J. W., W. M. Mitchell, A. Patel, T. A. Kopajtic, F. I. Carrol, A. H. Lewin, P. Abraham, and M. J. Kuhar. 1992. High-affinity binding of [^{125}I]RTI-55 to dopamine and serotonin transporters in rat brain. *Synapse.* 12:27–36.

Campbell, A. M., P. D. Kessler, and D. M. Fambrough. 1992. The alternative carboxyl termini of avian cardiac and brain sarcoplasmic reticulum/endoplasmic reticulum Ca^{2+}-ATPases are on opposite sides of the membrane. *Journal of Biological Chemistry.* 267:9321–9325.

Clark, J. A., A. Y. Deutch, P. Z. Gallipoli, and S. G. Amara. 1992. Functional expression and CNS distribution of a β-alanine-sensitive neuronal GABA transporter. *Neuron.* 9:337–348.

De Paulis, T., D. Kelder, S. B. Ross, and N. E. Stjernstrom. 1978. On the topology of the norepinephrine transport carrier in rat hypothalamus: the site of action of tricyclic uptake inhibitors. *Molecular Pharmacology.* 14:596–606.

Fremeau, R. T., M. G. Caron, and R. D. Blakely. 1992. Molecular cloning and expression of a high affinity L-proline transporter expressed in putative glutamatergic pathways of rat brain. *Neuron.* 8:915–926.

Frielle, T., K. W. Daniel, M. G. Caron, and R. J. Lefkowitz. 1988. Structural basis of β-adrenergic receptor subtype specificity studied with chimeric β₁/β₂-adrenergic receptors. *Proceedings of the National Academy of Sciences, USA.* 85:9494–9498.

Fuller, R. W., and D. T. Wong. 1990. Serotonin uptake and serotonin uptake inhibition. *Annals of the New York Academy of Sciences.* 600:68–80.

Guastella, J., N. Brecha, C. Weigmann, H. A. Lester, and N. Davidson. 1992. Cloning, expression, and localization of a rat brain high-affinity glycine transporter. *Proceedings of the National Academy of Sciences, USA.* 89:7189–7193.

Guastella, J., N. Nelson, H. Nelson, L. Czyzk, S. Keynan, M. C. Meidel, N. Davidson, H. A. Lester, and B. Kanner. 1990. Cloning and expression of a rat brain GABA transporter. *Science.* 249:1303–1306.

Hoffman, B. J., E. Mezey, and M. J. Brownstein. 1991. Cloning of a serotonin transporter affected by antidepressants. *Science.* 254:579–580.

Kilty, J. E., K. Lorang, and S. G. Amara. 1991. Cloning and expression of a cocaine-sensitive rat dopamine transporter. *Science.* 254:578–579.

Kitayama, S., S. Shimada, H. Xu, L. Markham, D. M. Donovan, and G. R. Uhl. 1992. Dopamine transporter site-directed mutations differentially alter substrate transport and cocaine binding. *Proceedings of the National Academy of Sciences, USA.* 89:7782–7785.

Kobilka, B. K., T. S. Kobilka, K. Daniel, J. W. Regan, M. G. Caron, and R. J. Lefkowitz. 1988. Chimeric α₂-, β₂-adrenergic receptors: delineation of domains involved in effector coupling and ligand binding specificity. *Science.* 240:650–656.

Koe, B. K. 1976. Molecular geometry of inhibitors of the uptake of catecholamines and serotonin in synaptosomal preparations of rat brain. *Journal of Pharmacology and Experimental Therapeutics.* 199:649–661.

Kozak, M. 1986. Point mutations define a sequence flanking the AUG initiator codon that modulates translation by eukaryotic ribosomes. *Cell.* 44:283–292.

Lee, S.-L., and B. L. Fanburg. 1986. Serotonin uptake by bovine pulmonary artery endothelial cells in culture. I. Characterization. *American Journal of Physiology.* 250 (*Cell Physiology* 19):C761–C765.

Li, M., Y. N. Jan, and L. Y. Jan. 1992. Specification of subunit assembly by the homophilic amino-terminal domain of the Shaker potassium channel. *Science.* 257:1225–1230.

Lingjarede, O. 1971. Uptake of serotonin in blood platelets in vitro. I. The effects of chloride. *Acta Physiologica Scandinavica.* 81:75–83.

Mabjeesh, N. J., and B. I. Kanner. 1992. Neither amino nor carboxyl termini are required for function of the sodium- and chloride-coupled γ-aminobutyric acid transporter from rat brain. *Journal of Biological Chemistry.* 267:2563–2568.

MacDonald, R. J., G. H. Swift, A. E. Przybyla, and J. M. Chirgwin. 1987. Isolation of RNA using guanidinium salts. *Methods in Enzymology.* 152:219–227.

Masters, S. B., K. A. Sullivan, R. T. Miller, B. Beiderman, N. G. Lopez, J. Ramachandran, and H. R. Bourne. 1988. Carboxyl terminal domain of Gₛₐ specifies coupling of receptors to stimulation of adenylyl cyclase. *Science.* 241:448–451.

Maxwell, R. A., R. M. Ferris, and J. E. Burcsu. 1976. Structural requirements for inhibition of noradrenaline uptake by phenethlamine derivatives, desipramine, cocaine, and other compounds. *In* The Mechanism of Neuronal and Extraneuronal Transport of Catecholamines. D. M. Paton, editor. Raven Press, New York. 95–153.

Maxwell, R. A., and H. L. White. 1978. Tricyclic and monamine oxidase inhibitor antidepressants: structure-activity relationships. *In* Handbook of Psychopharmacology. L. L. Iversen, S. D. Iversen, and S. H. Snyder, editors. Plenum Publishing Corp., New York. 83–155.

Mayser, W., H. Betz, and P. Schloss. 1991. Isolation of cDNAs encoding a novel transmitter of the neurotransmitter transporter gene family. *FEBS Letters.* 295:203–206.

Pacholczyk, T., R. D. Blakely, and S. G. Amara. 1991. Expression cloning of a cocaine- and antidepressant-sensitive human noradrenaline transporter. *Nature.* 350:350–354.

Ross, S. B., and A. L. Renyi. 1967. Accumulation of tritiated 5-hydroxytryptamine in brain slices. *Life Sciences.* 6:1407–1415.

Rudnick, G. 1977. Active transport of 5-hydroxytrptamine by plasma membrane vesicles isolated from human blood platelets. *Journal of Biological Chemistry.* 252:2170–2174.

Saiki, R. K., D. H. Gelfand, S. Stoffel, S. J. Scharf, R. Higuchi, G. T. Horn, K. B. Mullis, and H. A. Erlich. 1988. Primer-directed enzymatic amplification of DNA with a thermostable DNA polymerase. *Science.* 239:487–494.

Sanger, F., S. Nicklen, and A. R. Coulson. 1977. DNA sequencing with chain-terminating inhibitors. *Proceedings of the National Academy of Sciences, USA.* 74:5463–5467.

Shimada, S., S. Kitayama, C.-L. Lin, A. Patel, E. Nanthakumar, P. Gregor, M. Kuhar, and G. Uhl. 1991. Cloning and expression of a cocaine-sensitive dopamine transporter complementary DNA. *Science.* 254:576–578.

Smith, K. E., L. A. Borden, P. R. Hartig, T. Branchek, and R. L. Weinshank. 1992. Cloning and expression of a glycine transporter reveal colocalization with NMDA receptors. *Neuron.* 8:927–935.

Stocker, M., O. Pongs, M. Hoth, S. H. Heinemann, W. Stühmer, K.-H. Schröter, and J. P. Ruppersberg. 1991. Swapping of functional domains in voltage-gated K$^+$ channels. *Proceedings of the Royal Society of London.* 245:101–107.

Talvenheimo, J., P. J. Nelson, and G. Rudnick. 1979. Mechanism of imipramine inhibition of platelet 5-hydroxytryptamine transport. *Journal of Biological Chemistry.* 254:4631–4635.

Wall, S. C., R. W. Innis, and G. Rudnick. 1993. Binding of the cocaine analog [^{125}I]-2β-carbometoxy-3β-(4-iodophenyl)tropane (B-CIT) to serotonin and dopamine transporters: different ionic requirements for substrate and β-CIT binding. *Molecular Pharmacology.* In press.

Yamauchi, A., S. Uchida, H. M. Kwon, A. S. Preston, R. B. Robey, A. Garcia-Perez, M. B. Burg, and J. S. Handler. 1992. Cloning of a Na$^+$- and Cl$^-$-dependent betaine transporter that is regulated by hypertonicity. *Journal of Biological Chemistry.* 267:649–652.

Yan, H., J. Schlessinger, and M. V. Chao. 1991. Chimeric NGF-EGF receptors define domains responsible for neuronal differentiation. *Science.* 252:561–563.

Chapter 23

Identification of a New Family of Proteins Involved in Amino Acid Transport

Matthias A. Hediger, Yoshikatsu Kanai, Wen-Sen Lee, and Rebecca G. Wells

Department of Medicine, Renal Division, Brigham and Women's Hospital and Harvard Medical School, Boston, Massachusetts 02115

Molecular Biology and Function of Carrier Proteins © 1993 by The Rockefeller University Press

Introduction

In epithelial cells of the kidney and the small intestine, transport of amino acids proceeds via specialized transporters (Hopfer, 1987; Mircheff et al., 1982). Net absorption of amino acids from the lumen into the blood depends on active ion-coupled transporters located in the brush border membrane and facilitated transporters located in the basolateral membrane. Active transport systems are known to be coupled either to the inwardly directed Na^+, Cl^-, or H^+ gradients or to the outwardly directed K^+ gradient (Berry and Rector, 1991). For example, system A, which transports alanine and glycine, and system ASC, which prefers alanine, serine, and cysteine, are coupled to the inwardly directed Na^+ gradient (McCormick and Johnstone, 1988). The recently cloned acidic amino acid transport system $X_{A,G}^-$ is coupled to both the inwardly directed Na^+ and the outward directed K^+ gradients (Heinz et al., 1988; Kanai and Hediger, 1992). Transporters of β-amino acids are coupled to the inwardly directed Na^+ and Cl^- gradients (Wolff and Kinne, 1988). In contrast, system L, which has a preference for branched chain and aromatic amino acids (leucine, isoleucine, valine, phenylalanine), is not Na^+ dependent (Tate et al., 1989).

Clues to the specificity of amino acid transport systems have been provided by studies of patients with inborn errors of transport. For example, dibasic amino aciduria (lysinuric protein intolerance), which is characterized by impaired renal and intestinal absorption of lysine, arginine, and ornithine, but not cystine, suggests that a system exists that is specific for the transport of dibasic amino acids (Simell, 1989). A characteristic of this defect is impaired growth and reduced plasma ornithine levels, which in the liver affects the urea cycle. The disease may be caused by a defect in the Na^+-independent transport system for lysine, arginine, and ornithine known as system y^+. In the mouse, system y^+ has recently been characterized and found to possess 14 predicted membrane spanning regions (Kim et al., 1991; Wang et al., 1991). Oocyte expression studies confirmed that this protein mediates the uptake of dibasic amino acids but not cystine. Whether this transporter is defective in dibasic amino aciduria remains to be determined.

Since both cystine and dibasic amino acids are malabsorbed in the kidney and jejunum of patients with the autosomal recessive disease cystinuria, it is assumed that these amino acids are transported via a single kidney- and intestine-specific transporter that is different from y^+ (Segal and Thier, 1989). There is disagreement as to the Na^+ dependence of this transporter and to the existence of a second high K_m renal transport system for cystine alone. Another genetic defect of proximal tubule transport is Hartnup disease, which is a characterized by deficient transport of neutral amino acids but not of amino acids and glycine (Jepson, 1978). Investigation of these inborn errors of transport requires information on the molecular structure of the transporters involved. Expression cloning has been used to isolate kidney- and intestine-specific cDNAs from rat (clone D2, Wells and Hediger, 1992; NAA-Tr, Tate et al., 1992) and rabbit (rBAT, Bertran et al., 1992*a*) that stimulate the transport of amino acids when expressed in *Xenopus* oocytes. D2 and rBAT are 82% identical at the amino acid sequence level and both clones induce the high affinity uptake into oocytes of a previously unreported broad spectrum of amino acids including cystine and dibasic and neutral amino acids. Clone NAA-Tr has a coding sequence identical to that of D2 but was reported not to transport dibasic amino acids.

We review here our data on the expression cloning and characterization of clone D2 (Wells and Hediger, 1992), which encodes a type II–like membrane protein. We discuss our findings with 4F2 heavy chain, a sequence homologue of D2, which we show also induces amino acid transport when expressed in *Xenopus* oocytes, but which has different transport characteristics (Wells et al., 1992; Bertran et al., 1992*b*). We propose that D2 and 4F2 are members of a novel gene family of proteins that are involved in the activation of amino acid transport through a previously undescribed mechanism.

Expression Cloning and Isolation of Clone D2

Expression studies using *Xenopus* oocytes were performed to examine whether expression cloning could be used to clone cystine transporter cDNAs. Injection of mRNA from rat kidney and jejunum into oocytes resulted in a substantially greater uptake of ^{14}C-L-cystine compared with water-injected controls (Fig. 1 *A*). The uptake for kidney was ~25-fold more than that for jejunum. Size-fractionation of the kidney mRNA led to the identification of an RNA size range at 1.8–2.5 kb that strongly induced cystine uptake (Fig. 1 *A, inset*). This size range was used to construct a unidirectional cDNA library using the expression vector pBluescript (Stratagene, La Jolla, CA). The library was screened using the techniques that evolved during the expression cloning of the Na$^+$/glucose cotransporter from rabbit small intestine (Hediger et al., 1987). Progressively smaller pools were analyzed until a clone (D2) that induced cystine uptake 200–400 times above that of control oocytes was isolated (Fig. 1 *A*; see also Wells and Hediger, 1992). Hybrid depletion of D2 from kidney mRNA by incubation with an antisense oligonucleotide to the 5′ end of D2 resulted in a decrease in cystine uptake to near background, suggesting that D2 is responsible for the majority of the cystine uptake induced by total mRNA (Fig. 1 *B*). Northern blot analysis of rat kidney and intestine RNA showed that clone D2 hybridized intensely to a band at 2.2 kb and weakly to another at 4.4 kb. No hybridization was apparent for rat RNA from heart, liver, lung, spleen, brain, skeletal muscle, and pancreas.

Expression studies of oocytes injected with D2 cRNA demonstrated that, in addition to inducing the uptake of cystine, D2 also induces the uptake of dibasic and, interestingly, neutral amino acids (Fig. 1 *C*). A transport system of cystine and dibasic and neutral amino acids has not been previously described for the kidney, although some previous inhibition studies that addressed cystine uptake in perfused tubules or brush border membrane vesicles have included data that suggested a more broad substrate specificity for renal and jejunal dibasic amino acid transporters than is generally assumed (Foreman et al., 1980; Biber et al., 1983).

Inhibition studies using D2 cRNA-injected oocytes showed that cystine and dibasic or neutral amino acids in excess inhibited the uptake of ^{14}C-labeled cystine, neutral amino acids, and dibasic amino acids (Wells and Hediger, 1992). This suggested that the uptake for these different classes of amino acids proceeds via the same transporter. Uptake of both neutral and dibasic amino acids is stereospecific for L-amino acids. The K_m for L-cystine is 63.7 μM, which is in accord with high affinity cystine transport (Voelkl and Silbernagl, 1982; Schafer and Watkins, 1984).

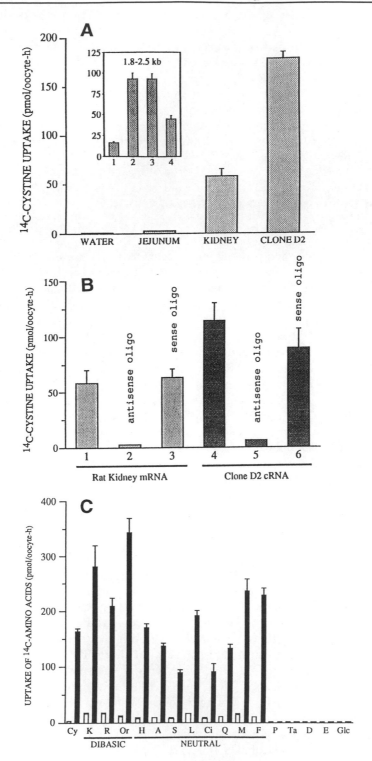

D2-induced amino acid transport is mostly (60–100%) Na$^+$ independent. The D2-induced uptake, however, is able to generate a high concentration gradient (up to 100-fold for cystine), suggesting that the driving force for transport may be provided by the translocation of other solutes.

Predicted Structure of the D2 Protein

The size of the D2 cDNA is 2279 nucleotides and corresponds to the message size observed in Northern analysis (Wells and Hediger, 1992). The deduced amino acid sequence is 683 amino acids long and there are seven potential N-glycosylation sites, all toward the COOH-terminal end. In vitro translation studies in the presence of pancreatic microsomal membranes indicate that at least some of these sites are N-glycosylated. An internal hydrophobic region at the N-terminal end which is predicted to be membrane spanning makes it likely that D2 codes for a type II membrane glycoprotein with a hydrophilic cytoplasmic NH$_2$ terminus, a single membrane-spanning domain, and an extracellular COOH terminus (Fig. 2). This structure is unlike that of most previously cloned channels and transporters, which have multiple membrane-spanning regions. A potassium channel with a single transmembrane region, however, has been recently cloned (Takumi et al., 1988). We have identified a leucine zipper motif (Buckland and Wild, 1989) toward the COOH terminus (residues 548–569) of the D2 protein, which raises the possibility that the protein undergoes dimerization (e.g., association to a second subunit).

Sequence Homologies to the D2 Protein

Search of protein sequence databases demonstrated significant similarity between the proposed extracellular COOH-terminal domain of the D2 protein and a family of carbohydrate-metabolizing enzymes (Wells and Hediger, 1992). These proteins include α-amylases and α-glucosidases from both prokaryotic and eukaryotic organ-

---

Figure 1. Expression cloning. (*A*) Screening of a rat kidney cDNA library for cystine transport results in the isolation of clone D2. The uptakes of ^{14}C-labeled L-cystine (2.2 µM) into *Xenopus* oocytes injected with water, rat kidney cortex, jejunum mRNA, or clone D2 cRNA are shown. All uptakes were performed in the presence of 100 mM Na$^+$. The inset shows peak uptake induced by injection of a 1.8–2.5-kb size fraction of mRNA. Each column represents the mean uptake of five to eight oocytes and the standard error is indicated by the error bars. (*B*) D2 is responsible for most if not all of the cystine uptake induced by rat kidney cortex mRNA. The hybrid depletion of D2 mRNA by incubation with an antisense oligonucleotide corresponding to the 5' end of clone D2 is shown. The mean of the uptake into water-injected oocytes has been subtracted. Columns *1* and *4* show uptake into oocytes injected with rat kidney mRNA and with clone D2 cRNA, respectively. Columns *2* and *5* show that uptake of both is decreased to 4–5% of baseline after incubation with an antisense oligonucleotide; incubation with a sense oligonucleotide (columns *3* and *6*) has a minimal effect. (*C*) D2 induces the transport of cystine as well as dibasic and neutral amino acids but not proline, taurine, acidic amino acids, or glucose. D2 cRNA-injected oocytes (*filled bars*) take up dibasic and neutral amino acids only. Open columns represent water-injected oocytes. *Cy,* L-cystine; *Or,* L-ornithine; *Ci,* L-citrulline (Ci); *Ta,* taurine; *glc;* D-glucose. All other amino acids used are indicated by the single letter code and are L-amino acids.

isms (Fig. 2). Sequence identities between the D2 amino acid sequence and those of members of this family are as high as 41%. D2 shares conserved residues with the members of this family. Of note are asparagine and aspartic acid residues postulated in porcine pancreatic amylase either to be part of the catalytic site (Asp 312 of D2) or to serve as Ca^{2+}-binding sites (Asn 211 and Asp 281 of D2; see Fig. 2). There is also significant sequence similarity (29% identity) to the human and mouse 4F2 heavy chain cell surface antigen (Quackenbush et al., 1987; Teixeira et al., 1987). This antigen is a type II membrane glycoprotein, as has been postulated for the D2 protein, and has a transmembrane domain at a position similar to D2.

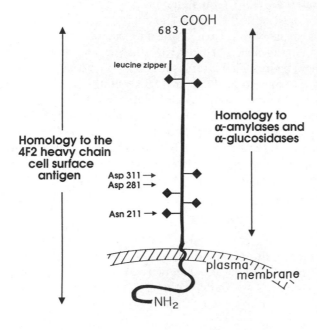

D2 MEMBRANE GLYCOPROTEIN

Figure 2. Predicted structure of the rat kidney and intestine specific D2 protein. The D2 protein is predicted to be a type II membrane glycoprotein with a hydrophilic cytoplasmic NH_2 terminus, a single membrane-spanning domain, and an extracellular COOH terminus. There are seven potential N-glycosylation sites, indicated by diamonds. The leucine zipper is marked by a vertical line. The COOH-terminal domain is homologous to a family of carbohydrate-metabolizing enzymes, which includes α-amylases and α-glucosidases from both prokaryotic and eukaryotic organisms. D2 shares conserved residues with the members of this family: Asn 211 and Asp 281 are predicted to correspond to the proposed Ca^{2+}-binding ligands Asn 100 and Asp 167 porcine pancreatic amylase. Asp 311 is predicted to correspond to a catalytic aspartic acid residue of porcine pancreatic amylase (see Wells and Hediger, 1992). There is significant similarity of the D2 protein to the human and mouse 4F2 heavy chain cell surface antigen (29% amino acid sequence identity) which is, like D2, a type II membrane glycoprotein.

Examination of the Function of the 4F2 Antigen Heavy Chain

The 4F2 cell surface antigen is a 125-kD disulfide-linked heterodimer composed of an 85-kD glycosylated heavy chain and a 41-kD nonglycosylated light chain (Teixeira et al., 1987). The antigen is associated with cell activation and tumor cell growth. It was originally identified by the production of a mouse monoclonal antibody (mAb4F2) against the human T cell tumor line HSB-2. Northern blot analysis with a mouse heavy chain cDNA probe has demonstrated that 4F2 heavy chain expression is widespread in mouse tissues. 4F2 expression is induced during the process of cellular activation, and remains at constant levels in exponentially growing cells. The function of the 4F2 proteins, however, has been unclear despite nearly a decade of research.

The similarity between the D2 and 4F2 amino acid sequences suggested that the two proteins might have similar functions. We, as well as Murer and colleagues, injected 4F2 heavy chain cRNA into oocytes and determined that it also induces the uptake of dibasic and neutral amino acids up to threefold over controls (Bertran et al., 1992*b*; Wells et al., 1992). The substrate specificity and Na^+ dependence of transport is different from that induced by cRNA from clone D2, but has some similarity to that of system y^+ (see Kim et al., 1991; Wang et al., 1991). The K_m of 4F2-induced arginine uptake is 43 μM. This is similar to the K_m values for water-injected oocytes (intrinsic activity, 48 μM), D2-injected oocytes (74 μM), and y^+-injected oocytes (70 μM), and is representative of high affinity transport.

Comparison of D2-, 4F2-, and Y$^+$-induced Amino Acid Transport

The properties of amino acid transport induced by D2, 4F2, and y^+ were compared using *Xenopus* oocyte expression studies as shown in Fig. 3. The different properties of the expressed transport systems are summarized below.

D2

D2 and its rabbit counterpart induce the high affinity uptake of dibasic and neutral amino acids as well as cystine (see also Fig. 1). The transport is >60% independent of Na^+.

4F2

4F2 induces uptake of dibasic and neutral amino acids but, in contrast to D2, this does not include cystine. The uptake of the neutral amino acids leucine and methionine is >80% Na^+ dependent, whereas the uptake of dibasic amino acids is Na^+ independent (Fig. 3 *B*). Likewise, the inhibition of arginine but not leucine by an excess of neutral amino acids (including homoserine) is Na^+ dependent (Fig. 3 *C*). This pattern of Na^+ dependence displayed by 4F2 was unexpected and has not been noted before for an amino acid transport system.

Methyl-amino-isobutyric acid (MeAIB), a specific substrate of the Na^+-dependent system A, has no effect on 4F2-induced arginine uptake (Fig. 3 *D*), thus ruling out the possibility that 4F2 activates system A. Interestingly, cystine, which is not taken up by 4F2 cRNA-injected oocytes, causes a 50% decrease in 4F2-induced arginine uptake but has no effect on leucine uptake (Fig. 3 *D*).

y$^+$

System y^+ mediates the Na^+-independent uptake of dibasic amino acids but not cystine (Fig. 3 *A*). In contrast to 4F2, y^+ does not mediate transport of the neutral amino acids leucine and methionine. Other neutral amino acids such as homoserine, however, are transported by system y^+ but only in the presence of Na^+. In comparison, 4F2 shows a much wider spectrum of Na^+-dependent neutral amino acid uptake than system y^+.

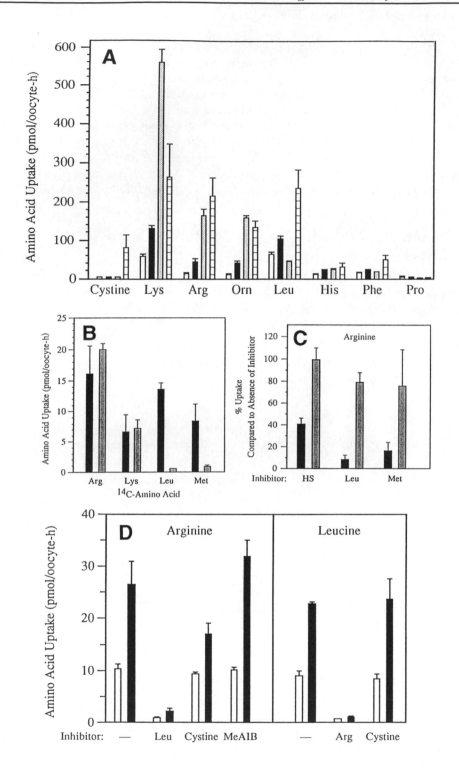

Functional Models for D2 and 4F2

The single transmembrane domains of D2 and 4F2 make these proteins unlike previously cloned transporters (e.g., the y^+ transporter), all of which have multiple membrane-spanning regions. The sequence similarity between D2 and glucosidases suggests that D2 may be a carbohydrate-metabolizing enzyme. We have been unable, however, to demonstrate α-amylase or maltase activity in cellular fractions from D2-injected oocytes. D2 and 4F2 might function as regulators of transporters, perhaps by acting as regulatory subunits of hetcrodimeric transporters. The 4F2 heavy chain exists as a heterodimer attached to a smaller, nonglycosylated light chain whose structure and function are unknown. If D2 and 4F2 function as regulatory subunits, it will be important to establish the structures of the transporters they are associated with. In oocytes, these proteins may associate with endogeneous transporters to induce uptake. D2-induced amino acid transport, however, has different characteristics from intrinsic amino acid transport in oocytes, suggesting that activation of an intrinsic oocyte transporter by D2 requires significant upregulation of that transporter or a change in its substrate specificity. The mouse blastocyst transporter $b^{o,+}$ is a Na^+-independent transporter with a broad substrate specificity for neutral and dibasic amino acids, and there is some evidence for weak $b^{o,+}$ activity in oocytes (Van Winkle et al., 1988).

4F2- and system y^+-induced amino acid transport share some characteristics (see Fig. 3), although there are significant differences. For example, the sodium requirement for 4F2-induced transport of dibasic and neutral amino acids and that described for system y^+ is different. This suggests that in animal tissues, 4F2 is

---

Figure 3. D2, 4F2, and $y^{\pm}$-injected oocytes demonstrate different patterns of amino acid uptake. (*A*) Comparison of D2, 4F2, and $y^{\pm}$-induced amino acid transport. Oocytes injected with water or with cRNA from the rat D2, mouse y^+ (Kim et al., 1991), or human 4F2 heavy chain (Quackenbush et al., 1987) were assayed 3 d (cystine, Lys, Leu) or 4 d (all others) after injection for the uptake of 100 μM (cystine, Lys, Leu) or 15 μM (all others) radiolabeled amino acids. All incubations were 1 h and were performed in the presence of 100 mM Na^+. Error bars represent the standard error. *Open bars,* water-injected; *filled bars,* 4F2-injected; *shaded bars,* y^+-injected; *horizontally striped bars,* D2-injected. (*B*) 4F2-induced uptake of leucine and methionine but not arginine and lysine is sodium dependent. Oocytes were injected with water or 4F2 cRNA and assayed 3 d later for uptake of 15 μM ^{14}C-arginine, lysine, leucine, or methionine in the presence of 100 mM Na^+ (*filled bars*) or 100 mM NMDG (*shaded bars*). The data shown represent uptake into 4F2-injected oocytes minus uptake into water-injected oocytes. (*C*) Inhibition of 4F2-induced arginine uptake by neutral amino acids is sodium dependent. Oocytes were injected with water or 4F2-cRNA and assayed 3 d after injection in the presence of either 100 mM Na^+ (*filled bars*) or 100 mM NMDG (*shaded bars*) for uptake of 15 μM ^{14}C-arginine alone or in the presence of 2 mM of an unlabeled inhibitor amino acid. Uptake into water-injected oocytes was subtracted and the percentage of uptake in the presence of an inhibitor was calculated; uptake in the absence of any inhibitor was considered to be 100%. *HS,* homoserine. (*D*) MeAIB, a specific substrate of system A, has no effect on the 4F2-induced arginine uptake; cystine causes a 50% decrease in arginine uptake but has no effect on leucine uptake. Oocytes injected with water or 4F2-cRNA were assayed 3 d after injection for uptake of 15 μM ^{14}C-labeled arginine (*left*) or leucine (*right*) alone and in the presence of 2 mM (0.5 mM for cystine) of an unlabeled amino acid, as indicated. All uptakes were performed in the presence of 100 mM Na^+. *Open bars,* water-injected oocytes; *filled bars,* 4F2-cRNA-injected oocytes.

310   *Molecular Biology and Function of Carrier Proteins*

unlikely to be an activator of system y^+. It is conceivable, however, that 4F2 activates an amino acid transport system that is similar to system y^+.

Our data on D2- and 4F2-induced amino acid transport do not rule out the possibility that these type II membrane glycoproteins are themselves transporters. The homology of the large extracellular domains of D2 and 4F2 to glucosidases, which bind to specific oligosaccharides, could be taken as an indication that this domain represents a specific binding site; for example, it could form a binding site for amino acids. The functional transporters may consist of several D2 or 4F2 polypeptides whose single membrane-spanning domains oligomerize to form homooligomeric pores that translocate amino acids.

Understanding the mechanisms involved in D2- and 4F2-induced amino acid transport may lead to important insights into the Na^+ dependency and substrate specificity of amino acid transport systems. Further experiments involving expression of D2 and 4F2 in transformed cell lines in addition to oocytes will clarify the role of this new family of proteins in amino acid transport.

Localization of the D2 mRNA in the Rat Kidney Proximal Tubule S3 Segment

To examine whether D2 is involved in high affinity transport of cystine in the proximal straight tubule (the latter part of S2 and all of S3), we have carried out combined in situ hybridization and immunocytochemistry experiments and determined the site of expression of D2 mRNA in the kidney tubule (Kanai et al., 1992). Hybridization of ^{35}S-labeled D2-antisense mRNA to a whole kidney section is shown in Fig. 4. The low magnification view shows dense hybridization to tubules in the outer stripe of the outer medulla, whereas the signal was absent from the inner stripe of the outer medulla. The signal was also detected over tubules in medullary rays, projecting into the cortex. D2 antisense RNA hybridized to the same tubular segments in the outer stripe of the outer medulla which were immunopositive with the anti-ecto-ATPase antibody that is specific for the S3 segment. These studies demonstrated that D2 mRNA is strongly expressed in the S3 segment of the rat kidney proximal tubule. Weak staining was also observed in the S1 and S2 segments. The signal was absent in all other parts of the kidney. This S3 specific expression of D2 mRNA coincides with the site of high affinity transport of cystine and other amino acids.

Potential Role of D2 in Cystinuria

Except for neutral amino acid transport, D2 exhibits characteristics and a tissue-specific expression similar to those previously described for the transporter predicted to be defective in cystinuria (Segal et al., 1977; Segal and Thier, 1989). A defect of the D2 protein could explain the impaired cystine and dibasic amino acid excretion seen in cystinuria, given potential compensation of neutral amino acid uptake by other neutral amino acid transporters (e.g., system L). Our results from hybrid depletion experiments, which suggest that D2 is responsible for most if not all of the renal cystine transport, are consistent with the potential role of D2 in cystinuria. A defect of the D2 protein could conceivably be involved in Hartnup disease, which is characterized by impaired reabsorption of neutral amino acids. As a first step toward

Figure 4. Localization of the D2 mRNA in kidney proximal tubule S3 segments. Low power, bright field micrograph showing the pattern of hybridization of the D2 antisense cRNA probe (^{35}S-labeled) in 5-μm cryostat sections of paraformaldehyde-fixed rat kidney. A strong signal is present in tubules from the outer stripe of the outer medulla, whereas the signal is absent from the inner stripe of the outer medulla. The positive tubules show a pattern of distribution that is characteristic of the S3 segment of the proximal tubule. This was confirmed by immunostaining adjacent sections using antibodies specific for different segments of the proximal tubule (Kanai et al., 1992).

studying the role of D2 in these diseases we have cloned, sequenced, and character-ized a human D2 cDNA (Lee et al., 1993). The availability of this sequence will allow us to examine the D2 gene in patients with renal aminoacidurias.

Conclusion

Our data suggest that D2- and 4F2-induced amino acid uptakes most likely represent the activities of separate transport systems with some characteristics of systems y^+ and $b^{o,+}$. We conclude that these type II membrane glycoproteins represent a new family of proteins which induce amino acid transport with distinct characteristics, possibly functioning as transport activators or regulatory subunits of transporters. The study of additional subunits associated with these proteins may lead to further insights into their in vivo function. The 4F2 light chain, for example, may represent a system y^+–like transport protein and its study may provide important insight into the role of 4F2 during cell activation and tumor cell growth. In addition, the availability of a human D2 clone will facilitate studies into the potential role of members of the D2 family in normal and abnormal amino acid transport such as occurs in cystinuria.

References

Berry, C. A., and F. C. Rector, Jr. 1991. Renal transport of glucose, amino acids, chloride, and water. *In* The Kidney. 4th ed. B. M. Brenner and F. C. Rector, Jr., editors. W. B. Saunders Company, Philadelphia. 245–282.

Bertran, J., S. Magagnin, A. Werner, D. Markovich, J. Biber, X. Testar, A. Zorzano, L. C. Kühn, M. Palacin, and H. Murer. 1992*a*. Stimulation of system y^+-like amino acid transport by the heavy chain of human 4F2 surface antigen in *Xenopus laevis* oocytes. *Proceedings of the National Academy of Sciences, USA.* 89:5606–5610.

Bertran, J., A. Werner, M. L. Moore, G. Stange, D. Markovich, J. Biber, X. Testar, A. Zorzano, M. Palacin, and H. Murer. 1992*b*. Expression cloning of a cDNA from rabbit kidney cortex that induces a single transport system for cystine and dibasic and neutral amino acids. *Proceedings of the National Academy of Sciences, USA.* 89:5601–5605.

Biber, J., G. Stange, B. Steiger, and H. Murer. 1983. Transport of L-cystine by rat renal brush border membrane vesicles. *Pflügers Archiv.* 396:335–341.

Buckland, R., and F. Wild. 1989. Leucine zipper motif extends. *Nature.* 338:547.

Foreman, J. W., S.-M. Hwang, and S. Segal. 1980. Transport interactions of cystine and dibasic amino acids in isolated rat renal tubules. *Metabolism Clinical and Experimental.* 29:53–61.

Hediger, M. A., M. J. Coady, T. S. Ikeda, and E. M. Wright. 1987. Expression cloning and cDNA sequencing of the Na^+/glucose cotransporter. *Nature.* 330:379–381.

Heinz, E., D. L. Sommerfeld, and K. H. Kinne. 1988. Electrogenicity of sodium/glutamate cotransport in rabbit renal brush-border membranes: a reevaluation. *Biochimica et Biophysica Acta.* 937:300–308.

Hopfer, U. 1987. Membrane transport mechanisms for hexoses and amino acids in the small intestine. *In* Physiology of the Gastrointestinal Tract. 2nd ed. L. R. Johnson, editor. Raven Press, New York. 1499–1524.

Jepson, J. B. 1978. Hartnup disease. *In* The Metabolic Basis of Inherited Disease. 4th ed. J. B. Sandbury, J. B. Wyngaarden, and D. S. Fredrickson, editors. McGraw-Hill Book Company, New York. 1563.

Kanai, Y., and M. A. Hediger. 1992. Primary structure and functional characterization of a high-affinity glutamate transporter. *Nature.* 360: 467–471.

Kanai, Y., M. G. Stelzner, W.-S. Lee, R. G. Wells, D. Brown, and M. A. Hediger. 1992. Expression of mRNA (D2) encoding a protein involved in amino acid transport in S3 proximal tubule. *American Journal of Physiology (Renal Fluid and Electrolyte Physiology).* 263:F1087–F1093.

Kim, J. W., E. I. Closs, L. M. Albritton, and J. M. Cunningham. 1991. Transport of cationic amino acids by the mouse ecotropic retrovirus receptor. *Nature.* 352:725–728.

Lee, W.-S., R. W. Wells, R. Sabbag, T. K. Mohandas, and M. A. Hediger. 1993. Cloning and chromosomal localization of a human kidney cDNA involved in cystine and dibasic and neutral amino acid transport. *Journal of Clinical Investigation.* In press.

McCormick, J. I., and R. M. Johnstone. 1988. Simple and effective purification of a Na$^+$-dependent amino acid transport system from Ehrlich ascites cell plasma membrane. *Proceedings of the National Academy of Sciences, USA.* 85:7877–7881.

Mircheff, A. K., I. Kippen, B. Hirayama, and E. M. Wright. 1982. Delineation of sodium-stimulated amino acid transport in rabbit kidney brush border vesicles. *Journal of Membrane Biology.* 64:113–122.

Quackenbush, E., M. Clabby, K. M. Gottesdiener, J. Barbosa, N. H. Jones, J. L. Strominger, S. Speck, and J. M. Leiden. 1987. Molecular cloning of complementary DNAs encoding the heavy chain of the human 4F2 cell-surface antigen: a type II membrane glycoprotein involved in normal and neoplastic cell growth. *Proceedings of the National Academy of Sciences, USA.* 84:6526–6530.

Schafer, J. A., and M. L. Watkins. 1984. Transport of L-cystine in isolated perfused proximal straight tubules. *Pflügers Archiv.* 401:143–151.

Segal, S., P. D. McNamara, and L. M. Pepe. 1977. Transport interaction of cystine and dibasic amino acids in renal brush border vesicles. *Science.* 197:169–170.

Segal, S., and S. O. Thier. 1989. Cystinuria. *In* The Metabolic Basis of Inherited Disease. 6th ed. C. R. Scriver, A.L. Beaudet, W. S. Sly, and D. Valle, editors. McGraw-Hill Book Company, New York. 2479–2496.

Simell, O. 1989. Lysinuric protein intolerance and other cationic aminoacidurias. *In* The Metabolic Basis of Inherited Disease. 6th ed. C. R. Scriver, A. L. Beaudet, W. S. Sly, and D. Valle, editors. McGraw-Hill Book Company, New York. 2497–2513.

Takumi, T., H. Ohkubo, and S. Nakanishi. 1988. Cloning of a membrane protein that induces a slow voltage-gated potassium current. *Science.* 242:1042–1045.

Tate, S. S., R. Urade, T. V. Getchell, and S. Udenfriend. 1989. Expression of the mammalian Na$^+$-independent L system amino acid transporter in *Xenopus* oocytes. *Archives of Biochemistry and Biophysics.* 275:591–596.

Tate, S. S., N. Yan, and S. Udenfriend. 1992. Expression cloning of a Na$^+$-independent neutral amino acid transporter from rat kidney. *Proceedings of the National Academy of Sciences, USA.* 89:1–5.

Teixeira, S., S. Di Grandi, and L. C. Kühn. 1987. Primary structure of the human 4F2 antigen heavy chain predicts a transmembrane protein with a cytoplasmic NH$_2$ terminus. *Journal of Biological Chemistry.* 262:9574–9580.

Van Winkle, L. J., A. L. Campione, and J. M. Gorman. 1988. Na$^+$-independent transport of basic and zwitterionic amino acids in mouse blastocysts by shared system and by processes which distinguish between these substrates. *Journal of Biological Chemistry.* 263:3150–3163.

Voelkl, H., and S. Silbernagl. 1982. Mutual inhibition of L-cystine/L-cysteine and other neutral amino acids during tubular reabsorption. *Pflügers Archiv.* 359:190–195.

Wang, H., M. P. Kavanaugh, R. A. North, and D. Kabat. 1991. Cell-surface receptor for ecotropic murine retroviruses is a basic amino-acid transporter. *Nature.* 352:729–731.

Wells, R. G., and M. A. Hediger. 1992. Cloning of a rat kidney cDNA that stimulates dibasic and neutral amino acid transport and has sequence similarity to glucosidases. *Proceedings of the National Academy of Sciences, USA.* 89:5596–5600.

Wells, R. G., W.-S. Lee, Y. Kanai, J. M. Leiden, and M. A. Hediger. 1992. The 4F2 antigen heavy chain induces uptake of neutral and dibasic amino acids in *Xenopus* oocytes. *Journal of Biological Chemistry.* 267:15285–15288.

Wolff, N. A., and R. Kinne. 1988. Taurine transport by rabbit kidney brush-border membranes: coupling to sodium, chloride, and membrane potential. *Journal of Membrane Biology.* 102:131–139.

List of Contributors

Franklin S. Abrams, Department of Biochemistry and Cell Biology, State University of New York at Stony Brook, Stony Brook, New York

B. D. Adair, Department of Molecular Biophysics and Biochemistry, Yale University, New Haven, Connecticut

Giovanna Ferro-Luzzi Ames, Department of Molecular and Cell Biology, University of California at Berkeley, Berkeley, California

Matthew P. Anderson, Departments of Internal Medicine and Physiology and Biophysics, Howard Hughes Medical Institute, University of Iowa College of Medicine, Iowa City, Iowa

Jon Beckwith, Department of Microbiology and Molecular Genetics, Harvard Medical School, Boston, Massachusetts

Carol Berkower, Department of Cell Biology and Anatomy, Johns Hopkins University School of Medicine, Baltimore, Maryland

Randy D. Blakely, Department of Anatomy and Cell Biology, Emory University School of Medicine, Atlanta, Georgia

Dana Boyd, Department of Microbiology and Molecular Genetics, Harvard Medical School, Boston, Massachusetts

A. Brünger, Department of Molecular Biophysics and Biochemistry, Yale University, New Haven, Connecticut

Ellen Buschman, Department of Biochemistry, McGill University, Montreal, Quebec, Canada

Nancy Carrasco, Department of Molecular Pharmacology, Albert Einstein College of Medicine, Bronx, New York

Joseph R. Casey, MRC Group in Membrane Biology, Department of Biochemistry, University of Toronto, Toronto, Ontario, Canada

Laurent Counillon, Centre de Biochimie-Centre National de la Recherche Scientifique, Université de Nice, Nice, France

Ge Dai, Department of Molecular Pharmacology, Albert Einstein College of Medicine, Bronx, New York

J. Deisenhofer, Howard Hughes Medical Institute and Department of Biochemistry, University of Texas Southwestern Medical Center, Dallas, Texas

Max Dolder, Maurice E. Müller Institute for High-Resolution Electron Microscopy at the Biocenter, University of Basel, Basel, Switzerland

Rhonda L. Dunten, Howard Hughes Medical Institute, Departments of Physiology and Microbiology & Molecular Genetics, Molecular Biology Institute, University of California Los Angeles, Los Angeles, California

Andreas Engel, Maurice E. Müller Institute for High-Resolution Electron Microscopy at the Biocenter, University of Basel, Basel, Switzerland

D. M. Engelman, Department of Molecular Biophysics and Biochemistry, Yale University, New Haven, Connecticut

J. M Flanagan, Department of Molecular Biophysics and Biochemistry, Yale University, New Haven, Connecticut

Philippe Gros, Department of Biochemistry, McGill University, Montreal, Quebec, Canada

Karl M. Hager, Department of Physiology, UCLA School of Medicine, Los Angeles, California

Akihiro Hazama, Department of Physiology, UCLA School of Medicine, Los Angeles, California

Matthias A. Hediger, Department of Medicine, Renal Division, Brigham and Women's Hospital and Harvard Medical School, Boston, Massachusetts

Andreas Hefti, Maurice E. Müller Institute for High-Resolution Electron Microscopy at the Biocenter, University of Basel, Basel, Switzerland

Bruce Hirayama, Department of Physiology, UCLA School of Medicine, Los Angeles, California

J. F. Hunt, Department of Molecular Biophysics and Biochemistry, Yale University, New Haven, Connecticut

Georg Jander, Department of Microbiology and Molecular Genetics, Harvard Medical School, Boston, Massachusetts

Jean Xin Jiang, Department of Biochemistry and Cell Biology, State University of New York at Stony Brook, Stony Brook, New York

H. R. Kaback, Howard Hughes Medical Institute, Departments of Physiology and Microbiology & Molecular Genetics, Molecular Biology Institute, University of California Los Angeles, Los Angeles, California

Stephen M. Kaminsky, Department of Molecular Pharmacology, Albert Einstein College of Medicine, Bronx, New York

Yoshikatsu Kanai, Department of Medicine, Renal Division, Brigham and Women's Hospital and Harvard Medical School, Boston, Massachusetts

Baruch I. Kanner, Department of Biochemistry, Hadassah Medical School, The Hebrew University, Jerusalem, Israel

Martin Klingenberg, Institute for Physical Biochemistry, University of Munich, Munich, Federal Republic of Germany

Ron R. Kopito, Department of Biological Sciences, Stanford University, Stanford, California

Carolina Landolt-Marticorena, MRC Group in Membrane Biology, Department of Biochemistry, University of Toronto, Toronto, Ontario, Canada

Gerard Leblanc, Laboratoire J. Maetz, Département de Biologie Cellulaire et Moléculaire, Commissariat à l'Energie Atomique, Villefranche-sur-mer, France

Wen-Sen Lee, Department of Medicine, Renal Division, Brigham and Women's Hospital and Harvard Medical School, Boston, Massachusetts

M. A. Lemmon, Department of Molecular Biophysics and Biochemistry, Yale University, New Haven, Connecticut

Orlie Levy, Department of Molecular Pharmacology, Albert Einstein College of Medicine, Bronx, New York

Zhaoping Li, Cardiovascular Research Laboratory and the Departments of Medicine and Physiology, University of California, Los Angeles, School of Medicine, Los Angeles, California

Ann E. Lindsey, Department of Biological Sciences, Stanford University, Stanford, California

Erwin London, Department of Biochemistry and Cell Biology, State University of New York at Stony Brook, Stony Brook, New York

Donald D. F. Loo, Department of Physiology, UCLA School of Medicine, Los Angeles, California

Peter C. Maloney, Department of Physiology, Johns Hopkins Medical School, Baltimore, Maryland

Susan Michaelis, Department of Cell Biology and Anatomy, Johns Hopkins University School of Medicine, Baltimore, Maryland

Kim R. Moore, Department of Anatomy and Cell Biology, Emory University School of Medicine, Atlanta, Georgia

Debora A. Nicoll, Cardiovascular Research Laboratory and the Departments of Medicine and Physiology, University of California, Los Angeles, School of Medicine, Los Angeles, California

Lucie Parent, Department of Molecular Physiology and Biophysics, Baylor College of Medicine, Houston, Texas

Kenneth D. Philipson, Cardiovascular Research Laboratory and the Departments of Medicine and Physiology, University of California, Los Angeles, School of Medicine, Los Angeles, California

Thierry Pourcher, Laboratoire J. Maetz, Département de Biologie Cellulaire et Moléculaire, Commissariat à l'Energie Atomique, Villefranche-sur-mer, France

Jacques Pouyssegur, Centre de Biochimie-Centre National de la Recherche Scientifique, Université de Nice, Nice, France

Will Prinz, Department of Microbiology and Molecular Genetics, Harvard Medical School, Boston, Massachusetts

Yan Qian, Program in Neuroscience, Emory University School of Medicine, Atlanta, Georgia

Reinhart A. F. Reithmeier, MRC Group in Membrane Biology, Department of Biochemistry, University of Toronto, Toronto, Ontario, Canada

Stephan Ruetz, Department of Biochemistry, McGill University, Montreal, Quebec, Canada

Miklós Sahin-Tóth, Howard Hughes Medical Institute, Departments of Physiology and Microbiology & Molecular Genetics, Molecular Biology Institute, University of California Los Angeles, Los Angeles, California

Carolina Salvador, Department of Molecular Pharmacology, Albert Einstein College of Medicine, Bronx, New York

Vivian E. Sarabia, MRC Group in Membrane Biology, Department of Medicine, University of Toronto, Toronto, Ontario, Canada

Stephane Supplisson, Department of Physiology, UCLA School of Medicine, Los Angeles, California

Domenico Tortorella, Department of Biochemistry and Cell Biology, State University of New York at Stony Brook, Stony Brook, New York

Beth Traxler, Department of Microbiology and Molecular Genetics, Harvard Medical School, Boston, Massachusetts

H. Treutlein, Department of Molecular Biophysics and Biochemistry, Yale University, New Haven, Connecticut

Eric Turk, Department of Physiology, UCLA School of Medicine, Los Angeles, California

Nancy D. Ulbrandt, Department of Biochemistry and Cell Biology, State University of New York at Stony Brook, Stony Brook, New York

Thomas Walz, Maurice E. Müller Institute for High-Resolution Electron Microscopy at the Biocenter, University of Basel, Basel, Switzerland

Jing Wang, MRC Group in Membrane Biology, Department of Medicine, University of Toronto, Toronto, Ontario, Canada

Rebecca G. Wells, Whitehead Institute for Biomedical Research, Cambridge, Massachusetts

Michael J. Welsh, Departments of Internal Medicine and Physiology and Biophysics, Howard Hughes Medical Institute, University of Iowa College of Medicine, Iowa City, Iowa

T. Hastings Wilson, Department of Physiology, Harvard Medical School, Boston, Massachusetts

Ernest M. Wright, Department of Physiology, UCLA School of Medicine, Los Angeles, California

Marie-Louise Zani, Laboratoire J. Maetz, Département de Biologie Cellulaire et Moléculaire, Commissariat à l'Energie Atomique, Villefranche-sur-mer, France

J. Zhang, Department of Molecular Biophysics and Biochemistry, Yale University, New Haven, Connecticut

Subject Index

Kidney
  mdr gene expression, 106
  sodium/hydrogen antiporters, 174
Kinetics
  ADP/ATP carrier, mitochondrial, 207–208
  anion exchanger binding, 196
  melibiose permease, 214–215
  sodium/hydrogen antiporters, 173–174, 177

β-Lactamase gene fusion systems, 27
  comparison with alkaline phosphatase, 31–32
Lactose permease of *Escherichia coli*, 1–9
  intramembrane charged residues, 4–6
  structure, 2–4
LacY carrier, 157
Leishmania tarentolae, *mdr* homologue, 98–100
Lipids
  bilayers, hydrogen bond stability, 12, 13
  fluorescence quenching, 56
  and membrane-inserted diphtheria toxin, 49
Liver, *mdr* gene expression, 106

Major histocompatibility complex (MHC), *mdr*
    homologue, 100
MalF protein, alkaline phosphatase fusion, 28–
    30, 34
Maltose, TM transport, and P-gp segment ho-
    mology, 98
Maltose permease, 82
Marchantia polymorpha, *mdr* homologue, 100
Mating pheromone
  yeast *STE6* gene, complementation by mouse
    mdr3, 105–108
  See also Yeast, STE6 protein.
mdr
  chimeric genes, 101–102
  cloned genes, and P-gp polypeptides, 97–98
  function in transfected cells, 100–101
  super gene family, 98–100
mdr1
  NB domains, 103–1014
  TM domain 11, 104–105
mdr3
  complementation of STE6 mutation, 105–106
Melibiose permease of *Escherichia coli*, 213–227
  acidic residues in hydrophobic core, 217
  Asp to Cys or Asn mutations, 217–221
  Asp to Glu mutations, 221–223
  cation coordination, ASP role, 223–224
  functional properties, 214–215
  identification and secondary structure, 215–216
  point mutations altering cationic selectivity, 216
Membrane-bound complex, histidine permease,
    82–85
Membrane proteins
  evolution of, 147–160
    bacterial proton pumps, 151–152
    carriers

common structure of superfamily, 155–158
    gene duplication and fusion, 153–155
    origins, 153
    transmembrane α helices, 153
  cell volume regulation by active transport,
    148–149
    osmotic crisis of early cells, 148
    proton circulation in early cells, 149
    transport link to metabolism, proton pump,
    149–150
  glycophorin A dimerization, 11–21
  high-resolution three-dimensional structure,
    39–43
  lactose permease, 1–9
  topology, and gene fusion. *See* Gene fusion.
  translocation, bacterial toxin models, 45–61
Mitochondria
  ADP/ATP carrier, 201–212
    cardiolipin interaction, 205, 208
    folding, 205–207
    gating mechanisms, 208–210
    nucleotide binding site, 207
    site-directed mutagenesis, 207–208
    structure, 203–205
  uncoupling protein, 157
mtp, *mdr* homologue, 100
Multidrug resistance (MDR)
  gene duplication and fusion, evolutionary pat-
    tern, 155
  transport protein, functional domains, 95–117
    chimeric genes, 101–103
    cloned *mdr* genes and P-gp polypeptides,
    97–98
    complementation of yeast STE6 mutation by
    mouse *mdr3*, 105–106
    mdr gene in normal tissues, 106
    mdr super gene family, 98–100
    in transfected cells, 100–101
    mutational analysis
      of NB domains, 103–104
      of TM domain 11, 104–105
  See also Periplasmic permeases.
Mutagenesis
  ADP/ATP carrier, mitochondrial, 207–208
  glycophorin A dimerization, 14–17
  lac permease, intramembrane charged resi-
    dues, 4–6
  melibiose permease, 216–221
  NB domains of *mdr1*, 103–104
  sodium/calcium exchanger, 189–190
  STE6 protein, 132–133
  TM domain 11 of *mdr1* and *mdr3*, 104–105

NHE1 isoform. *See* Sodium/hydrogen exchanger
    isoforms.
Nucleotide binding (NB)
  ADP/ATP carrier, mitochondrial, 207